VOLUME ONE HUNDRED AND FORTY THREE

INTERNATIONAL REVIEW OF NEUROBIOLOGY

Imaging in Movement Disorders:
Imaging Applications in
Non-Parkinsonian and Other
Movement Disorders

INTERNATIONAL REVIEW OF NEUROBIOLOGY

VOLUME 143

SERIES EDITOR

VOLUME ONE HUNDRED AND FORTY THREE

INTERNATIONAL REVIEW OF NEUROBIOLOGY

Imaging in Movement Disorders:
Imaging Applications in
Non-Parkinsonian and Other
Movement Disorders

Edited by

MARIOS POLITIS, MD MSc DIC PhD FRCP FEAN
Professor of Neurology & Neuroimaging,
Director, Neurodegeneration Imaging Group,
Institute of Psychiatry, Psychology and Neuroscience,
King's College London, United Kingdom

ACADEMIC PRESS

An imprint of Elsevier

Academic Press is an imprint of Elsevier
50 Hampshire Street, 5th Floor, Cambridge, MA 02139, United States
525 B Street, Suite 1650, San Diego, CA 92101, United States
The Boulevard, Langford Lane, Kidlington, Oxford OX5 1GB, United Kingdom
125 London Wall, London, EC2Y 5AS, United Kingdom

First edition 2018

Notices
Knowledge and best practice in this field are constantly changing. As new research and experience broaden our understanding, changes in research methods, professional practices, or medical treatment may become necessary.

Practitioners and researchers must always rely on their own experience and knowledge in evaluating and using any information, methods, compounds, or experiments described herein. In using such information or methods they should be mindful of their own safety and the safety of others, including parties for whom they have a professional responsibility.

To the fullest extent of the law, neither the Publisher nor the authors, contributors, or editors, assume any liability for any injury and/or damage to persons or property as a matter of products liability, negligence or otherwise, or from any use or operation of any methods, products, instructions, or ideas contained in the material herein.

ISBN: 978-0-12-815420-5
ISSN: 0074-7742

For information on all Academic Press publications
visit our website at https://www.elsevier.com/books-and-journals

Working together
to grow libraries in
developing countries

www.elsevier.com • www.bookaid.org

Publisher: Zoe Kruze
Acquisition Editor: Sam Mahfoudh
Editorial Project Manager: Teresa Pons-Ferrer
Production Project Manager: Abdulla Sait
Cover Designer: Christian J. Bilbow

Typeset by SPi Global, India

CONTENTS

CONTRIBUTORS

Daniela Berg
Department of Neurology, Christian-Albrechts-University of Kiel, Kiel; Department of Neurodegeneration, Hertie-Institute for Clinical Brain Research, University of Tuebingen, Tuebingen, Germany

Luis Pedro Faria de Abreu
Departamento de Neurociências do Hospital de Santa Maria; Assistente Convidado da Faculdade de Medicina de Lisboa, Universidade de Lisboa, Lisbon, Portugal

Edoardo Rosario de Natale
Neurodegeneration Imaging Group, Maurice Wohl Clinical Neuroscience Institute, Institute of Psychiatry, Psychology and Neuroscience (IoPPN), King's College London, London, United Kingdom

Mark J. Edwards
Neuroscience Research Centre, Institute of Molecular and Clinical Sciences, St George's University of London, London, United Kingdom

Christos Ganos
Department of Neurology, Charité, University Medicine Berlin, Berlin, Germany

Davide Martino
Department of Clinical Neurosciences, University of Calgary; Hotchkiss Brain Institute, Calgary, AB, Canada

Mario Mascalchi
Meyer Children Hospital; Department of Experimental and Clinical Biomedical Sciences "Mario Serio", University of Florence, Florence, Italy

Gennaro Pagano
Neurodegeneration Imaging Group, Maurice Wohl Clinical Neuroscience Institute, Institute of Psychiatry, Psychology and Neuroscience (IoPPN), King's College London, London, United Kingdom

Giuseppe Plazzi
IRCCS Istituto delle Scienze Neurologiche di Bologna; Department of Biomedical and Neuromotor Sciences, Unit of Neurology, University of Bologna, Bologna, Italy

Marios Politis
Neurodegeneration Imaging Group, Maurice Wohl Clinical Neuroscience Institute, Institute of Psychiatry, Psychology and Neuroscience (IoPPN), King's College London, London, United Kingdom

Giovanni Rizzo
IRCCS Istituto delle Scienze Neurologiche di Bologna; Department of Biomedical and Neuromotor Sciences, Unit of Neurology, University of Bologna, Bologna, Italy

Kristina Simonyan
Department of Otolaryngology, Massachusetts Eye and Ear Infirmary; Department of Neurology, Massachusetts General Hospital; Harvard Medical School, Boston, MA, United States

Tiago Teodoro
Neuroscience Research Centre, Institute of Molecular and Clinical Sciences, St George's University of London, London, United Kingdom; Instituto de Medicina Molecular, Faculdade de Medicina de Lisboa, Universidade de Lisboa, Lisbon, Portugal

Alessandra Vella
Nuclear Medicine, "Le Scotte" University Hospital, Siena, Italy

Heather Wilson
Neurodegeneration Imaging Group, Maurice Wohl Clinical Neuroscience Institute, Institute of Psychiatry, Psychology and Neuroscience (IoPPN), King's College London, London, United Kingdom

Yulia Worbe
Centre de Référence National Maladie Rare 'Syndrome Gilles de la Tourette; Sorbonne Université, UMR S 1127, CNRS UMR 7225, ICM; Départment de Physiologie, Hôpital Saint-Antoine, Paris, France

Rezzak Yilmaz
Department of Neurology, Christian-Albrechts-University of Kiel, Kiel, Germany

PREFACE

Magnetic resonance imaging (MRI), position emission tomography (PET), single photon emission computed tomography (SPECT) are noninvasive brain imaging techniques, which allow the *in vivo* investigation of brain function. These techniques have played a pivotal role in expanding our understanding of brain structure, function, and behavior, as well as the basis of mechanisms underlying movement disorders. Neuroimaging techniques are rapidly growing in number and complexity with huge potential to unlock the determinants that influences neural mechanisms in physiological conditions and subsequently in pathological conditions. Worldwide, researchers are employing these neuroimaging tools to study the physiopathological changes within the brain of living humans. The combination of these techniques with a fine clinical evaluation of the patients provides an invaluable opportunity to identify the relationships between clinical presentation and basic neuroscience; helping to reduce the distance between bench and bed-side. Novel neuroimaging techniques are becoming increasingly translated into clinical practice and may be used as biomarkers to monitor disease progression and treatment efficacy of new neuroprotective drugs.

This book brings together the fundamentals of neuroimaging methodology with direct applications across the spectrum of movement disorders and dementias, and the lessons learned from neuroimaging tools in the context of movement disorders, including Dystonia, Essential Tremor, Restless Leg Syndrome, Tourette's Syndrome and Tic Disorders, Functional Movement Disorders, Cerebellar Disorders, and REM Sleep Behavior Disorder. Addition sections cover structural, functional, and molecular imaging of Parkinson's Disease Cognitive Impairment and Dementia with Lewy Bodies, and novel Imaging Applications in Movement Disorders, including Transcranial Sonography, Imaging Brain Networks, and imaging transplantation. Principles of Hybrid PET–MRI, physics and instrumentation, normal distribution of radiopharmaceuticals, and protocols central to the field are described in detail.

This book is an up-to-date and comprehensive textbook addressing all aspects of neuroimaging techniques with direct applications in Non-Parkinsonian Movement Disorders and Dementias. Therefore, it is an indispensable educational and research reference of value to movement disorder specialists, physicians, researchers, and students. I hope the readers find these articles informative, engaging, and helpful in their practices. I wish to express my sincere gratitude to all the contributing authors, all their efforts to create this exciting book are immensely appreciated.

MARIOS POLITIS

The Neurodegeneration Imaging Group, King's College London, London, United Kingdom

ACKNOWLEDGMENTS

I am very grateful to Heather Wilson and Gennaro Pagano, Neurodegeneration Imaging Group of the King's College London, who are very bright and talented emerging scientists for their editorial assistance and their help with development of these volume series.

CHAPTER ONE

Neuroimaging Applications in Dystonia

Kristina Simonyan[1]

Department of Otolaryngology, Massachusetts Eye and Ear Infirmary, Boston, MA, United States
Department of Neurology, Massachusetts General Hospital, Boston, MA, United States
Harvard Medical School, Boston, MA, United States
[1]Corresponding author: e-mail address: kristina_simonyan@meei.harvard.edu

Contents

Abstract

Dystonia is a neurological disorder characterized by involuntary, repetitive movements. Although the precise mechanisms of dystonia development remain unknown, the diversity of its clinical phenotypes is thought to be associated with multifactorial pathophysiology, which is linked not only to alterations of brain organization, but also environmental stressors and gene mutations. This chapter will present an overview of the pathophysiology of isolated dystonia through the lens of applications of major neuroimaging methodologies, with links to genetics and environmental factors that play a prominent role in symptom manifestation.

Dystonia is a neurological movement disorder, which is characterized by sustained or intermittent muscle contractions, causing abnormal, often repetitive movements, postures, or both (Albanese et al., 2013). Isolated dystonia (formerly known as primary), where dystonic symptoms are the

1

only disease manifestation, is a rare disorder with a prevalence of about 3–30 per 100,000 in the general population (Asgeirsson, Jakobsson, Hjaltason, Jonsdottir, & Sveinbjornsdottir, 2006; de Carvalho Aguiar & Ozelius, 2002; Nutt, Muenter, Melton, Aronson, & Kurland, 1988). It typically develops spontaneously and progresses into a chronic debilitating condition, oftentimes severely impacting not only the freedom of movements but also various other aspects of patient's life, leading to continuous stress, social embarrassment, and even loss of employment. Dystonia comprises a large number of clinical syndromes, with adult-onset focal dystonia being the most common phenotype, followed by segmental dystonia, and much rare early (childhood)-onset generalized dystonia. Some forms of dystonia selectively affect higher-order motor control, resulting in impairments of unique patterns of highly learned behaviors, such as writing, speaking, or playing a musical instrument.

The exact pathophysiological mechanisms of isolated dystonia remain unknown. Based on the current state of knowledge, the diversity of clinical phenotypes of dystonia is thought to be associated with multifactorial pathophysiology, which is linked not only to alterations of brain organization, but also environmental stressors and gene mutations (Fig. 1). This chapter will present an overview of the pathophysiology of isolated dystonia through the lens of applications of major neuroimaging methodologies, with links to genetics and environmental factors that play a prominent role in symptom manifestation.

1. MULTI-MODAL NEUROIMAGING APPLICATION IN DYSTONIA

Rapid advances in brain imaging have had a tremendous impact on the evolution of our understanding of dystonia development. Narrowly considered for decades as a basal ganglia disorder due to the predilection for striatal lesions to trigger secondary (or combined) dystonias (Marsden, Obeso, Zarranz, & Lang, 1985; Vitek, 2002), neuroimaging studies addressing different aspects of brain organization have been transformative in changing our views about the complex pathophysiological mechanisms underlying this disorder and opening new perspectives for its objective diagnosis and therapeutic interventions. The early neuroimaging studies were heavily rooted in the use of positron emission tomography that captured brain glucose metabolism and cerebral blood flow alterations in dystonia patients compared to healthy subjects. Further advances in functional brain imaging

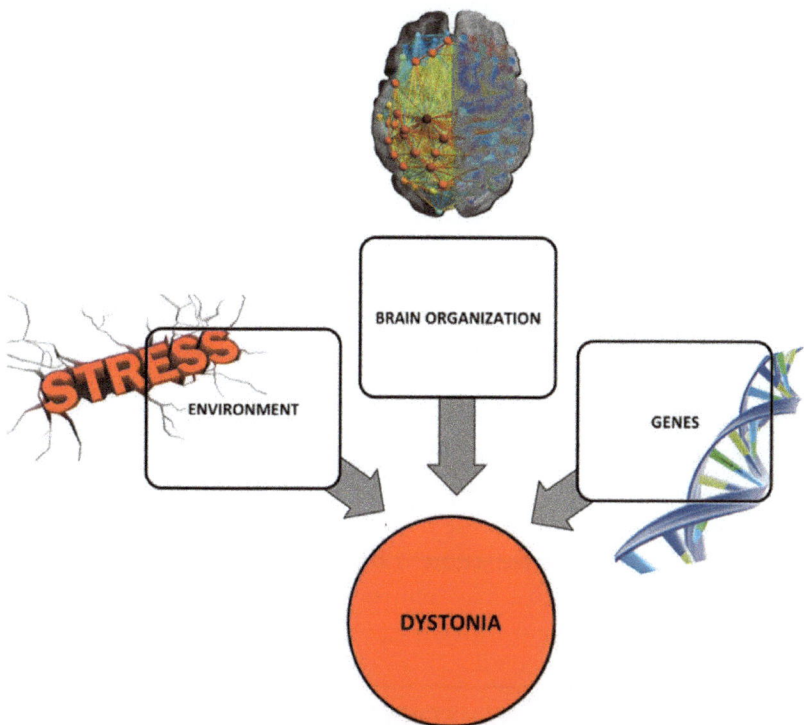

Fig. 1 Schematic representation of multifactorial pathophysiology of isolated dystonia with major factors including abnormal brain structural, functional and neurochemical organization, underlying gene mutations, and environmental triggers, a combination of which leads to the manifestation of dystonic symptoms.

using functional magnetic resonance imaging (MRI) and magnetoencephalography as well as structural imaging using diffusion weighted and high-resolution MR sequences provided deeper insights into brain regional alterations at both structural and functional levels. Finally, relatively recent applications of mathematical modeling of larger-scale brain networks revealed global neural abnormalities that were not recognized in initial studies, reinforcing the notion that isolated dystonia is a network disorder. In parallel, neuroreceptor mapping studies with positron emission tomography and magnetic resonance spectroscopy helped define the aberrations of dopaminergic and GABAergic neurotransmission across different forms of dystonia as one of the potential underlying causes of altered network function.

Recent studies have also attempted to link and understand the impact of gene mutations on the development of brain abnormalities in dystonia.

Starting with the first discovery of DYT1 (*TOR1A*) gene for early-onset generalized dystonia in 1997 (Ozelius et al., 1997), several studies have been designed to elucidate distinct brain disorganization in manifesting and non-manifesting mutation carriers. Further on, rapid advances in next-generation whole-exome sequencing led to identification of additional dystonia genes, including DYT6 (*THAP 1*), DYT24 (*ANO3*), DYT25 (*GNAL*), DYT4 (*TUBB4*), and DYT23 (*CIZ1*), with the last two still pending an independent confirmation (Klein, 2014). These mutations are known to predominantly cause segmental dystonias that affect adjacent body regions. Again, neuroimaging studies revealed important associations between some of these genetic factors and neural aberrancies.

Arguably, the least explored aspect in the pathophysiology of dystonia is the role of environmental variables, such as extended sun exposure for blepharospasm, neck trauma and surgery in cervical dystonia, viral infections in laryngeal dystonia, and overuse practice in focal hand dystonia and musician's dystonia. The current effort is directed to elucidating the possible environmental stressors that may influence the development of brain alterations and by this trigger the manifestation of dystonic symptoms.

2. STRUCTURAL NEUROIMAGING OF DYSTONIA

2.1 Gray Matter Structural Organization

The absence of gross structural abnormalities on conventional MRI is one of the clinical hallmarks of dystonia. In fact, it is often a criterion confirming the differential diagnosis of dystonia. On the other hand, detailed evaluation of high-resolution T1-weighted and diffusion-weighted MR sequences in numerous neuroimaging studies led to identification of *microstructural* cortical and subcortical alterations across all forms of isolated dystonia.

One of the first studies dates back to late 1990s, which applied MRI-based stereological volumetry to reveal a 10% increase of putaminal volume in patients with blepharospasm and focal hand dystonia (Black, Ongur, & Perlmutter, 1998). With the focused investigation on the striatum, this finding was contemporary and supportive of prevailing knowledge at the time that dystonia is a basal ganglia disorder due to intrinsic structural changes. However, further development of methodologies, such as voxel-based morphometry and cortical thickness analyses, allowed for evaluation of volumetric changes within and/or between patient and control groups based on statistical parametric mapping. Analytic pipelines for both methodologies use a refined and fully automated approach of segmentation of high-resolution T1-weighted brain images into gray matter, white matter, and cerebrospinal fluid. Voxel-based morphometry measures the quantity of tissue within a voxel and is dependent on local cortical thickness and/or cortical surface area. In contrast, cortical thickness measurements are based on determining the inner and outer cortical boundaries or surfaces and have a high sensitivity in detecting cortical changes, especially those prone to gene mutations. Both methodologies are complimentary, with voxel-based morphometry providing information about the volumetric organization of the entire brain and cortical thickness measurements being more limited to the cortical ribbon.

The use of these analytic methods made it apparent that putaminal volumetric changes are indeed a common feature of gray matter alterations across different forms of dystonia, including both focal and generalized phenotypes (Draganski et al., 2009; Draganski, Thun-Hohenstein, Bogdahn, Winkler, & May 2003; Etgen, Muhlau, Gaser, & Sander, 2006; Granert, Peller, Jabusch, Altenmuller, & Siebner, 2011; Obermann et al., 2007; Pantano et al., 2011; Simonyan & Ludlow, 2012) (Fig. 2A). In addition, these studies revealed that gray matter changes in focal dystonias further extend to the other divisions of the basal ganglia, most commonly including caudate nucleus and globus pallidus, as well as thalamus and cerebellum (Delmaire et al., 2005; Draganski et al., 2003; Egger et al., 2007; Filip et al., 2017; Obermann et al., 2007; Pantano et al., 2011; Simonyan & Ludlow, 2012; Waugh et al., 2016; Zeuner et al., 2015). More importantly, structural alterations were shown to involve some of the key cortical regions, responsible for sensorimotor processing, integration and motor execution, such as frontoparietal, supplementary motor and primary sensorimotor areas (Delmaire et al., 2005; Draganski et al., 2003; Egger et al., 2007; Garraux et al., 2004; Granert, Peller, Jabusch, et al., 2011; Horovitz, Gallea, Najee-Ullah, & Hallett, 2013; Martino et al., 2011; Pantano et al., 2011; Ramdhani et al., 2014; Simonyan & Ludlow, 2012). Among these, primary sensorimotor changes tended to localize to dystonia-affected body regions within the sensorimotor homunculus. For example, gray matter volumetric increases in patients with focal hand dystonia were observed in the hand area of sensorimotor cortex (Delmaire et al., 2005; Egger et al., 2007; Garraux et al., 2004), whereas in patients with laryngeal dystonia these were found within the larynx representation (Bianchi et al., 2017; Kostic et al., 2016; Simonyan & Ludlow, 2012), further showing relevance of cortical aberrations to the pathophysiology of distinct forms of dystonia.

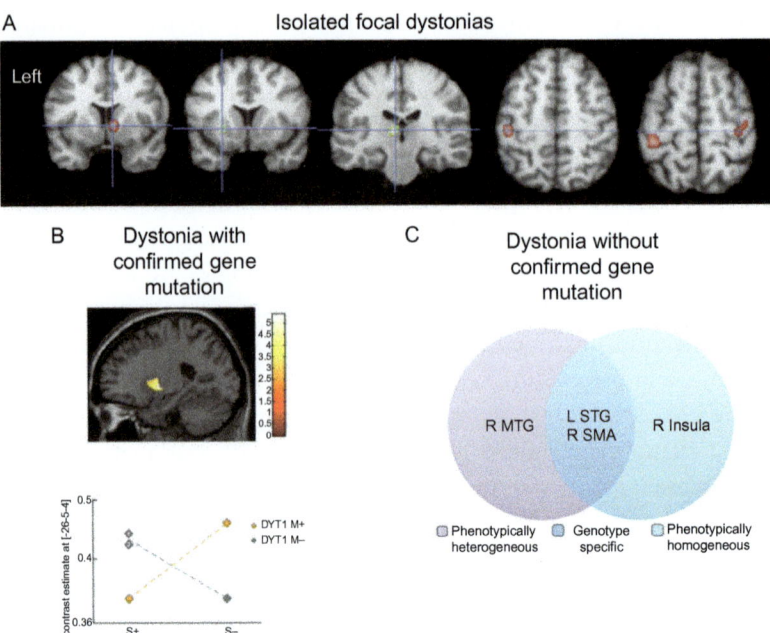

Fig. 2 Microstructural gray matter abnormalities in isolated dystonia. (A) Anatomic likelihood estimation meta-analysis of common gray matter volumetric abnormalities across focal dystonias. (B) Gray matter volumetric increases in bilateral putamen in non-manifesting DYT1 mutation carriers and non-DYT1 dystonia patients compared to manifesting DYT1 mutation carriers and healthy subjects. The plot shows crossover interaction between genotype and phenotype factors. DYT M+, mutation-positive dystonia; DYT M−, mutation negative dystonia. (C) Cortical thickness changes across different phenotypes and putative genotypes (without confirmed gene mutation but based on a family history of dystonia) in patients with laryngeal dystonia. MTG, middle temporal gyrus; STG, superior temporal gyrus; SMA, supplementary motor area; R, right; L, left. *Panel (A): Adapted from Zheng, Z., Pan, P., Wang, W., & Shang, H. (2012). Neural network of primary focal dystonia by an anatomic likelihood estimation meta-analysis of gray matter abnormalities.* Journal of the Neurological Sciences, 316(1–2), 51–55. https://doi.org/10.1016/j.jns.2012.01.032. *Panel (B): Adapted from Draganski, B., Schneider, S. A., Fiorio, M., Kloppel, S., Gambarin, M., Tinazzi, M., et al. (2009). Genotype-phenotype interactions in primary dystonias revealed by differential changes in brain structure.* NeuroImage, 47(4), 1141–1147. doi:10.1016/j.neuroimage.2009.03.057. *Panel (C): Adapted from Bianchi, S., Battistella, G., Huddlestone, H., Scharf, R., Fleysher, L., Rumbach, A. F., et al. (2017). Phenotype- and genotype-specific structural alterations in spasmodic dysphonia.* Movement Disorders, 32(4), 560–568. https://doi.org/10.1002/mds.26920.

In contrast to this body of literature in focal dystonias, examinations of gray matter organization in isolated dystonias with underlying gene mutations (hereditary dystonias) are scarce. There are only a few reports showing genotype–phenotype interactions at the level of structural alterations.

One study revealed that non-manifesting DYT1 mutation carriers and non-DYT1 dystonia patients have significantly increased bilateral putaminal volume compared to manifesting DYT1 mutation carriers and healthy subjects (Draganski et al., 2009) (Fig. 2B). Another study reported a volumetric increase of globus pallidus in patients with generalized dystonia, although the DYT1 status was not known in this cohort (Egger et al., 2007). As gene discovery for isolated focal dystonias remains stagnant and extremely challenging due, in part, to rare availability of large families, low penetrance, and variable expressivity, the design of neuroimaging studies has also been limited by somewhat arbitrary assignments of experimental cohorts based on a confirmed family history of dystonia only. Based on such approach, one recent study in patients with sporadic and familial forms of laryngeal dystonia identified phenotype-specific gray matter abnormalities in primary and associative areas of motor control that were distinct from genotype-specific changes within the cortical regions controlling sensory processing (Bianchi et al., 2017) (Fig. 2C).

Along with the progress in mapping and characterizing gray matter organization in dystonia, questions pertaining to the exact cellular mechanisms and causes of gray matter alterations as revealed by voxel-based morphometry and cortical thickness measurements remain poorly understood. It is possible that gray matter volumetric increases in some forms of dystonia, especially in musician's dystonia and other task-specific dystonias, may be due to abnormal motor training and overuse that generally lead to volumetric increases in related brain regions in healthy populations (Ceccarelli et al., 2009; Draganski et al., 2004; Granert, Peller, Gaser, et al., 2011; Quallo et al., 2009). Another possible mechanisms may involve the formation of abnormal new connections by dendritic spine growth and axonal remodeling, and/or the strengthening of existing synaptic connections (Chklovskii, 2004; Chklovskii, Mel, & Svoboda, 2004; Holtmaat, Wilbrecht, Knott, Welker, & Svoboda, 2006; Sur & Rubenstein, 2005). Future ultra-high-resolution imaging of gray matter, ideally coupled with postmortem neuropathology, may be able to reveal which of these causative mechanisms are responsible for observed neuroimaging changes across dystonias.

2.2 White Matter Structural Organization

Most of the knowledge about white matter organization in isolated dystonia comes from the use of diffusion tensor imaging, which is a form of diffusion-weighted imaging that examines axonal organization in brain tissue. It is based on the principal of random displacement of water molecules and differential quantification of their diffusivity in parallel and perpendicular directions along the axonal course. Diffusion imaging measures include anisotropy indices (e.g., fractional anisotropy) to quantify the directionality of water diffusivity, reflecting axonal integrity and tissue coherence. Other diffusivity indices (e.g., mean diffusivity, trace of diffusion tensor) examine the magnitude of water movement independent of the direction and provide information about the organization of the extracellular space and the intracellular water content. Similar to voxel-based morphometry and cortical thickness measurements, diffusivity measures are sensitive to microstructural white matter abnormalities not evident on conventional MRI.

In addition to the assessment of regional white matter organization, diffusion imaging methods allow for in vivo reconstruction of the white matter tracts using deterministic and probabilistic tractography. This is considered as a valid method for examination of white matter connectivity due to its capability to model fiber direction(s) per voxel (single with deterministic tractography and multiple with probabilistic tractography), providing realistic estimates of the wiring strength of gross white matter projections. However, tractography is still limited for mapping the connectivity between brain regions with high fiber complexity and uncertainty about the fiber directions, especially in the presence of dense local cortico–cortical connections. The likelihood of tractography is excessively higher in estimation of the tracts from the regions of interest to immediately surrounding areas than those in more distant brain regions, leading to the distance bias. Similarly, the tractography algorithm is prone to higher propagation of tracts to larger targets and even to generating false-positive tracts. Moreover, tractography by-passes monosynaptic connections and does not allow inferring directionality of the connection, which makes it difficult to assess the structural influence of one brain region upon another. While a powerful analytic method for in vivo examination of structural connectivity in humans, the results of diffusion tractography warrant a careful comparison to those obtained with neuroanatomical tract tracing methods in non-human primates, most importantly to avoid identification of spurious "novel" connectivity and its impairments in diseased populations.

Studies assessing regional white matter organization in isolated dystonia have been mostly focused on the use of various diffusivity measures between and within different groups of patients and healthy subjects. The general unifying finding across these studies is identification of reduced axonal integrity and increased water diffusivity, involving both cortical and sub-cortical structures along the cortico-striato-pallido-thalamic and cerebello-thalamo-cortical pathways (Blood et al., 2012; Bonilha et al., 2007; Colosimo et al., 2005; Delmaire et al., 2009; Fabbrini et al., 2008; Horovitz, Ford, Najee-Ullah, Ostuni, & Hallett, 2012; Prell et al., 2013; Simonyan et al., 2008) (Fig. 3A). Specificity of these alterations is relevant to the distinct forms of dystonia, such as reduced white matter integrity was reported in the posterior limb of internal capsule, including the corticospinal tract, in focal hand dystonia (Delmaire et al., 2009) and in the genu of internal capsule, involving

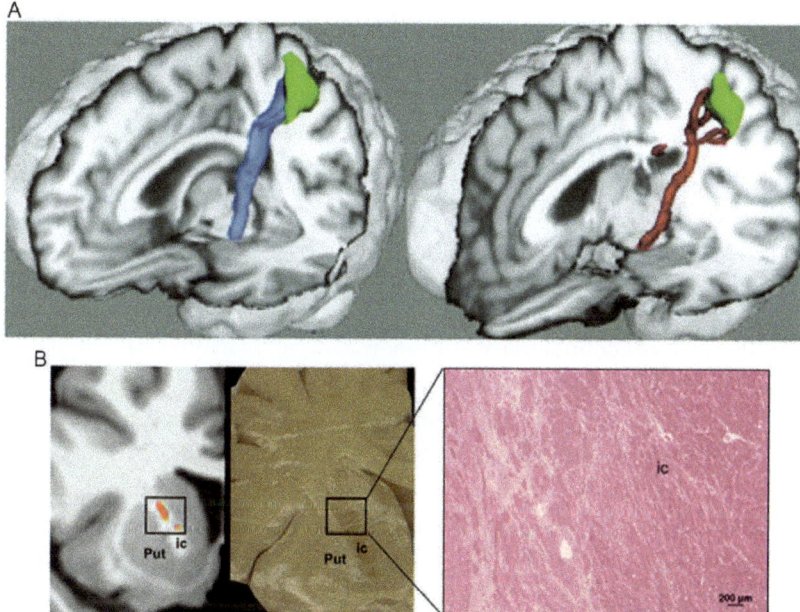

Fig. 3 (A) Decreased gray matter volume in the facial portion of the precentral gyrus and reduced volume of the corticobulbar tract in healthy subjects (left) and patients with blepharospasm (right). (B) Specificity of white matter changes depending on the phenotype of dystonia: reduced white matter integrity along the corticobulbar tract in laryngeal dystonia, with subsequent postmortem neuropathology confirming axonal degeneration and demyelination within the corticobulbar tract in laryngeal dystonia. Put, putamen; ic, internal capsule. *Panel (A): Adapted from Horovitz, S. G., Ford, A., Najee-Ullah, M. A., Ostuni, J. L., & Hallett, M. (2012). Anatomical correlates of blepharospasm. Translational Neurodegeneration, 1(1), 12. https://doi.org/10.1186/2047-9158-1-12. Panel (B): Adapted from Simonyan, K., Tovar-Moll, F., Ostuni, J., Hallett, M., Kalasınsky, V. F., Lewin-Smith, M. R., et al. (2008). Focal white matter changes in spasmodic dyspho-nia: A combined diffusion tensor imaging and neuropathological study. Brain, 131(Pt. 2), 447–459. https://doi.org/10.1093/brain/awm303.*

corticobulbar tract, in laryngeal dystonia (Simonyan et al., 2008). The latter finding has been substantiated by neuropathological evidence of decreased axonal density and myelin content, potentially underlying the change in water diffusivity (Simonyan et al., 2008) (Fig. 3B).

Among the hereditary forms of dystonia, both manifesting and non-manifesting DYT1 carriers showed reduced white matter integrity in subgyral white matter of sensorimotor cortex, albeit with greater abnormalities in manifesting than non-manifesting carriers (Carbon, Kingsley, et al., 2004; Carbon, Kingsley, Tang, Bressman, & Eidelberg, 2008). A similar finding was observed in the study of manifesting DYT6 carriers (Cheng et al., 2012). In addition, DYT1 carriers showed abnormal white matter organization of dorsal pontine tegmentum adjacent to the left superior cerebellar peduncle (Carbon et al., 2008). Diffusion tractography, examining the cerebellar involvement, identified reduced integrity in the cerebellar lobule VI adjacent to dentate nucleus in DYT1 and DYT6 manifesting carriers as well as reduced thalamo-cortical connectivity in non-manifesting dystonia gene carriers (Argyelan et al., 2009). On the other hand, substantial phenotypic variability of hereditary dystonia appears to be linked to reduced integrity of somatotopic projections related to asymptomatic body regions (Vo, Sako, Niethammer, et al., 2015). Similarly, non-manifesting carriers showed additionally reduced connectivity within the distal thalamo-cortical segment of the cerebello-thalamo-cortical tract, pointing to the protective neural mechanisms underlying the manifestation of hereditary dystonia. Corroborating these findings, an ex vivo study in the heterozygous DYT1 knock-in mouse model showed reduced connectivity of the cerebello-thalamic, thalamo-cortical and thalamo-striatal pathways in mutants compared to wild type (Ulug et al., 2011). Another study reported deficits of free water diffusivity and widespread increases in functional connectivity of the stratum with somatosensory cortex, thalamus, cerebellum and brainstem in the conditional knock-out mice of the DYT1 protein torsinA (DeSimone et al., 2017). As the cerebello-thalamo-cortical pathway facilitates intracortical inhibition via projections to interneurons in the sensorimotor cortex (Molinari, Filippini, & Leggio, 2002), these findings may suggest the presence of the intermediary pattern in non-manifesting carriers that may act as a buffer against the aberrant outflow from the proximal part of this pathway (Niethammer, Carbon, Argyelan, & Eidelberg, 2011).

Taken together, the use of volumetric and diffusion imaging helped establish the presence of microstructural alterations of gray and white matter in dystonia that are not restricted to the putamen but rather extend to surrounding subcortical and further cortical and cerebellar regions, predominantly within the basal ganglia-thalamo-cortical and cerebello-thalamo-cortical circuitries. Identification of structural alterations specific to the disorder genotype and phenotype has been important for providing a critical step toward future delineation of imaging markers and the potential targets for novel therapeutic interventions in this disorder.

3. FUNCTIONAL NEUROIMAGING OF DYSTONIA

3.1 Positron Emission Tomography

Early studies in dystonia relied heavily on the use of positron emission tomography to capture the cerebral blood flow and glucose metabolism in patient within-group and between-group comparisons with healthy subjects. Positron emission tomography is based on radioactivity emitted after a radiolabeled tracer is injected intravenously. The nucleus of the radioisotope emits a positron, which collides with an electron in the tissue and in the process converts mass to energy in the form of two photons. The tomograph uses scintillation crystals placed around the subject's head to detect these photons. The crystals absorb the photons, producing light that is converted into an electrical signal.

One of the most commonly used tracers to quantify brain activity via glucose metabolism in dystonia is $[^{18}F]$-fluorodeoxyglucose ($[^{18}F]FDG$), which is taken up by brain regions depending on metabolic needs. As regional metabolism is dependent on synaptic activity, changes in tracer uptake correlate with changes in regional neuronal activity. Using $[^{18}F]FDG$, several studies reported abnormal metabolism in basal ganglia, thalamus, cerebellum, and sensorimotor cortex in patients with focal dystonia (Esmaeli-Gutstein, Nahmias, Thompson, Kazdan, & Harvey, 1999; Hutchinson et al., 2000; Kerrison et al., 2003; Magyar-Lehmann et al., 1997; Suzuki et al., 2007). During induced sleep in patients with blepharospasm, hypometabolic activity was additionally found in the frontal eye field (Brodmann area 8), which is involved in planning of complex movements and is associated with supranuclear control of the eyelid opening (Hutchinson et al., 2000). Evaluation of the effectiveness of botulinum toxin treatment in blepharospasm showed an association with normalization of metabolic activity in the cerebellum and pons (Suzuki et al., 2007).

Studies in hereditary dystonias also identified distinct patterns of regional metabolic activity. Specifically, DYT1 carriers showed metabolic increases in the striatum and cerebellum; DYT6 carriers had metabolic reductions in the putamen, thalamus, cerebellar cortex, and upper brainstem; and DYT11 mutation carriers had both metabolic increases in the deep cerebellar nuclei and vermis and metabolic decreases in the medial prefrontal cortex and pre-supplementary motor area (Carbon & Eidelberg, 2009; Carbon, Su, et al., 2004). The findings in DYT1 patients were substantiated in the *Tor1a* het-erozygous knock-out mouse, which showed metabolic abnormalities in the striatum and cerebellum as well as sensorimotor cortex and subthalamic nucleus (Vo, Sako, Dewey, Eidelberg, & Ulug, 2015).

Another commonly used radiotracer in dystonia research is $[^{15}O]$ H_2O, which allowed the assessment of the regional cerebral blood flow during production of both symptomatic and asymptomatic tasks while in the scanner (Ali et al., 2006; Ceballos-Baumann et al., 1995; Ibanez, Sadato, Karp, Deiber, & Hallett, 1999; Lerner et al., 2004; Playford, Passingham, Marsden, & Brooks, 1998). Similar to glucose metabolism, the regional cerebral blood flow appeared to be abnormal in the basal ganglia, thalamus, cerebellum, and sensorimotor cortex in focal dystonia. In addition, wider spread abnormalities were identified involving prefrontal, posterior parietal and temporal regions necessary for motor planning and sensorimotor processing. Examination of the regional covariance pattern during writing and at rest in patients with writer's cramp further revealed reduced correla-tions between putamen and premotor cortex and between bilateral premotor cortex (Ibanez et al., 1999), suggesting that not only regional brain activity but also functional connectivity may be abnormal in dystonia. Defi-cient activity and connectivity of the preparatory cortical regions and basal ganglia pointed to a loss of inhibition during the generation of motor com-mands, likely arising from the striatal region.

3.2 Functional MRI During Task Production and Resting State

The development of functional MRI methodology in early 1990s (Kwong et al., 1992; Ogawa, Lee, Nayak, & Glynn, 1990) has completely changed the landscape of neuroimaging studies examining brain activity and func-tional connectivity in dystonia. Functional MRI is based on a blood oxygen-level dependent contrast, which captures the hemodynamic response by relying on differential magnetic susceptibility of oxyhemo-globin and deoxyhemoglobin due to changes during neuronal activity. Functional MRI has been used during various tasks, both eliciting and not eliciting dystonic movements. Similar to positron emission tomogra-phy and structural imaging studies, this line of research again confirmed the presence of alterations in basal ganglia, thalamus, cerebellum and sensorimotor cortex across different forms of isolated dystonia and hinted to abnormal integration of sensorimotor information flow within the basal ganglia-thalamo-cortical and cerebello-thalamo-cortical circuitries (e.g. Baker, Andersen, Morecraft, & Smith, 2003; Burciu et al., 2017; Butterworth et al., 2003; Haslinger, Altenmuller, Castrop, Zimmer, & Dresel, 2010; Hu, Wang, Liu, & Zhang, 2006; Islam et al., 2009; Kadota et al., 2010; Pujol et al., 2000; Simonyan & Ludlow, 2010). Additionally, some studies pointed to abnormal sensory processing by primary somatosensory cortex that may contribute to the pathophysiology of dystonia (Dresel, Haslinger, Castrop, Wohlschlaeger, & Ceballos-Baumann, 2006; Haslinger et al., 2010; Simonyan & Ludlow, 2010), whereas others mapped abnormal somatotopy of digit representation in primary somatosensory cortex and putamen in focal hand dystonia (Butterworth et al., 2003; Delmaire et al., 2005; Nelson, Blake, & Chen, 2009). Studies examining the effects of botulinum toxin injections on brain activity in patients with focal dystonia provided largely controversial results, showing either modulated or non-modulated brain activity following the treatment (Ali et al., 2006; Dresel et al., 2011; Haslinger et al., 2005; Nevrly et al., 2018). On the other hand, subclinical abnormalities in sensory discrimination were described as a mediational endophenotype of dystonia (Hutchinson et al., 2013) and linked to alterations in primary somatosensory and middle frontal cortices (Termsarasab et al., 2016). Importantly, differential associations between abnormal sensory discrimination and functional abnormalities were observed depending on the phenotype and genotype of dystonia, including greater cerebellar involvement in familial laryngeal dystonia cases and greater putaminal and cortical sensorimotor inclusion in different phenotypes of laryngeal dystonia (Termsarasab et al., 2016).

Another step forward in understanding the neuroimaging pathophysiology of dystonia came with the development of resting-state functional MRI, which relies on the measurement of low frequency physiological fluctuations in the BOLD signal, reflecting the functional brain organization during various activation states (Biswal, Yetkin, Haughton, & Hyde, 1995; Smith et al., 2009). The resting-state functional MRI approach proved to be useful in circumventing the challenges associated with the implementation of the task-related designs across different forms of dystonia, which exhibit distinct symptoms, thus making the choice of a single, commonly affected task production unfeasible. Furthermore, because the explanation of functional changes across distinct dystonia phenotypes and genotypes may sometimes be ambiguous due to a combination of motor and sensory components, examination of the resting-state activity and connectivity provides a more uniform and coherent understanding of neural alterations.

Resting-state functional MRI studies in focal hand dystonia found decreased connectivity in primary somatosensory region with concomitant increases in the putamen as well as functional decoupling of dorsal premotor cortex from parietal cortex (Delnooz, Helmich, Toni, & van de Warrenburg, 2012; Mohammadi et al., 2012). In cervical dystonia, alterations were found not only in sensorimotor but also in visual and executive control networks (Delnooz, Pasman, Beckmann, & van de Warrenburg, 2013), whereas changes in blepharospasm were related to the default-mode network (Yang et al., 2013). It was further shown that vulnerable connectivity of primary sensorimotor and inferior parietal cortices in laryngeal dystonia is tightly associated with the polygenic risk of dystonia, likely representing an endophenotypic imaging marker of this disorder; genes contributing to the polygenic score are involved in synaptic transmission and neuron development (Battistella, Fuertinger, Fleysher, Ozelius, & Simonyan, 2016; Putzel et al., 2018). Schematic knowledge of functional alterations in dystonia can be viewed in Fig. 4.

Using these findings, a few studies started to probe existing machine learning algorithms in attempt to characterize neural markers of dystonia. Classification algorithms represent a powerful tool for identification of single

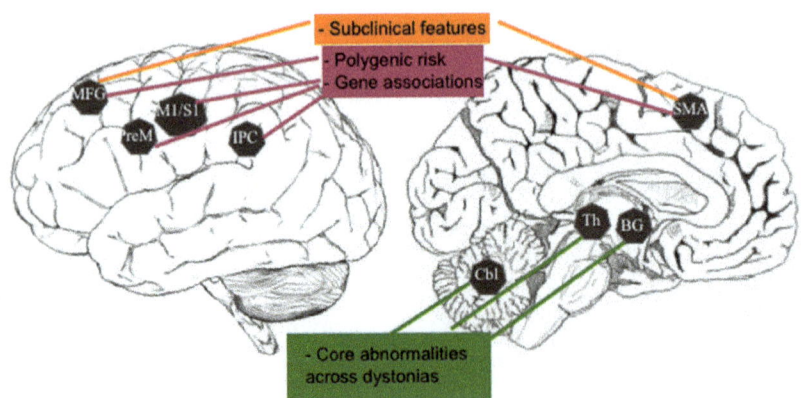

Fig. 4 Schematic representation of common abnormalities of brain metabolism and activation across different forms of dystonia and their relevance to the pathophysiology of this disorder. MGF, middle frontal gyrus; PreM, premotor cortex; M1/S1, primary sensorimotor cortex; IPC, inferior parietal cortex; SMA, supplementary motor area; BG, basal ganglia; Th, thalamus; Cbl, cerebellum.

traits or a combination of features that characterize and separate two or more classes of objects or subjects. Algorithmic classifiers using functional MRI voxel-wise time series have been successfully applied in several neurodegenerative disorders (Fornari, Maeder, Meuli, Ghika, & Knyazeva, 2012; Janousova, Schwarz, & Kasparek, 2015; Yourganov et al., 2014). The first study in dystonia used multivariate classification algorithm of linear discriminant analysis (LDA) based on the measures of abnormal functional resting-state connectivity in primary sensorimotor, premotor and inferior parietal regions, achieving 71% accuracy in classifying laryngeal dystonia and healthy controls (Battistella et al., 2016). It further improved its accuracy in classifying familial vs sporadic laryngeal patients at 81% and remained at the same 71% accuracy level when considering different (adductor and abductor) phenotypes. As such, this study used neuroimaging data to disambiguate focal dystonia from a normal state and differentiate the disorder based not on clinical evaluations of its symptoms but of neuroimaging heterogeneity, opening new avenues to the development of disorder-specific diagnostic biomarkers.

3.3 Dystonia as a Functional Network Disorder

Starting with the positron emission tomography studies in generalized dystonia that showed a presence of the abnormal metabolic network (Eidelberg et al., 1998, 1995; Niethammer et al., 2011), the overall concept of functional alterations not being limited to the basal ganglia, as historically proposed, has evolved into propositions of focal dystonia, too, to represent a functional network disorder (Lehericy, Tijssen, Vidailhet, Kaji, & Meunier, 2013; Neychev, Gross, Lehericy, Hess, & Jinnah, 2011; Quartarone & Hallett, 2013; Ramdhani & Simonyan, 2013; Zoons, Booij, Nederveen, Dijk, & Tijssen, 2011). This concept has been recently experimentally substantiated in a study that used a graph theoretical approach to probe the organization of large-scale functional networks across different forms of focal dystonia. Compared to healthy subjects, patients showed altered network architecture, which was characterized by an abnormal breakdown of the basal ganglia-thalamo-cerebellar community, a loss of pivotal regions of information transfer (hubs) in the premotor cortex, and a pronounced decline in sensorimotor and inferior parietal cortical connectivity (Battistella, Termsarasab, Ramdhani, Fuertinger, & Simonyan, 2017) (Fig. 5A). These findings pointed to the unified pathophysiological mechanism underlying different forms of dystonia due to common network alterations, while suggesting the concurrent presence of

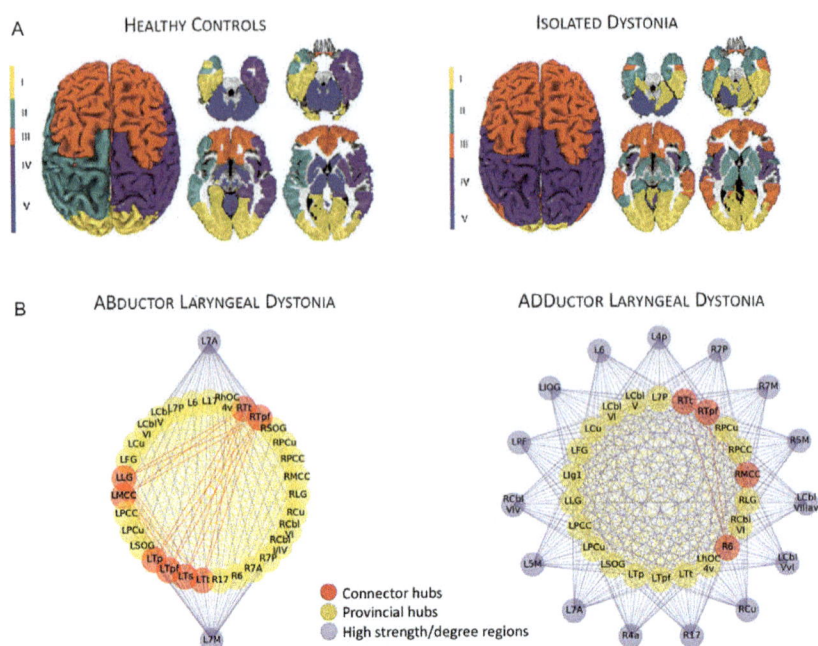

Fig. 5 (A) Large-scale network organization based on resting-state functional MRI group-averaged networks in healthy subjects and patients with different forms of focal isolated dystonia. This panel shows the regional distribution of neural communities based on the inter-regional coupling. Patients show disintegration of neural communities compared to healthy subjects. (B) Functional architecture of the neural network is influenced by the phenotype of dystonia; an example in abductor vs adductor forms of laryngeal dystonia. The panel depicts phenotype-specific abnormal distribution of connector and provincial hubs (regions of highest information transfer), which, as a result, establish a characteristic pattern of connectivity with non-hub regions across the entire brain. 6/17, area 6/17; 7A/7P, subdivisions of area 7; Cbl-I/IV/Cbl-V/Cbl-VI, cerebellar lobules I/IV/V/VI; Cu/PCu, cuneus/precuneus; FG, fusiform gyrus; Ig1, part Ig1 of the insula; LG, lingual gyrus; MCC/PCC, middle/posterior cingulate cortex; SOG, superior orbital gyrus; Tp/Tpf/Ts/Tt, parietal/prefrontal/somatosensory/temporal subdivisions of the thalamus; hOC4v, ventral part of area hOC4; L, left; R, right. *Panel (A): Adapted from Battistella, G., Termsarasab, P., Ramdhani, R. A., Fuertinger, S., & Simonyan, K. (2017). Isolated focal dystonia as a disorder of large-scale functional networks.* Cerebral Cortex, 27(2), 1203–1215. doi:10.1093/cercor/bhv313. *Panel (B): Adapted from Fuertinger, S., & Simonyan, K. (2017). Connectome-wide phenotypical and genotypical associations in focal dystonia.* The Journal of Neuroscience, 37(31), 7438–7449. https://doi.org/10.1523/JNEUROSCI.0384-17.2017.

pathophysiologically divergent mechanisms contributing to different forms of dystonia. In line with this assumption, another study using a similar analytic approach mapped marked differences in the topological organization of parietal regions between phenotypically different forms of laryngeal dystonia

(Fuertinger & Simonyan, 2017) (Fig. 5B). Moreover, the interface between sporadic genotype and most common adductor phenotype yielded distinct functional communities of interacting regions that were primarily governed by intramodular hub regions. On the other hand, the interface between less common familial genotype and abductor phenotype was associated with numerous long-ranging hub regions and an abnormal integration of left thalamus and basal ganglia.

Another aspect of network abnormality is how one region exerts its influences upon another. Although the current analytic methodology, dynamic causal modeling, has severe limitations such as the total number of regions to be realistically explored cannot exceed 4 or 5, rendering examination of whole brain connectivity not feasible, it has nevertheless been useful in showing malfunctioning intracortical connections between primary motor cortex and supplementary motor area as well as abnormal reciprocal excitatory connectivity in the cortico-cerebellar circuitry during performance of a motor task in patients with writer's cramp (Rothkirch et al., 2018). Future studies should focus on the development of new analytic tools to assess the mechanistic properties of abnormal functional network in dystonia by examining its effective connectivity.

4. NEURORECEPTOR MAPPING STUDIES IN DYSTONIA
4.1 Dopaminergic Neurotransmission

Despite the basal ganglia being at the epicenter of dystonia pathophysiology, until recently the neurochemical underpinning of these abnormalities remained poorly understood. The basal ganglia set the pattern for facilitation of voluntary movements and simultaneous inhibition of competing/interfering movements by balancing excitation and inhibition within the thalamo-cortical circuitry. This is achieved by a synergistic action of the net excitatory direct basal ganglia pathway, which predominantly expresses dopamine D_1 family receptors, and the net inhibitory basal ganglia pathway, which predominantly expresses dopamine D_2 family receptors (Gerfen, 1992, 2000; Surmeier, Yan, & Song, 1998). Endogenously released striatal dopamine influences direct and indirect pathways both separately and via bridging collaterals between the two pathways, allowing dynamic modulation of thalamo-cortical neurons for physiologically normal facilitation of movement initiated in the motor cortex (Fig. 6A).

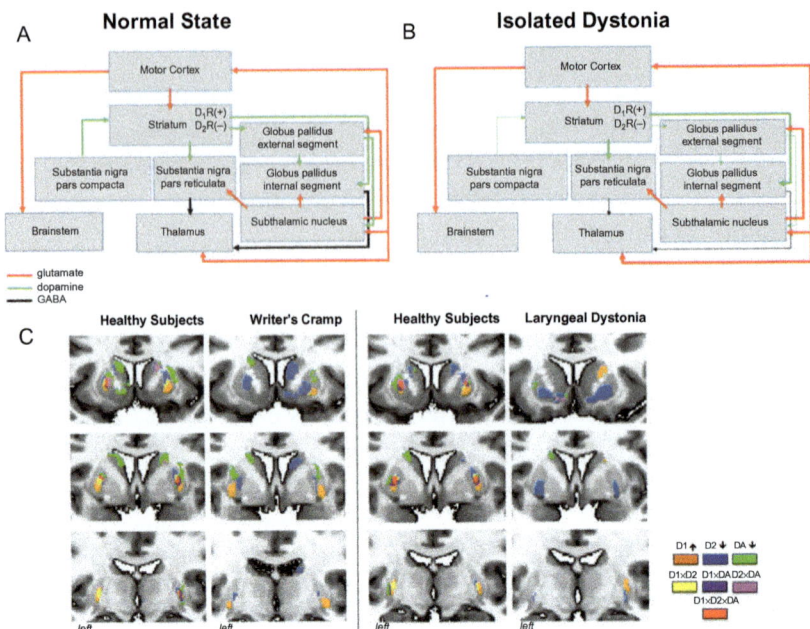

Fig. 6 Schematic representation of neurotransmission during (A) normal state and (B) in isolated dystonia. Striatal dopaminergic input from the substantia nigra, pars compacta, is weakened; dopaminergic neurotransmission is enhanced via the direct pathway and diminished via the indirect pathway. (C) Topological distribution of striatal dopaminergic function in healthy subjects and patients with writer's cramp and laryngeal dystonia. Within each patient and control group, a conjunction analysis was used to examine the overlap and distinct distribution between the significant clusters derived from three measures of dopaminergic function: D_1 receptor binding; D_2 receptor binding, and striatal phasic dopamine release during finger tapping (for the comparison with writer's cramp) and sentence production (for the comparison with laryngeal dystonia). Healthy subjects have a great degree of overlap between all three measures as well as smaller regions of distinct receptor distribution. This topology is reversed in patients with dystonia, where the overlap is largely diminished. The legend provides the color scheme for overlapping as well as distinct regions of receptor activation and dopamine release. DA, dopamine.

It has been suggested that abnormal dopamine levels may modulate striatal synaptic plasticity in dystonia (Breakefield et al., 2008; Hallett, 2004; Todd & Perlmutter, 1998), while increased D_1-mediated excitation and decreased D_2 receptor-mediated inhibition may alter the balance between the basal ganglia direct and indirect pathways, cause overall disinhibition of the thalamocortical circuitry, and contribute to dystonia muscle contractions during performance of fine motor tasks (Hallett, 1993).

Using PET or single-photon emission computed tomography with specialized radioligands to target striatal dopamine receptors, decreased striatal dopamine D_2 receptor binding at rest was found in patients with both focal and generalized forms of dystonia as well as non-manifesting DYT1 and DYT6 gene carriers (Asanuma et al., 2005; Berman, Hallett, Herscovitch, & Simonyan, 2013; Berger et al., 2007; Carbon et al., 2009; Horie et al., 2009; Horstink et al., 1997; Naumann et al., 1998; Perlmutter et al., 1997; Simonyan, Berman, Herscovitch, & Hallett, 2013). Among these, two studies leveraged the ability of endogenously released phasic dopamine to displace the bound radiotracer in order to assess dopaminergic function during symptomatic and asymptomatic tasks (Berman et al., 2013; Simonyan et al., 2013). It was reported that patients with focal hand dystonia and laryngeal dystonia exhibit abnormally decreased striatal phasic dopamine release during symptomatic task production (finger tapping and speaking, respectively), whereas the levels of striatal dopamine release during asymptomatic tasks (speaking and finger tapping, respectively) are abnormally increased. While there is no apparent neurodegeneration or cell loss within the basal ganglia, experimentally reduced striatal D_2 receptor binding was also observed in the dt^{sz} mutant hamster and associated with increased striatal dopamine release during the manifestation of dystonic episodes (Hamann & Richter, 2004; Nobrega, Richter, Tozman, Jiwa, & Loscher, 1996). Decreased neuroreceptor binding at rest is thought to reflect decreased D_2 receptor availability due to decreased receptor density and/or increased tonic dopamine levels in the synapses. This contributes to disinhibition within the indirect basal ganglia pathway and leads to an inability to suppress unwanted "nearby" motor contractions during the production of specific actions, a well-established abnormality in isolated dystonia. On the other hand, decreased phasic dopaminergic activity may represent a disorder-specific pathophysiological trait involved in generation of dystonic symptoms, whereas increased dopamine release during unaffected and unrelated motor tasks may be due to compensatory adaptation of the nigrostriatal dopaminergic system.

In terms of involvement of the direct basal ganglia pathway in dystonia, the recent study revealed increased availability of striatal dopamine D_1 receptors, suggesting hyperactivity of the direct pathway in patients with focal hand dystonia and laryngeal dystonia (Simonyan, Cho, Hamzehei Sichani, Rubien-Thomas, & Hallett, 2017). This dopaminergic alteration followed a well-known somatotopic organization of the striatum, with changes localized to the striatal hand and larynx representations, respectively.

Furthermore, dopaminergic dysfunctions involved both associative (anterior) and sensorimotor (posterior) striatal subdivisions, potentially having direct impact not only on hyperexcitability of motor cortex but also on parietal and prefrontal projection regions, which are responsible for the control of sensorimotor integration and preparation to motor execution and which are known to have abnormal activity and connectivity in dystonia. In fact, these cortical alterations may be a result of propagation of abnormal dopaminergic function via influencing beta oscillations within different striato-cortical loops.

When examining topological distribution of D_1 and D_2 receptor abnormalities, abnormal segregation of hyperfunctional direct and hypofunctional indirect pathways within the striatum became apparent, with negligible, if any, overlap between the two pathways (Fig. 6C). Moreover, the loss of overlap between the regions of dopamine D_1 and D_2 receptor availability and phasic dopamine release suggested complex disorganization of a nigro-striatal input.

Overall, these data showed that disorganization of striatal dopaminergic neurotransmission is of a global scale involving both the direct and indirect basal ganglia pathways and representing a common pathophysiological trait in dystonia.

4.2 GABAergic Neurotransmission

A loss of surround inhibition is considered as one of the main pathophysiological features of isolated dystonia (Quartarone & Hallett, 2013). Initial studies employing magnetic resonance spectroscopy to quantify GABAergic function in dystonia patients reported that GABA levels are significantly decreased in sensorimotor cortex and lentiform nucleus in patients with writer's cramp compared to healthy subjects (Levy & Hallett, 2002). This findings failed a replication in the follow up study that used a somewhat different analytic approach (Herath, Gallea, van der Veen, Horovitz, & Hallett, 2010). Subsequent positron emission studies with [^{11}C] flumazenil radiotracer produced more stable and reproducible results, demonstrating reduced binding of the ligand to $GABA_A$ receptors in primary sensorimotor, premotor, anterior cingulate, supplementary motor area, inferior parietal and insular cortex as well as caudate nucleus and cerebellum with some variations of affected regions across different forms of dystonia, including carriers of DYT1 mutation (Berman et al., 2018; Gallea et al., 2018; Garibotto et al., 2011; Simonyan, 2017). This GABAergic deficiency may result from the loss or decreased density of GABA$_A$ receptors, abnormal binding properties of these receptors, or decreased GABA synthesis due to a loss of inhibitory interneurons in dystonia. One study showed that decreased GABA$_A$ receptor availability in inferior parietal cortex is associated with increased gray matter volume (Simonyan, 2017) and increased brain activity (Gallea et al., 2018). Given its dysfunctional connectivity with sensorimotor regions and an association with the polygenic risk of dystonia (Gallea, Horovitz, Najee-Ullah, & Hallett, 2016; Putzel et al., 2018), inferior parietal cortex may be critical for setting off the disinhibition within the dystonic network.

Taken together, these studies re-evaluated the involvement of the basal ganglia circuitry in the pathophysiology of isolated dystonia as follows (Simonyan et al., 2017). Attenuated and topologically misplaced nigro-striatal dopamine release acts upon upregulated direct basal ganglia pathway and downregulated indirect pathway, which leads to overly excessive excitatory striatal output via the direct pathway in the presence of decreased inhibitory striatal output via the indirect pathway (Fig. 6B). This, collectively, disinhibits the thalamus and propagates to the motor cortex and other sensorimotor cortical regions, potentially underlying dissociations between activity in the striatum and sensorimotor cortex in the development of the dystonia-characteristic cortico-striatal loop.

5. CONTRIBUTION OF NEUROIMAGING TO UNDERSTANDING THE PATHOPHYSIOLOGY OF DYSTONIA

Based on advanced neuroimaging methodologies and analytic techniques, some unifying conclusions about neural alterations in isolated dystonia can be drawn. First, brain changes are not restricted to the basal ganglia; rather, these extend to other subcortical, cerebellar and sensorimotor cortical regions, forming a dysfunctional network with the major impairments within the striato-thalamo-cortical and cerebellar-thalamo-cortical pathways. Second, genotype and phenotype interactions appear to differentially impact brain network disorganization in dystonia, leading to distinct manifestations of the disorder. Third, altered dopaminergic and GABAergic neurotransmission underlies circuit abnormalities by influencing the balance between the basal ganglia direct and indirect pathways, propagating to disinhibition and hyperexcitability of cortical regions that are involved not only in the sensorimotor control but also important for the sensorimotor integration and preparation to the movement execution.

The progress in the field of neuroimaging methodologies continues having a direct impact on unraveling dystonic brain disorganization, piece-by-piece. With the further development of advanced imaging methodologies, the studies will be adequately powered and designed to provide yet lacking explanations of whether wide-ranging neural changes are causative, compensatory, or both in isolated dystonia. This, in turn, will be crucial for identification of novel criteria for enhanced and objective diagnosis of dystonia as well as for the development of new therapeutic and neurosurgical approaches to target these aberrations.

REFERENCES

Albanese, A., Bhatia, K., Bressman, S. B., Delong, M. R., Fahn, S., Fung, V. S., et al. (2013). Phenomenology and classification of dystonia: A consensus update. *Movement Disorders, 28*(7), 863–873. https://doi.org/10.1002/mds.25475.

Ali, S. O., Thomassen, M., Schulz, G. M., Hosey, L. A., Varga, M., Ludlow, C. L., et al. (2006). Alterations in CNS activity induced by botulinum toxin treatment in spasmodic dysphonia: An H215O PET study. *Journal of Speech, Language, and Hearing Research: JSLHR, 49*(5), 1127–1146. https://doi.org/10.1044/1092-4388(2006/081).

Argyelan, M., Carbon, M., Niethammer, M., Ulug, A. M., Voss, H. U., Bressman, S. B., et al. (2009). Cerebellothalamocortical connectivity regulates penetrance in dystonia. *The Journal of Neuroscience, 29*(31), 9740–9747. https://doi.org/10.1523/JNEUROSCI.2300-09.2009.

Asanuma, K., Ma, Y., Okulski, J., Dhawan, V., Chaly, T., Carbon, M., et al. (2005). Decreased striatal D2 receptor binding in non-manifesting carriers of the DYT1 dystonia mutation. *Neurology, 64*(2), 347–349. https://doi.org/10.1212/01.WNL.0000149764.34953.BF.

Asgeirsson, H., Jakobsson, F., Hjaltason, H., Jonsdottir, H., & Sveinbjornsdottir, S. (2006). Prevalence study of primary dystonia in Iceland. *Movement Disorders, 21*(3), 293–298. https://doi.org/10.1002/mds.20674.

Baker, R. S., Andersen, A. H., Morecraft, R. J., & Smith, C. D. (2003). A functional magnetic resonance imaging study in patients with benign essential blepharospasm. *Journal of Neuro-Ophthalmology, 23*(1), 11–15.

Battistella, G., Fuertinger, S., Fleysher, L., Ozelius, L. J., & Simonyan, K. (2016). Cortical sensorimotor alterations classify clinical phenotype and putative genotype of spasmodic dysphonia. *European Journal of Neurology, 23*(10), 1517–1527. https://doi.org/10.1111/ene.13067.

Battistella, G., Termsarasab, P., Ramdhani, R. A., Fuertinger, S., & Simonyan, K. (2017). Isolated focal dystonia as a disorder of large-scale functional networks. *Cerebral Cortex, 27*(2), 1203–1215. https://doi.org/10.1093/cercor/bhv313.

Berger, H. J., van der Werf, S. P., Horstink, C. A., Cools, A. R., Oyen, W. J., & Horstink, M. W. (2007). Writer's cramp: Restoration of striatal D2-binding after successful biofeedback-based sensorimotor training. *Parkinsonism & Related Disorders, 13*(3), 170–173. https://doi.org/10.1016/j.parkreldis.2006.09.003.

Berman, B. D., Hallett, M., Herscovitch, P., & Simonyan, K. (2013). Striatal dopaminergic dysfunction at rest and during task performance in writer's cramp. *Brain, 136,* 3645–3658. Pt. 12. https://doi.org/10.1093/brain/awt282.

Berman, B. D., Pollard, R. T., Shelton, E., Karki, R., Smith-Jones, P. M., & Miao, Y. (2018). GABAA receptor availability changes underlie symptoms in isolated cervical dystonia. *Frontiers in Neurology, 9,* 188. https://doi.org/10.3389/fneur.2018.00188.

Bianchi, S., Battistella, G., Huddlestone, H., Scharf, R., Fleysher, L., Rumbach, A. F., et al. (2017). Phenotype- and genotype-specific structural alterations in spasmodic dysphonia. *Movement Disorders, 32*(4), 560–568. https://doi.org/10.1002/mds.26920.

Biswal, B., Yetkin, F. Z., Haughton, V. M., & Hyde, J. S. (1995). Functional connectivity in the motor cortex of resting human brain using echo-planar MRI. *Magnetic Resonance in Medicine, 34*(4), 537–541.

Black, K. J., Ongur, D., & Perlmutter, J. S. (1998). Putamen volume in idiopathic focal dystonia. *Neurology, 51*(3), 819–824.

Blood, A. J., Kuster, J. K., Woodman, S. C., Kirlic, N., Makhlouf, M. L., Multhaupt-Buell, T. J., et al. (2012). Evidence for altered basal ganglia-brainstem connections in cervical dystonia. *PLoS One, 7*(2), e31654. https://doi.org/10.1371/journal.pone.0031654.

Bonilha, L., de Vries, P. M., Vincent, D. J., Rorden, C., Morgan, P. S., Hurd, M. W., et al. (2007). Structural white matter abnormalities in patients with idiopathic dystonia. *Movement Disorders, 22*(8), 1110–1116. https://doi.org/10.1002/mds.21295.

Breakefield, X. O., Blood, A. J., Li, Y., Hallett, M., Hanson, P. I., & Standaert, D. G. (2008). The pathophysiological basis of dystonias. *Nature Reviews. Neuroscience, 9*(3), 222–234. https://doi.org/10.1038/nrn2337.

Burciu, R. G., Hess, C. W., Coombes, S. A., Ofori, E., Shukla, P., Chung, J. W., et al. (2017). Functional activity of the sensorimotor cortex and cerebellum relates to cervical dystonia symptoms. *Human Brain Mapping, 38*(9), 4563–4573. https://doi.org/10.1002/hbm.23684.

Butterworth, S., Francis, S., Kelly, E., McGlone, F., Bowtell, R., & Sawle, G. V. (2003). Abnormal cortical sensory activation in dystonia: An fMRI study. *Movement Disorders, 18*(6), 673–682. https://doi.org/10.1002/mds.10416.

Carbon, M., & Eidelberg, D. (2009). Abnormal structure-function relationships in hereditary dystonia. *Neuroscience, 164*(1), 220–229. https://doi.org/10.1016/j.neuroscience.2008.12.041.

Carbon, M., Kingsley, P. B., Su, S., Smith, G. S., Spetsieris, P., Bressman, S., et al. (2004). Microstructural white matter changes in carriers of the DYT1 gene mutation. *Annals of Neurology, 56*(2), 283–286. https://doi.org/10.1002/ana.20177.

Carbon, M., Kingsley, P. B., Tang, C., Bressman, S., & Eidelberg, D. (2008). Microstructural white matter changes in primary torsion dystonia. *Movement Disorders, 23*(2), 234–239. https://doi.org/10.1002/mds.21806.

Carbon, M., Niethammer, M., Peng, S., Raymond, D., Dhawan, V., Chaly, T., et al. (2009). Abnormal striatal and thalamic dopamine neurotransmission: Genotype-related features of dystonia. *Neurology, 72*(24), 2097–2103. https://doi.org/10.1212/WNL.0b013e3181aa538f.

Carbon, M., Su, S., Dhawan, V., Raymond, D., Bressman, S., & Eidelberg, D. (2004). Regional metabolism in primary torsion dystonia: Effects of penetrance and genotype. *Neurology, 62*(8), 1384–1390.

Ceballos-Baumann, A. O., Passingham, R. E., Warner, T., Playford, E. D., Marsden, C. D., & Brooks, D. J. (1995). Overactive prefrontal and underactive motor cortical areas in idiopathic dystonia. *Annals of Neurology, 37*(3), 363–372. https://doi.org/10.1002/ana.410370313.

Ceccarelli, A., Rocca, M. A., Pagani, E., Falini, A., Comi, G., & Filippi, M. (2009). Cognitive learning is associated with gray matter changes in healthy human individuals: A tensor-based morphometry study. *NeuroImage, 48*(3), 585–589. https://doi.org/10.1016/j.neuroimage.2009.07.009.

Cheng, F. B., Wan, X. H., Feng, J. C., Ma, L. Y., Hou, B., Feng, F., et al. (2012). Subcellular distribution of THAP1 and alterations in the microstructure of brain white matter in DYT6 dystonia. *Parkinsonism & Related Disorders*, *18*(8), 978–982. https://doi.org/10.1016/j.parkreldis.2012.05.008.

Chklovskii, D. B. (2004). Synaptic connectivity and neuronal morphology: Two sides of the same coin. *Neuron*, *43*(5), 609–617. https://doi.org/10.1016/j.neuron.2004.08.012.

Chklovskii, D. B., Mel, B. W., & Svoboda, K. (2004). Cortical rewiring and information storage. *Nature*, *431*(7010), 782–788. https://doi.org/10.1038/nature03012.

Colosimo, C., Pantano, P., Calistri, V., Totaro, P., Fabbrini, G., & Berardelli, A. (2005). Diffusion tensor imaging in primary cervical dystonia. *Journal of Neurology, Neurosurgery, and Psychiatry*, *76*(11), 1591–1593. https://doi.org/10.1136/jnnp.2004.056614.

de Carvalho Aguiar, P. M., & Ozelius, L. J. (2002). Classification and genetics of dystonia. *Lancet Neurology*, *1*(5), 316–325.

Delmaire, C., Krainik, A., Tezenas du Montcel, S., Gerardin, E., Meunier, S., Mangin, J. F., et al. (2005). Disorganized somatotopy in the putamen of patients with focal hand dystonia. *Neurology*, *64*(8), 1391–1396. https://doi.org/10.1212/01.WNL.0000158424.01299.76.

Delmaire, C., Vidailhet, M., Wassermann, D., Descoteaux, M., Valabregue, R., Bourdain, F., et al. (2009). Diffusion abnormalities in the primary sensorimotor pathways in writer's cramp. *Archives of Neurology*, *66*(4), 502–508. https://doi.org/10.1001/archneurol.2009.8.

Delnooz, C. C., Helmich, R. C., Toni, I., & van de Warrenburg, B. P. (2012). Reduced parietal connectivity with a premotor writing area in writer's cramp. *Movement Disorders*, *27*(11), 1425–1431. https://doi.org/10.1002/mds.25029.

Delnooz, C. C., Pasman, J. W., Beckmann, C. F., & van de Warrenburg, B. P. (2013). Task-free functional MRI in cervical dystonia reveals multi-network changes that partially normalize with botulinum toxin. *PLoS One*, *8*(5), e62877. https://doi.org/10.1371/journal.pone.0062877.

DeSimone, J. C., Pappas, S. S., Febo, M., Burciu, R. G., Shukla, P., Colon-Perez, L. M., et al. (2017). Forebrain knock-out of torsinA reduces striatal free-water and impairs whole-brain functional connectivity in a symptomatic mouse model of DYT1 dystonia. *Neurobiology of Disease*, *106*, 124–132. https://doi.org/10.1016/j.nbd.2017.06.015.

Draganski, B., Gaser, C., Busch, V., Schuierer, G., Bogdahn, U., & May, A. (2004). Neuroplasticity: Changes in grey matter induced by training. *Nature*, *427*(6972), 311–312. https://doi.org/10.1038/427311a.

Draganski, B., Schneider, S. A., Fiorio, M., Kloppel, S., Gambarin, M., Tinazzi, M., et al. (2009). Genotype-phenotype interactions in primary dystonias revealed by differential changes in brain structure. *NeuroImage*, *47*(4), 1141–1147. https://doi.org/10.1016/j.neuroimage.2009.03.057.

Draganski, B., Thun-Hohenstein, C., Bogdahn, U., Winkler, J., & May, A. (2003). "Motor circuit" gray matter changes in idiopathic cervical dystonia. *Neurology*, *61*(9), 1228–1231.

Dresel, C., Bayer, F., Castrop, F., Rimpau, C., Zimmer, C., & Haslinger, B. (2011). Botulinum toxin modulates basal ganglia but not deficient somatosensory activation in orofacial dystonia. *Movement Disorders*, *26*(8), 1496–1502. https://doi.org/10.1002/mds.23497.

Dresel, C., Haslinger, B., Castrop, F., Wohlschlaeger, A. M., & Ceballos-Baumann, A. O. (2006). Silent event-related fMRI reveals deficient motor and enhanced somatosensory activation in orofacial dystonia. *Brain*, *129*(Pt. 1), 36–46. https://doi.org/10.1093/brain/awh665.

Egger, K., Mueller, J., Schocke, M., Brenneis, C., Rinnerthaler, M., Seppi, K., et al. (2007). Voxel based morphometry reveals specific gray matter changes in primary dystonia. *Movement Disorders*, *22*(11), 1538–1542. https://doi.org/10.1002/mds.21619.

Eidelberg, D., Moeller, J. R., Antonini, A., Kazumata, K., Nakamura, T., Dhawan, V., et al. (1998). Functional brain networks in DYT1 dystonia. *Annals of Neurology*, *44*(3), 303–312. https://doi.org/10.1002/ana.410440304.

Eidelberg, D., Moeller, J. R., Ishikawa, T., Dhawan, V., Spetsieris, P., Przedborski, S., et al. (1995). The metabolic topography of idiopathic torsion dystonia. *Brain, 118*, 1473–1484. Pt. 6.

Esmaeli-Gutstein, B., Nahmias, C., Thompson, M., Kazdan, M., & Harvey, J. (1999). Positron emission tomography in patients with benign essential blepharospasm. *Ophthalmic Plastic & Reconstructive Surgery, 15*(1), 23–27.

Etgen, T., Muhlau, M., Gaser, C., & Sander, D. (2006). Bilateral grey-matter increase in the putamen in primary blepharospasm. *Journal of Neurology, Neurosurgery, and Psychiatry, 77*(9), 1017–1020. https://doi.org/10.1136/jnnp.2005.087148.

Fabbrini, G., Pantano, P., Totaro, P., Calistri, V., Colosimo, C., Carmellini, M., et al. (2008). Diffusion tensor imaging in patients with primary cervical dystonia and in patients with blepharospasm. *European Journal of Neurology, 15*(2), 185–189. https://doi.org/10.1111/j.1468-1331.2007.02034.x.

Filip, P., Gallea, C., Lehericy, S., Bertasi, E., Popa, T., Marecek, R., et al. (2017). Disruption in cerebellar and basal ganglia networks during a visuospatial task in cervical dystonia. *Movement Disorders, 32*(5), 757–768. https://doi.org/10.1002/mds.26930.

Fornari, E., Maeder, P., Meuli, R., Ghika, J., & Knyazeva, M. G. (2012). Demyelination of superficial white matter in early Alzheimer's disease: A magnetization transfer imaging study. *Neurobiology of Aging, 33*(2), 428.e7–19. https://doi.org/10.1016/j.neurobiolaging.2010.11.014.

Fuertinger, S., & Simonyan, K. (2017). Connectome-wide phenotypical and genotypical associations in focal dystonia. *The Journal of Neuroscience, 37*(31), 7438–7449. https://doi.org/10.1523/JNEUROSCI.0384-17.2017.

Gallea, C., Herath, P., Voon, V., Lerner, A., Ostuni, J., Saad, Z., et al. (2018). Loss of inhibition in sensorimotor networks in focal hand dystonia. *NeuroImage. Clinical, 17*, 90–97. https://doi.org/10.1016/j.nicl.2017.10.011.

Gallea, C., Horovitz, S. G., Najee-Ullah, M., & Hallett, M. (2016). Impairment of a parieto-premotor network specialized for handwriting in writer's cramp. *Human Brain Mapping, 37*(12), 4363–4375. https://doi.org/10.1002/hbm.23315.

Garibotto, V., Romito, L. M., Elia, A. E., Soliveri, P., Panzacchi, A., Carpinelli, A., et al. (2011). In vivo evidence for GABA(A) receptor changes in the sensorimotor system in primary dystonia. *Movement Disorders, 26*(5), 852–857. https://doi.org/10.1002/mds.23553.

Garraux, G., Bauer, A., Hanakawa, T., Wu, T., Kansaku, K., & Hallett, M. (2004). Changes in brain anatomy in focal hand dystonia. *Annals of Neurology, 55*(5), 736–739. https://doi.org/10.1002/ana.20113.

Gerfen, C. R. (1992). The neostriatal mosaic: Multiple levels of compartmental organization. *Trends in Neurosciences, 15*(4), 133–139.

Gerfen, C. R. (2000). Molecular effects of dopamine on striatal-projection pathways. *Trends in Neurosciences, 23*(10 Suppl), S64–S70.

Granert, O., Peller, M., Gaser, C., Groppa, S., Hallett, M., Knutzen, A., et al. (2011). Manual activity shapes structure and function in contralateral human motor hand area. *NeuroImage, 54*(1), 32–41. https://doi.org/10.1016/j.neuroimage.2010.08.013.

Granert, O., Peller, M., Jabusch, H. C., Altenmuller, E., & Siebner, H. R. (2011). Sensorimotor skills and focal dystonia are linked to putaminal grey-matter volume in pianists. *Journal of Neurology, Neurosurgery, and Psychiatry, 82*(11), 1225–1231. https://doi.org/10.1136/jnnp.2011.245811.

Hallett, M. (1993). Physiology of basal ganglia disorders: An overview. *The Canadian Journal of Neurological Sciences, 20*(3), 177–183.

Hallett, M. (2004). Dystonia: Abnormal movements result from loss of inhibition. *Advances in Neurology, 94*, 1–9.

Hamann, M., & Richter, A. (2004). Striatal increase of extracellular dopamine levels during dystonic episodes in a genetic model of paroxysmal dyskinesia. *Neurobiology of Disease, 16*(1), 78–84. https://doi.org/10.1016/j.nbd.2004.01.005.

Haslinger, B., Altenmuller, E., Castrop, F., Zimmer, C., & Dresel, C. (2010). Sensorimotor overactivity as a pathophysiologic trait of embouchure dystonia. *Neurology, 74*(22), 1790–1797. https://doi.org/10.1212/WNL.0b013e3181e0f784.

Haslinger, B., Erhard, P., Dresel, C., Castrop, F., Roettinger, M., & Ceballos-Baumann, A. O. (2005). "Silent event-related" fMRI reveals reduced sensorimotor activation in laryngeal dystonia. *Neurology, 65*(10), 1562–1569. https://doi.org/10.1212/01.wnl.0000184478.59063.db.

Herath, P., Gallea, C., van der Veen, J. W., Horovitz, S. G., & Hallett, M. (2010). In vivo neurochemistry of primary focal hand dystonia: A magnetic resonance spectroscopic neurometabolite profiling study at 3T. *Movement Disorders, 25*(16), 2800–2808. https://doi.org/10.1002/mds.23306.

Holtmaat, A., Wilbrecht, L., Knott, G. W., Welker, E., & Svoboda, K. (2006). Experience-dependent and cell-type-specific spine growth in the neocortex. *Nature, 441*(7096), 979–983. https://doi.org/10.1038/nature04783.

Horie, C., Suzuki, Y., Kiyosawa, M., Mochizuki, M., Wakakura, M., Oda, K., et al. (2009). Decreased dopamine D receptor binding in essential blepharospasm. *Acta Neurologica Scandinavica, 119*(1), 49–54. https://doi.org/10.1111/j.1600-0404.2008.01053.x.

Horovitz, S. G., Ford, A., Najee-Ullah, M. A., Ostuni, J. L., & Hallett, M. (2012). Anatomical correlates of blepharospasm. *Translational Neurodegeneration, 1*(1), 12. https://doi.org/10.1186/2047-9158-1-12.

Horovitz, S. G., Gallea, C., Najee-Ullah, M., & Hallett, M. (2013). Functional anatomy of writing with the dominant hand. *PLoS One, 8*(7), e67931. https://doi.org/10.1371/journal.pone.0067931.

Horstink, C. A., Praamstra, P., Horstink, M. W., Berger, H. J., Booij, J., & Van Royen, E. A. (1997). Low striatal D2 receptor binding as assessed by [123I]IBZM SPECT in patients with writer's cramp. *Journal of Neurology, Neurosurgery, and Psychiatry, 62*(6), 672–673.

Hu, X. Y., Wang, L., Liu, H., & Zhang, S. Z. (2006). Functional magnetic resonance imaging study of writer's cramp. *Chinese Medical Journal, 119*(15), 1263–1271.

Hutchinson, M., Kimmich, O., Molloy, A., Whelan, R., Molloy, F., Lynch, T., et al. (2013). The endophenotype and the phenotype: Temporal discrimination and adult-onset dystonia. *Movement Disorders, 28*(13), 1766–1774. https://doi.org/10.1002/mds.25676.

Hutchinson, M., Nakamura, T., Moeller, J. R., Antonini, A., Belakhlef, A., Dhawan, V., et al. (2000). The metabolic topography of essential blepharospasm: A focal dystonia with general implications. *Neurology, 55*(5), 673–677.

Ibanez, V., Sadato, N., Karp, B., Deiber, M. P., & Hallett, M. (1999). Deficient activation of the motor cortical network in patients with writer's cramp. *Neurology, 53*(1), 96–105.

Islam, T., Kupsch, A., Bruhn, H., Scheurig, C., Schmidt, S., & Hoffmann, K. T. (2009). Decreased bilateral cortical representation patterns in writer's cramp: A functional magnetic resonance imaging study at 3.0 T. *Neurological Sciences, 30*(3), 219–226. https://doi.org/10.1007/s10072-009-0045-7.

Janousova, E., Schwarz, D., & Kasparek, T. (2015). Combining various types of classifiers and features extracted from magnetic resonance imaging data in schizophrenia recognition. *Psychiatry Research, 232*(3), 237–249. https://doi.org/10.1016/j.pscychresns.2015.03.004.

Kadota, H., Nakajima, Y., Miyazaki, M., Sekiguchi, H., Kohno, Y., Amako, M., et al. (2010). An fMRI study of musicians with focal dystonia during tapping tasks. *Journal of Neurology, 257*(7), 1092–1098. https://doi.org/10.1007/s00415-010-5468-9.

Kerrison, J. B., Lancaster, J. L., Zamarripa, F. E., Richardson, L. A., Morrison, J. C., Holck, D. E., et al. (2003). Positron emission tomography scanning in essential blepharospasm. *American Journal of Ophthalmology, 136*(5), 846–852.

Klein, C. (2014). Genetics in dystonia. *Parkinsonism & Related Disorders, 20*(Suppl. 1), S137–S142. https://doi.org/10.1016/S1353-8020(13)70033-6.

Kostic, V. S., Agosta, F., Sarro, L., Tomic, A., Kresojevic, N., Galantucci, S., et al. (2016). Brain structural changes in spasmodic dysphonia: A multimodal magnetic resonance imaging study. *Parkinsonism & Related Disorders, 25*, 78–84. https://doi.org/10.1016/j.parkreldis.2016.02.003.

Kwong, K. K., Belliveau, J. W., Chesler, D. A., Goldberg, I. E., Weisskoff, R. M., Poncelet, B. P., et al. (1992). Dynamic magnetic resonance imaging of human brain activity during primary sensory stimulation. *Proceedings of the National Academy of Sciences of the United States of America, 89*(12), 5675–5679.

Lehericy, S., Tijssen, M. A., Vidailhet, M., Kaji, R., & Meunier, S. (2013). The anatomical basis of dystonia: Current view using neuroimaging. *Movement Disorders, 28*(7), 944–957. https://doi.org/10.1002/mds.25527.

Lerner, A., Shill, H., Hanakawa, T., Bushara, K., Goldfine, A., & Hallett, M. (2004). Regional cerebral blood flow correlates of the severity of writer's cramp symptoms. *NeuroImage, 21*(3), 904–913. https://doi.org/10.1016/j.neuroimage.2003.10.019.

Levy, L. M., & Hallett, M. (2002). Impaired brain GABA in focal dystonia. *Annals of Neurology, 51*(1), 93–101.

Magyar-Lehmann, S., Antonini, A., Roelcke, U., Maguire, R. P., Missimer, J., Meyer, M., et al. (1997). Cerebral glucose metabolism in patients with spasmodic torticollis. *Movement Disorders, 12*(5), 704–708. https://doi.org/10.1002/mds.870120513.

Marsden, C. D., Obeso, J. A., Zarranz, J. J., & Lang, A. E. (1985). The anatomical basis of symptomatic hemidystonia. *Brain, 108*, 463–483. Pt. 2.

Martino, D., Di Giorgio, A., D'Ambrosio, E., Popolizio, T., Macerollo, A., Livrea, P., et al. (2011). Cortical gray matter changes in primary blepharospasm: A voxel-based morphometry study. *Movement Disorders, 26*(10), 1907–1912. https://doi.org/10.1002/mds.23724.

Mohammadi, B., Kollewe, K., Samii, A., Beckmann, C. F., Dengler, R., & Munte, T. F. (2012). Changes in resting-state brain networks in writer's cramp. *Human Brain Mapping, 33*(4), 840–848. https://doi.org/10.1002/hbm.21250.

Molinari, M., Filippini, V., & Leggio, M. G. (2002). Neuronal plasticity of interrelated cerebellar and cortical networks. *Neuroscience, 111*(4), 863–870.

Naumann, M., Pirker, W., Reiners, K., Lange, K. W., Becker, G., & Brucke, T. (1998). Imaging the pre- and postsynaptic side of striatal dopaminergic synapses in idiopathic cervical dystonia: A SPECT study using [123I] epidepride and [123I] beta-CIT. *Movement Disorders, 13*(2), 319–323. https://doi.org/10.1002/mds.870130219.

Nelson, A. J., Blake, D. T., & Chen, R. (2009). Digit-specific aberrations in the primary somatosensory cortex in writer's cramp. *Annals of Neurology, 66*(2), 146–154. https://doi.org/10.1002/ana.21626.

Nevrly, M., Hlustik, P., Hok, P., Otruba, P., Tudos, Z., & Kanovsky, P. (2018). Changes in sensorimotor network activation after botulinum toxin type A injections in patients with cervical dystonia: A functional MRI study. *Experimental Brain Research, 236*(10), 2627–2637. https://doi.org/10.1007/s00221-018-5322-3.

Neychev, V. K., Gross, R. E., Lehericy, S., Hess, E. J., & Jinnah, H. A. (2011). The functional neuroanatomy of dystonia. *Neurobiology of Disease, 42*(2), 185–201. https://doi.org/10.1016/j.nbd.2011.01.026.

Niethammer, M., Carbon, M., Argyelan, M., & Eidelberg, D. (2011). Hereditary dystonia as a neurodevelopmental circuit disorder: Evidence from neuroimaging. *Neurobiology of Disease, 42*(2), 202–209. https://doi.org/10.1016/j.nbd.2010.10.010.

Nobrega, J. N., Richter, A., Tozman, N., Jiwa, D., & Loscher, W. (1996). Quantitative autoradiography reveals regionally selective changes in dopamine D1 and D2 receptor binding in the genetically dystonic hamster. *Neuroscience, 71*(1), 927–937.

Nutt, J. G., Muenter, M. D., Melton, L. J., 3rd, Aronson, A., & Kurland, L. T. (1988). Epidemiology of dystonia in Rochester, Minnesota. *Advances in Neurology, 50*, 361–365.

Obermann, M., Yaldizli, O., De Greiff, A., Lachenmayer, M. L., Buhl, A. R., Tumczak, F., et al. (2007). Morphometric changes of sensorimotor structures in focal dystonia. *Movement Disorders*, *22*(8), 1117–1123. https://doi.org/10.1002/mds.21495.

Ogawa, S., Lee, T. M., Nayak, A. S., & Glynn, P. (1990). Oxygenation-sensitive contrast in magnetic resonance image of rodent brain at high magnetic fields. *Magnetic Resonance in Medicine*, *14*(1), 68–78.

Ozelius, L. J., Hewett, J. W., Page, C. E., Bressman, S. B., Kramer, P. L., Shalish, C., et al. (1997). The early-onset torsion dystonia gene (DYT1) encodes an ATP-binding protein. *Nature Genetics*, *17*(1), 40–48. https://doi.org/10.1038/ng0997-40.

Pantano, P., Totaro, P., Fabbrini, G., Raz, E., Contessa, G. M., Tona, F., et al. (2011). A transverse and longitudinal MR imaging voxel-based morphometry study in patients with primary cervical dystonia. *AJNR. American Journal of Neuroradiology*, *32*(1), 81–84. https://doi.org/10.3174/ajnr.A2242.

Perlmutter, J. S., Stambuk, M. K., Markham, J., Black, K. J., McGee-Minnich, L., Jankovic, J., et al. (1997). Decreased [18F]spiperone binding in putamen in idiopathic focal dystonia. *The Journal of Neuroscience*, *17*(2), 843–850.

Playford, E. D., Passingham, R. E., Marsden, C. D., & Brooks, D. J. (1998). Increased activation of frontal areas during arm movement in idiopathic torsion dystonia. *Movement Disorders*, *13*(2), 309–318. https://doi.org/10.1002/mds.870130218.

Prell, T., Peschel, T., Kohler, B., Bokemeyer, M. H., Dengler, R., Gunther, A., et al. (2013). Structural brain abnormalities in cervical dystonia. *BMC Neuroscience*, *14*, 123. https://doi.org/10.1186/1471-2202-14-123.

Pujol, J., Roset-Llobet, J., Rosines-Cubells, D., Deus, J., Narberhaus, B., Valls-Sole, J., et al. (2000). Brain cortical activation during guitar-induced hand dystonia studied by functional MRI. *NeuroImage*, *12*(3), 257–267. https://doi.org/10.1006/nimg.2000.0615.

Putzel, G. G., Battistella, G., Rumbach, A. F., Ozelius, L. J., Sabuncu, M. R., & Simonyan, K. (2018). Polygenic risk of spasmodic dysphonia is associated with vulnerable sensorimotor connectivity. *Cerebral Cortex*, *28*(1), 158–166. https://doi.org/10.1093/cercor/bhw363.

Quallo, M. M., Price, C. J., Ueno, K., Asamizuya, T., Cheng, K., Lemon, R. N., et al. (2009). Gray and white matter changes associated with tool-use learning in macaque monkeys. *Proceedings of the National Academy of Sciences of the United States of America*, *106*(43), 18379–18384. https://doi.org/10.1073/pnas.0909751106.

Quartarone, A., & Hallett, M. (2013). Emerging concepts in the physiological basis of dystonia. *Movement Disorders*, *28*(7), 958–967. https://doi.org/10.1002/mds.25532.

Ramdhani, R. A., Kumar, V., Velickovic, M., Frucht, S. J., Tagliati, M., & Simonyan, K. (2014). What's special about task in dystonia? A voxel-based morphometry and diffusion weighted imaging study. *Movement Disorders*, *29*(9), 1141–1150. https://doi.org/10.1002/mds.25934.

Ramdhani, R. A., & Simonyan, K. (2013). Primary dystonia: Conceptualizing the disorder through a structural brain imaging lens. *Tremor and Other Hyperkinetic Movements (New York, N. Y.)*, *3*.

Rothkirch, I., Granert, O., Knutzen, A., Wolff, S., Govert, F., Pedersen, A., et al. (2018). Dynamic causal modeling revealed dysfunctional effective connectivity in both, the cortico-basal-ganglia and the cerebello-cortical motor network in writers' cramp. *NeuroImage. Clinical*, *18*, 149–159. https://doi.org/10.1016/j.nicl.2018.01.015.

Simonyan, K. (2017). Inferior parietal cortex as a hub of loss of inhibutuib and maladaptive plasticity. In *Paper presented at the Annual Meeting of Americal Academy of Neurology, Boston*.

Simonyan, K., Berman, B. D., Herscovitch, P., & Hallett, M. (2013). Abnormal striatal dopaminergic neurotransmission during rest and task production in spasmodic dysphonia. *The Journal of Neuroscience*, *33*(37), 14705–14714. https://doi.org/10.1523/JNEUROSCI.0407-13.2013.

Simonyan, K., Cho, H., Hamzehei Sichani, A., Rubien-Thomas, E., & Hallett, M. (2017). The direct basal ganglia pathway is hyperfunctional in focal dystonia. *Brain, 140*(12), 3179–3190. https://doi.org/10.1093/brain/awx263.

Simonyan, K., & Ludlow, C. L. (2010). Abnormal activation of the primary somatosensory cortex in spasmodic dysphonia: An fMRI study. *Cerebral Cortex, 20*(11), 2749–2759. https://doi.org/10.1093/cercor/bhq023.

Simonyan, K., & Ludlow, C. L. (2012). Abnormal structure-function relationship in spasmodic dysphonia. *Cerebral Cortex, 22*(2), 417–425. https://doi.org/10.1093/cercor/bhr120.

Simonyan, K., Tovar-Moll, F., Ostuni, J., Hallett, M., Kalasinsky, V. F., Lewin-Smith, M. R., et al. (2008). Focal white matter changes in spasmodic dysphonia: A combined diffusion tensor imaging and neuropathological study. *Brain, 131*(Pt. 2), 447–459. https://doi.org/10.1093/brain/awm303.

Smith, S. M., Fox, P. T., Miller, K. L., Glahn, D. C., Fox, P. M., Mackay, C. E., et al. (2009). Correspondence of the brain's functional architecture during activation and rest. *Proceedings of the National Academy of Sciences of the United States of America, 106*(31), 13040–13045. https://doi.org/10.1073/pnas.0905267106.

Sur, M., & Rubenstein, J. L. (2005). Patterning and plasticity of the cerebral cortex. *Science, 310*(5749), 805–810. https://doi.org/10.1126/science.1112070.

Surmeier, D. J., Yan, Z., & Song, W. J. (1998). Coordinated expression of dopamine receptors in neostriatal medium spiny neurons. *Advances in Pharmacology, 42*, 1020–1023.

Suzuki, Y., Mizoguchi, S., Kiyosawa, M., Mochizuki, M., Ishiwata, K., Wakakura, M., et al. (2007). Glucose hypermetabolism in the thalamus of patients with essential blepharospasm. *Journal of Neurology, 254*(7), 890–896. https://doi.org/10.1007/s00415-006-0468-5.

Termsarasab, P., Ramdhani, R. A., Battistella, G., Rubien-Thomas, E., Choy, M., Farwell, I. M., et al. (2016). Neural correlates of abnormal sensory discrimination in laryngeal dystonia. *NeuroImage. Clinical, 10*, 18–26. https://doi.org/10.1016/j.nicl.2015.10.016.

Todd, R. D., & Perlmutter, J. S. (1998). Mutational and biochemical analysis of dopamine in dystonia: Evidence for decreased dopamine D2 receptor inhibition. *Molecular Neurobiology, 16*(2), 135–147. https://doi.org/10.1007/BF02740641.

Ulug, A. M., Vo, A., Argyelan, M., Tanabe, L., Schiffer, W. K., Dewey, S., et al. (2011). Cerebellothalamocortical pathway abnormalities in torsinA DYT1 knock-in mice. *Proceedings of the National Academy of Sciences of the United States of America, 108*(16), 6638–6643. https://doi.org/10.1073/pnas.1016445108.

Vitek, J. L. (2002). Pathophysiology of dystonia: A neuronal model. *Movement Disorders, 17*(Suppl. 3), S49–S62.

Vo, A., Sako, W., Dewey, S. L., Eidelberg, D., & Ulug, A. M. (2015). 18FDG-microPET and MR DTI findings in Tor1a+/− heterozygous knock-out mice. *Neurobiology of Disease, 73*, 399–406. https://doi.org/10.1016/j.nbd.2014.10.020.

Vo, A., Sako, W., Niethammer, M., Carbon, M., Bressman, S. B., Ulug, A. M., et al. (2015). Thalamocortical connectivity correlates with phenotypic variability in dystonia. *Cerebral Cortex, 25*(9), 3086–3094. https://doi.org/10.1093/cercor/bhu104.

Waugh, J. L., Kuster, J. K., Levenstein, J. M., Makris, N., Multhaupt-Buell, T. J., Sudarsky, L. R., et al. (2016). Thalamic volume is reduced in cervical and laryngeal dystonias. *PLoS One, 11*(5), e0155302. https://doi.org/10.1371/journal.pone.0155302.

Yang, J., Luo, C., Song, W., Chen, Q., Chen, K., Chen, X., et al. (2013). Altered regional spontaneous neuronal activity in blepharospasm: A resting state fMRI study. *Journal of Neurology, 260*(11), 2754–2760. https://doi.org/10.1007/s00415-013-7042-8.

Yourganov, G., Schmah, T., Churchill, N. W., Berman, M. G., Grady, C. L., & Strother, S. C. (2014). Pattern classification of fMRI data: Applications for analysis of spatially distributed cortical networks. *NeuroImage, 96*, 117–132. https://doi.org/10.1016/j.neuroimage.2014.03.074.

Zeuner, K. E., Knutzen, A., Granert, O., Gotz, J., Wolff, S., Jansen, O., et al. (2015). Increased volume and impaired function: The role of the basal ganglia in writer's cramp. *Brain and Behavior: A Cognitive Neuroscience Perspective, 5*(2), e00301. https://doi.org/10.1002/brb3.301.

Zoons, E., Booij, J., Nederveen, A. J., Dijk, J. M., & Tijssen, M. A. (2011). Structural, functional and molecular imaging of the brain in primary focal dystonia—A review. *NeuroImage, 56*(3), 1011–1020. https://doi.org/10.1016/j.neuroimage.2011.02.045.

CHAPTER TWO

Neuroimaging Applications in Restless Legs Syndrome

Giovanni Rizzo*,†,1, Giuseppe Plazzi*,†

*IRCCS Istituto delle Scienze Neurologiche di Bologna, Bologna, Italy
†Department of Biomedical and Neuromotor Sciences, Unit of Neurology, University of Bologna, Bologna, Italy
[1]Corresponding author: e-mail address: g.rizzo@unibo.it

Contents

Abstract

Neuroimaging studies provide information useful to understand the pathophysiology of restless legs syndrome. Molecular PET and SPECT imaging findings mainly supported dysfunction of dopaminergic pathways involving not only the nigrostriatal but also mesolimbic pathways. Magnetic resonance imaging (MRI) studies have used different techniques. Studies using iron-sensitive sequences supported the presence of a regionally variable low brain iron content, mainly at the level of substantia nigra and thalamus. The search for brain structural or microstructural abnormalities by voxel-based morphometry, diffusion tensor imaging or cortical thickness analysis has reported none or variable findings in restless legs syndrome patients, most of them in regions belonging to sensorimotor and limbic/nociceptive networks. Functional MRI studies have substantially demonstrated activation or connectivity changes in the same networks. Magnetic resonance spectroscopy studies showed metabolic changes in the thalamus, which is a hub of these networks. In summary, neuroimaging findings in restless legs syndrome support the presence of reduction of brain iron content, of dysfunction of mesolimbic and nigrostriatal dopaminergic pathways, and of abnormalities at level of limbic/nociceptive and sensorimotor networks.

International Review of Neurobiology, Volume 143
ISSN 0074-7742
https://doi.org/10.1016/bs.irn.2018.09.012

31

1. INTRODUCTION

Restless legs syndrome is a common sensorimotor disorder. Its prevalence in the general population is of 10–12% (with a range of 5–20%, depending on the study), increasing with age and higher in women than in men (Allen et al., 2003; Trenkwalder, Paulus, & Walters, 2005). In European and American populations, about 2–3% of adults suffer from clinically significant symptoms (Allen et al., 2005). Clinical presentation is characterized by an irresistible urge to move the legs, associated with unpleasant paraesthesias in the legs and sometimes in the arms. These sensations occur at rest, mainly in the evening or at night, are relieved by movement, and are typically attenuated by dopaminergic drugs (Trenkwalder et al., 2005). Most patients have also periodic limb movements (PLMs) in sleep and wakefulness and may complain of insomnia and/or hypersomnia (Allen et al., 2003; Trenkwalder et al., 2005). In 70–80% of cases it is an idiopathic disorder without apparent cause, whereas for the remainder it has been described as a symptomatic syndrome, associated with pregnancy, uremia, iron depletion, polyneuropathy, spinal disorders, and rheumatoid arthritis (Bassetti, Mauerhofer, Gugger, Mathis, & Hess, 2001; Trenkwalder et al., 2005), although these conditions are probably more correctly considered "risk factors" (Zucconi & Ferini-Strambi, 2004). The International Restless Legs Syndrome Study Group (IRLSSG) established the clinical diagnostic criteria in 1995 and reviewed them in 2003 and 2014 (Allen et al., 2003, 2014; Walters, 1995).

The pathophysiology of restless legs syndrome is poorly understood. Clinical, neurophysiological, and pharmacological observations point toward an involvement of central nervous structures and networks, although the areas involved are somewhat uncertain, with both the dopaminergic system and iron metabolism being implicated.

Several imaging studies have focused on the evaluation of the central nervous system in restless legs syndrome patients. Different imaging techniques have been used, each one evaluating different aspects putatively involved in the pathophysiology of this disorder. A lot of information has been provided, with some discrepancies in the findings and divergences in the interpretations offered. Notwithstanding, most of the data can be merged in a convergent interpretative model.

These imaging studies include molecular imaging investigations, by using positron emission tomography (PET) and single positron emission computed tomography (SPECT), mainly focusing on the dopaminergic pathway, and with few studied focusing on different neurotransmitter systems as serotoninergic or opioid pathways. Furthermore, an increasing number of magnetic resonance imaging (MRI) studies became available, employing various techniques such as iron-sensitive MRI methods, voxel-based morphometry, magnetic resonance spectroscopy, diffusion weighted sequences, and functional MRI using a task or during the resting state.

In this chapter, we reviewed all these studies reporting their findings as completely as possible, attempting to integrate all current imaging study results into a convergent pathophysiological interpretation.

2. IMAGING STUDIES IN RESTLESS LEGS SYNDROME PATIENTS

2.1 Molecular Imaging Studies

Molecular imaging uses tracers labeled with radioactive isotopes to study the density of particular receptors or the regional cerebral blood flow and metabolism in specific areas. Some studies used PET or SPECT imaging in restless legs syndrome patients. One study has used [^{18}F]fluorodeoxyglucose PET to measure cerebral metabolism in six restless legs syndrome patients. No abnormal regional metabolic uptake was found, although all patients were free of restless legs syndrome symptoms during the PET scan examination (Trenkwalder et al., 1999). Conversely, a SPECT study evaluated regional cerebral blood flow in two patients with familial painful restless legs syndrome during the state of pain induced by immobility, using [^{99m}Tc]HMPAO-SPECT, and found reduced regional cerebral blood flow in the caudate nuclei and increased regional cerebral blood flow in the thalami and anterior cingulate with increasing pain (San Pedro et al., 1998).

All other studies focused on the dopaminergic system, mainly at level of both presynaptic and postsynaptic compartment of the nigro-striatal pathway. As for the presynaptic compartment, SPECT studies evaluated the binding of dopamine to their specific transporters (DAT) located in the presynaptic terminals of the neurons (Eisensehr et al., 2001; Kim et al., 2012; Lin et al., 2016; Linke et al., 2004; Michaud, Soucy, Chabli, Lavigne, & Montplaisir, 2002; Tribl et al., 2002), detecting no change in most cases (Eisensehr et al., 2001; Linke et al., 2004; Michaud et al., 2002; Tribl et al., 2002). Two studies reported DAT dysfunction. Specifically, one found a significantly decreased striatal [^{99m}Tc]TRODAT-1 binding in early restless legs syndrome patients (Lin et al., 2016), while the other an increased striatal DAT density at the level of the caudate and posterior putamen in old patients with moderately severe restless legs syndrome by using [^{123}I]βCIT (Kim et al., 2012). As for PET, three studies evaluated the presynaptic dopaminergic compartment by using [^{18}F]dopa, disclosing no change in one (Trenkwalder et al., 1999) and a mild reduction of striatal uptake in the other two (Ruottinen et al., 2000; Turjanski, Lees, & Brooks, 1999). In a more recent study, the authors used a PET ligand for DAT, [^{11}C]-D-*threo*-methylphenidate (Earley et al., 2011). They scanned patients for 90 min straight after tracer infusion in order to evaluate membrane-bound DAT rather than total cellular DAT as in the SPECT studies (which employ scans after a 24 h delay), disclosing a decreased binding potential in putamen and caudate but not in the ventral striatum of restless legs syndrome subjects.

As regards the postsynaptic dopamine (D2) receptors, three old SPECT studies (Staedt et al., 1993, 1995a,b) on "nocturnal myoclonus syndrome" patients, in the majority associated with restless legs syndrome, detected reduced [^{123}I]IBZM striatal binding as well as one study on restless legs syndrome patients (Michaud et al., 2002). No change in [^{123}I]IBZM binding was reported by other three studies (Eisensehr et al., 2001; Kim et al., 2012; Tribl et al., 2002). Most of the PET studies used [^{11}C]raclopride to study postsynaptic D2 receptors in restless legs syndrome patients. Two studies reported a reduced striatal uptake (Earley et al., 2013; Turjanski et al., 1999). In one of them, the authors evaluated not only the striatal D2 receptor binding potentials per se, which were reduced, but also the density of receptors on the membrane (βmax) and the receptor–ligand dissociation constant or receptor affinity (K_d), which were unchanged, together interpreted as increased synaptic dopamine (Earley et al., 2013). Differently, a third PET study reported a significantly lower mean magnitude of [^{11}C]raclopride binding potential in the restless legs syndrome group at level of the mesolimbic dopamine region (D2/D3 receptors), i.e., nucleus accumbens and caudate, but not in the nigrostriatal dopamine region (Oboshi et al., 2012). Furthermore, the mesolimbic binding correlated negatively with clinical severity scores and positively with the degree of improvement after dopaminergic treatment (Oboshi et al., 2012). The results of a further PET study supported a limbic dopaminergic impairment in restless legs syndrome, although with opposite results (Cervenka et al., 2006). In this study, the authors disclosed a higher [^{11}C]raclopride binding potential in limbic and associative striatal subregions, and higher [^{11}C]FLB 457 (a higher affinity D2 radioligand) binding potential in medial and posterior thalamus, anterior cingulate cortex and insulae (Cervenka et al., 2006). All these brain structures serve the medial nociceptive system, involved in the affective-motivational component of pain. An involvement of the nociceptive system was also suggested by negative correlation between restless legs syndrome severity and the binding levels of the non-selective opioid receptor radioligand $[^{11}C]$diprenorphine in orbitofrontal and anterior cingulate cortex reported by different authors (von Spiczak et al., 2005). Few data are available for different neurotransmission system, as the serotoninergic system, which is strictly related to the limbic network. One SPECT study used $[^{123}I]βCIT$ tracer to study serotonergic neurotransmission in restless legs syndrome, evaluating the availability of serotonin transporter (SERT) in the pons and medulla (Jhoo et al., 2010). SERT availability, although similar in restless legs syndrome and control groups, negatively correlated with the disease severity in restless legs syndrome patients (Jhoo et al., 2010).

2.2 Iron-Sensitive MRI Studies

Several MRI techniques are available to measure non-heme iron content in brain tissue. Indeed, paramagnetic iron proportionally increases proton transverse relaxation rates. The most iron sensitive parameters are T2* or T2', and T2 to a lesser extent. Relaxometry is frequently used to evaluate the different relaxation rates R2 (1/T2), R2* (1/T2*), and R2' (1/T2'=R2*−R2). Other techniques, i.e., phase imaging, susceptibility weighted imaging and quantitative susceptibility mapping, evaluate differences in tissue magnetic susceptibility, which correlates directly with iron content (Haacke et al., 2005).

A number of MRI studies used iron-sensitive sequences and quantitative analyses of the data derived from them to evaluate brain iron level in restless legs syndrome subjects, in order to test the hypothesis of reduced iron content in the brain of these patients. In the early works of this type, the authors assessed regional brain iron concentration in restless legs syndrome patients by R2' measurement, defined as "iron index" (Allen, Barker, Wehrl, Song, & Earley, 2001; Earley, Barker, Horska, & Allen, 2006). They reported reduced R2' values at the level of the substantia nigra only in the early-onset restless legs syndrome patients (<45 years), supporting the presence of a lower iron content in this region and in this subgroup of patients. Conversely, a recent study reported a significantly lower iron index in the substantia nigra of patients with late-onset restless legs syndrome compared with controls, but not in patients with early-onset restless legs syndrome, using R2′ measurement by 3 T MRI (Moon et al., 2014). These results were in line with those of a previous study that disclosed increased T2 relaxation, compatible with low iron content, in the substantia nigra pars compacta and not in the pars reticolata, in patients with late onset restless legs syndrome (Margariti et al., 2012). Another group reported higher mean T2 values, suggesting lower iron content, of multiple brain regions, i.e., caudate head and medial, dorsal, and ventral thalamus, but not substantia nigra, in restless legs syndrome patients, without differentiating between early-onset and late-onset cases (Godau, Klose, Di Santo, Schweitzer, & Berg, 2008).

Subsequent studies used different approaches of analysis of susceptibility-weighted sequences. One study analyzed phase imaging disclosing significantly higher phase values in the restless legs syndrome patients compared with healthy controls at the level of the substantia nigra, putamen, pallidum, and particularly the thalamus, indicating diffuse but regionally variable low brain iron content in idiopathic restless legs syndrome patients, without differentiating between early-onset and late-onset cases (Rizzo et al., 2013).

A recent 7 T MRI study applied quantitative susceptibility mapping to assess possible brain iron deficiency in patients with restless legs syndrome, mostly with an early-onset disease (Li et al., 2016). Compared with healthy controls, restless legs syndrome patients showed significantly decreased magnetic susceptibility in the thalamus and dentate nucleus, but not in the substantia nigra. However, magnetic susceptibility in substantia nigra correlated with the periodic limb movements in sleep index, with significantly lower susceptibility values in restless legs syndrome patients with severe motor signs (periodic limb movements in sleep ≥ 100 times/h), providing evidence of a possible link between brain iron deficiency in restless legs syndrome and periodic limb movements in sleep (Li et al., 2016) (Fig. 1A–C).

Not all published MRI studies reported data supporting lower brain iron content in restless legs syndrome. Indeed, one study using T2* gradient echo MRI sequences did not find significant differences in regional signal intensity of 12 brain regions, i.e., substantia nigra, pallidum, caudate head, thalamus, occipital white matter, and frontal white matter bilaterally (Knake et al., 2010). A further study even showed a decreased T2 relaxation time in the right internal globus pallidus and the subthalamic nucleus of untreated patients with early-onset restless legs syndrome, indicating increased iron content, without any change in the substantia nigra (Astrakas et al., 2008).

Clinical inhomogeneity or technical considerations could explained such variability in the results. First, from the clinical point of view the groups of

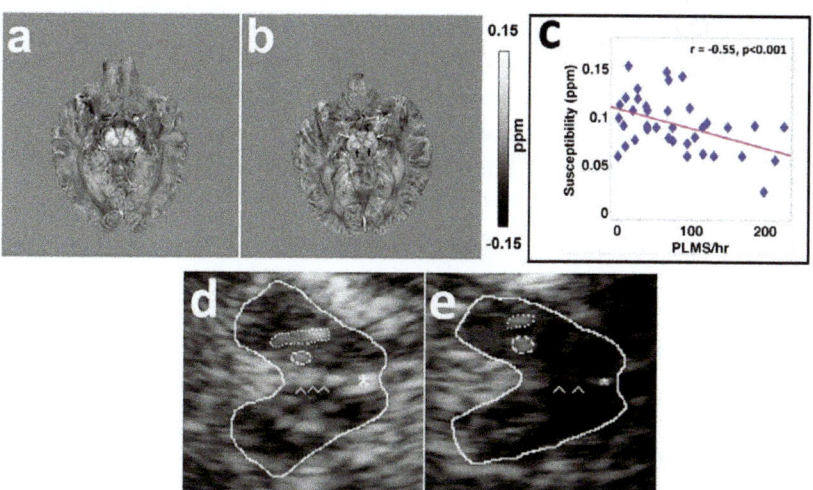

Fig. 1 Quantitative susceptibility maps of a normal control (A) and a restless legs syndrome patient with severe PLMs (B) showing the substantia nigra and red nucleus acquired at 7 T. Higher intensity in the image indicates higher iron content. Note the lower intensity of substantia nigra (white arrows) and red nucleus (black arrows) of restless legs syndrome patient (B) compared to normal control (A). (C) Correlation between magnetic susceptibility measured in the substantia nigra and PLMS measure in RLS patients. Ultrasound images of the mesencephalic brainstem of a control (D) and a restless legs syndrome patient (E). There is a line drawn around the hypoechogenic brainstem. Within the ipsilateral brainstem, the substantia nigra (dotted line) and the red nucleus (dashed line) can be delineated, and the continuous brainstem midline (raphe, arrows) and the aqueduct (asterisk) can be seen. Note the co-occurrence of a marked hypoechogenicity of the substantia nigra and of the raphe in the RLS patient, whereas the red nucleus appears normally echogenic. *Panels (A) and (B): From Rizzo, G., Li, X., Galantucci, S., Filippi, M., Cho, Y. W. (2017). Brain imaging and networks in restless legs syndrome.* Sleep Medicine, 31, 39–48. *Panel (C): From Li, X., Allen, R. P., Earley, C. J., Liu, H., Cruz, T. E., Edden, R. A. E., et al. (2016). Brain iron deficiency in idiopathic restless legs syndrome measured by quantitative magnetic susceptibility at 7 tesla.* Sleep Medicine, 22, 75–82. *Panels (D) and (E): From Godau, J., Wevers, A. K., Gaenslen, A., Di Santo, A., Liepelt, I., Gasser, et al. (2008). Sonographic abnormalities of brainstem structures in restless legs syndrome.* Sleep Medicine, 9(7), 782–789.

studied patients were different across the studies, in terms of age, age at onset, disease duration, disease severity, severity of periodic limb movements in sleep, treatment. Furthermore, it is possible that only a portion of restless legs syndrome patients has actually a low brain iron content as a pathogenic trigger. Second, the differences in the field of the scanner, in the sequences used, in the method of analysis are surely important. This concept is clearly supported by the results of a study that aimed at assessing the relationship between the different relaxometry methods and different region of interest approaches using each of these methods on a single population of controls and restless legs syndrome subjects (Moon et al., 2015). The authors found lower iron content in several brain regions of the patients, but such involvement changed with the different method applied (Moon et al., 2015).

A number of studies provided data supporting the low brain iron content hypothesis in restless legs syndrome using a different technique, transcranial B-Mode sonography, reporting hypoechogenicity of the substantia nigra (Godau, Klose, et al., 2008; Godau, Schweitzer, Liepelt, Gerloff, & Berg, 2007, Godau, Wevers, et al., 2008; Ryu, Lee, & Baik, 2011; Schmidauer et al., 2005), red nucleus and brainstem raphe (Godau, Wevers, et al., 2008) of restless legs syndrome patients (Fig. 1D and E). Substantia nigra hypoechogenicity inversely correlated with T2 values (Godau, Klose, et al., 2008) and was interpreted as secondary to iron deficiency.

Globally, all these findings, although with some discrepancies, support the hypothesis of reduced iron content in several regions of the brain of restless legs syndrome patients.

2.3 Magnetic Resonance Spectroscopy Studies

Magnetic resonance spectroscopy permits measurement of the concentration of specific biochemical compounds in the brain in precisely defined regions guided by MR imaging. Proton magnetic resonance spectroscopy is the most used in clinical practice, and detect a number of metabolites including N-acetyl-aspartate-containing compounds (markers for neuronal health, viability and number), choline containing compounds (major constituents of the membranes), creatine-phosphocreatine (whose peak is relatively stable and commonly used as a concentration reference), glutamate and glutamine (linked to excitatory neurotransmission), myoinositol (a glial marker), scyllo-inositol (closely coupled with mI amount), lactate (the end product of anaerobic glycolysis) and lipids. Using high field strengths, it is possible to detect additional metabolites at short TEs, as the inhibitory neurotransmitter gamma-aminobutyric acid and the antioxidant glutathione, and to separate glutamate and glutamine (Oz et al., 2014)

Three studies have used proton magnetic resonance spectroscopy in restless legs syndrome patients (Allen, Barker, Horska, & Earley, 2013; Rizzo, Tonon, et al., 2012; Winkelman, Schoerning, Platt, & Jensen, 2014) (Fig. 2). The first two studies used 1.5 T MRI scanners and both evaluated the thalamus, although with a different localization of the volume of

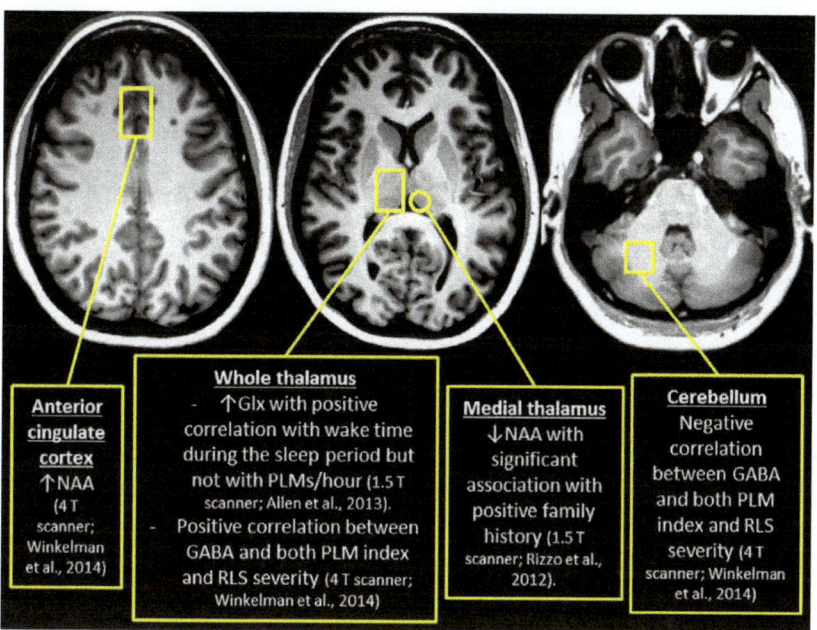

Anterior cingulate cortex
↑NAA (4 T scanner; Winkelman et al., 2014)

Whole thalamus
- ↑Glx with positive correlation with wake time during the sleep period but not with PLMs/hour (1.5 T scanner; Allen et al., 2013).
- Positive correlation between GABA and both PLM index and RLS severity (4 T scanner; Winkelman et al., 2014)

Medial thalamus
↓NAA with significant association with positive family history (1.5 T scanner; Rizzo et al., 2012).

Cerebellum
Negative correlation between GABA and both PLM index and RLS severity (4 T scanner; Winkelman et al., 2014)

Fig. 2 Volumes of interest used in the ^{1}H-MRS studies in restless legs syndrome patients and corresponding results.

interest (Allen et al., 2013; Rizzo, Tonon, et al., 2012). As the thalamus is a hub of multiple brain networks, studying its different subregions implies studying different brain networks. One study aimed to evaluate the metabolism of the medial portion of the thalamus, as a part of the limbic and nociceptive system, in idiopathic restless legs syndrome patients (Rizzo, Tonon, et al., 2012), disclosing a significantly reduced N-acetyl aspartate/creatine ratio and N-acetyl-aspartate concentrations in patients compared with healthy controls. Lower N-acetyl-aspartate concentrations were significantly associated with a family history of restless legs syndrome. In the same subjects, the thalamus was evaluated by diffusion tensor imaging, voxel-based morphometry, volumetric and shape analysis without detecting differences between the two groups. Accordingly, the results suggested an involvement of medial thalamic nuclei of a functional nature rather than a neuronal loss, supporting the hypothesis that a limbic system dysfunction plays a role in the pathophysiology of idiopathic restless legs syndrome (Rizzo, Tonon, et al., 2012). In the second proton magnetic resonance spectroscopy study investigating the thalamus of restless legs syndrome patients, the region of interest was localized at level of the whole right structure (Allen et al., 2013). A thalamic increase of the glutamate-glutamine/creatine ratio in patients compared to controls was disclosed, which correlated with the wake time during the sleep period and all other restless legs syndrome-related polysomnographic sleep variables except for periodic limb movements in sleep/hour (Allen et al., 2013). The authors suggested that glutamatergic dysfunction could underlie the arousal sleep disturbance and not the periodic limb movements in sleep, putatively more related to the dopaminergic system involvement, supporting the general hypothesis of increased glutamatergic activity in restless legs syndrome, which together with dopaminergic dysfunction may lead to dual mechanisms potentially underlying the clinical abnormalities seen in restless legs syndrome (Allen et al., 2013).

A further proton magnetic resonance spectroscopy study used a high field (4 T) MRI scanner, localizing the volumes of interest at level of the whole thalamus, the dorsal anterior cingulate cortex and the cerebellum of restless legs syndrome patients and controls subjects (Winkelman et al., 2014). Levels of metabolites were no different between two groups in the thalamus and the cerebellum, whereas N-acetyl-aspartate levels in the anterior cingulate cortex were higher in restless legs syndrome than in controls. Furthermore, gamma-aminobutyric acid levels positively correlated with both periodic limb movement indices and restless legs syndrome severity in the thalamus and negatively with both of these measures in the cerebellum in restless legs syndrome subjects. The authors interpreted the correlations found for the thalamic and cerebellar metabolites as a cerebellar–thalamic interaction modulating the intensity of restless legs syndrome sensory and motor symptoms (Winkelman et al., 2014). The increased anterior cingulate N-acetyl-aspartate levels could be due to an overactivation of this structure possibly associated with the affective components of the painful symptoms in this disorder (Winkelman et al., 2014).

2.4 Structural MRI Studies

A number of methods have been developed to search for brain structural changes. Most of these use high-resolution 3D T1-weighted MRI scans. Voxel based morphometry is able to detect subtle volume changes in the human brain by performing a voxel-by-voxel comparison of the regional attenuation of brain gray and white matter intensity across groups of subjects. Volumetric and shape analyses allow the evaluation of the volume and the shape of specific subcortical structures. Cortical thickness measurement investigates differences in cortical thickness across subjects. A further structural MRI technique is diffusion tensor imaging, which is sensitive to the random thermal movement of water molecules (Brownian motion) in neural tissues and provides quantitative measurements of brain microstructure via the diffusivity values (mean diffusivity, fractional anisotropy, axial diffusivity, and radial diffusivity, which are all parameters sensitive to neuronal and/or glial integrity).

Magnetic resonance studies using quantitative structural techniques have evaluated restless legs syndrome patients with the aim to search for subtle structural changes providing variable and partially discordant results (Table 1). Most of these studies used voxel-based morphometry approach, differing in some methodological aspects as the scanner field, the software used for the image analyses, and the statistical analyses, specifically concerning the correction for multiple comparisons. Most voxel-based morphometry studies reported no change (Belke et al., 2015; Celle, Roche, Peyron, et al., 2010; Comley et al., 2012; Margariti et al., 2012; Rizzo, Manners, et al., 2012). The other studies reported small and variable abnormalities: (i) bilateral gray matter increases in the pulvinar (Etgen et al., 2005); (ii) regional decreases of gray matter volume in the bilateral primary somatosensory cortex and left-sided primary motor areas (Unrath et al., 2007); (iii) a slightly increased gray matter density in the ventral hippocampus and middle orbitofrontal gyrus (Hornyak et al., 2007); (iv) reduced white matter volumes in small areas of the genu of the corpus callosum, anterior cingulum, and precentral gyrus (Connor et al., 2011); (v) regional decreases of gray matter volume in the left hippocampal gyrus, both parietal lobes, medial frontal areas and cerebellum (Chang et al., 2015).

A number of studies applied diffusion tensor imaging analysis using a voxel-based or tract-based approach (Belke et al., 2015; Chang et al., 2014; Lindemann et al., 2016; Rizzo, Manners, et al., 2012; Unrath et al., 2008; Zhuo et al., 2017). Also these studies reported variable white matter abnormalities in restless legs syndrome patients including: (i) multiple bilateral subcortical areas of significantly reduced fractional anisotropy, mainly close to the primary and associated motor and somatosensory cortices (Unrath et al., 2008); (ii) decreased fractional anisotropy in the genu of the corpus callosum and the inferior frontal gyrus white matter (Chang et al., 2014); (iii) altered fractional anisotropy in subcortical white matter, mainly in the temporal regions, the internal capsule, the pons, and the cerebellum (Belke et al., 2015). Two studies did not detect changes (Rizzo, Manners, et al., 2012; Zhuo et al., 2017), although in one of these, applying an uncorrected threshold of significance the analyses revealed subtle abnormalities, i.e., reduced

Table 1 Review of the Structural MRI Studies in RLS Patients Voxel-Based Morphometry (VBM)

Author	Cohort Studied	Technique	Results
Etgen et al. (2005)	51 patients and 51 controls, divided in two sub-groups of 28/28 [a] and 23/23 [b] from different centres Gender (M/F): 7/21 [a] and 6/17 [b] Age (years): 53.3±8.0 [a] and 59.3±10.1 [b] Disease duration (years): 19.2±14.3 [a] and 11.4±11.1 [b] IRLSSGRS score: 29.5±13.4 [a] and 28.8±5.9 [b] Positive family history in 9 (32%) [a] and 12 (43%) [b] The most patients on dopaminergic treatment	VBM using SPM1 software 1.5 T scanner	GM increase in pulvinar bilaterally (statistical test: GLM; threshold at $P<0.001$ uncorrected)
Hornyak et al. (2007)	14 patients and 14 controls Gender (M/F): 5/9 Age (years): 49.6±12.5 Disease duration (years): 16±12 IRLSSGRS score: 25.6±6.7 Positive family history in 9 (64%) 3 patients with minimal dopaminergic treatment exposure	VBM using SPM2 software 1.5 T scanner	GM decrease in left hippocampus and middle orbitofrontal gyrus (threshold at $P<0.05$ corrected for multiple comparisons)

Study	Sample	Method	Results
Unrath, Juergling, Schork, and Kassubek (2007)	63 patients and 40 controls Gender (M/F): 18/45 Age (years): 63.7 ± 11.4 Disease duration (years): 16 ± 12 IRLSSGRS score: 27.4 ± 7.1 Positive family history in 29 (46%) 57 (90.5%) patients on dopaminergic treatment	VBM using SPM2 software 1.5 T scanner	GM decreased in the bihemispheric primary somatosensory cortex, which additionally extended into left-sided primary motor areas (threshold at $P < 0.05$ corrected for small-volume. No significant results at $P < 0.05$ corrected for multiple comparisons)
Celle, Roche, Peyron, et al. (2010)	17 patients and 54 controls Gender (M/F): 1/16 Age (years): 65.9 ± 0.8 Disease duration not reported IRLSSGRS score 15.7 ± 4.8 Data on family history not reported No patient on dopaminergic treatment	VBM using SPM2 software 1.5 T scanner	No significant results at $P < 0.05$ corrected for multiple comparisons. (GM increase in left inferior occipital and left calcarine cortex and GM decrease in right superior temporal cortex at $P < 0.001$ uncorrected)
Comley et al. (2012)	16 patients and 16 controls Gender (M/F): 8/8 Age (years): 55 ± 7 Disease duration (years): 27 ± 12 IRLSSGRS score: 18.5 ± 3.9 Positive family history in 7 (56%) No patient on dopaminergic treatment	VBM using FSL–VBM software 1.5 T scanner	No significant results at $P < 0.05$ corrected for multiple comparisons
Connor et al. (2011)	23 patients and 23 controls Gender (M/F): 5/15 Age (years): 55.9 ± 13 Disease duration not reported IRLSSGRS score 24.4 ± 6.1 Data on family history not reported 6 (26%) patients on dopaminergic treatment	VBM using SPM5 software 3.0 T scanner	Reductions of WM volumes in small areas of the genu of the corpus callosum, anterior cingulum and precentral gyrus (threshold at $P < 0.001$ uncorrected)

Continued

Table 1 Review of the Structural MRI Studies in RLS Patients—cont'd
Voxel-Based Morphometry (VBM)

Author	Cohort Studied	Technique	Results
Rizzo, Tonon, et al. (2012)	20 patients and 20 controls Gender (M/F): 7/13 Age (years): 50±9 Disease duration (years): 10±9 IRLSSGRS score: 22±7 Positive family history in 8 (40%) 8 (40%) patients on dopaminergic treatment	VBM using SPM8 and FSL–VBM software 1.5 T scanner	No significant results at $P<0.05$ corrected and uncorrected for multiple comparisons
Margariti et al. (2012)	11 patients and 11 controls Gender (M/F): 2/9 Age (years): 55.3±8.4 Disease duration (years): 17.5±14.05 IRLSSGRS score: 18.4±4.8 Data on family history not reported No patient on dopaminergic treatment	VBM using SPM5 software 1.5 T scanner	No significant results at $P<0.05$ corrected for multiple comparisons
Belke et al. (2015)	12 patients and 12 controls Gender (M/F): 3/9 Age (years): 58.5±8.35 Disease duration (years): 12.1±8.7 IRLSSGRS score not reported (6 pts. with CGI=4and 6 pts. with CGI=5) Positive family history in 4 (33.3%) 8 (66.6%) patients on dopaminergic treatment	VBM using FSL–VBM software 1.5 T scanner	No significant results at $P<0.05$ corrected for multiple comparisons

Chang et al. (2015)	46 patients and 46 controls Gender (M/F): 14/32 Age (years): 55.9 ± 11.4 Disease duration (months): 123.9 ± 101.4 IRLSSGRS score: 26.9 ± 7.0 Data on family history not reported 13 (28.3%) patients on dopaminergic treatment	VBM using SPM8 VBM8 toolbox 3 T scanner	Reduced volume in some areas of the left hippocampal gyrus, both parietal lobes, medial frontal areas, lateral temporal areas and cerebellum (uncorrected, $P < 0.001$)

Diffusion tensor imaging (DTI)

Unrath, Muller, Ludolph, Riecker, and Kassubek (2008)	45 patients and 30 controls Gender (M/F): 13/32 Age (years): 59.9 ± 10.2 Disease duration (years): 21.6 ± 18.7 IRLSSGRS score: 28.1 ± 6.6 Positive family history in 14 (31.1%) 39 (86.7%) patients on dopaminergic treatment	Voxel-wise analysis of DTI using TIFT software 1.5 T scanner	Multiple subcortical areas of reduced FA bihemispherically close to the primary and associate motor and somatosensory cortices, in the right thalamus (posterior ventral lateral nucleus), in the motor projectional fibers and adjacent to the left anterior cingulum (threshold at $P < 0.05$ corrected for multiple comparisons)
Rizzo, Tonon, et al. (2012)	20 patients and 20 controls Gender (M/F): 7/13 Age (years): 50 ± 9 Disease duration (years): 10 ± 9 IRLSSGRS score: 22 ± 7 Positive family history in 8 (40%) 8 (40%) patients on dopaminergic treatment	Voxel-wise analysis of DTI using SPM8 and TBSS software 1.5 T scanner	No significant results at $P < 0.05$ corrected for multiple comparisons. (SPM analysis of DTI at $P < 0.001$ uncorrected: decreased FA and increased RD in right peridentate WM of restless legs syndrome patients, correlation between MD and AD values with IRLSSGRS score in left sensory-motor WM and splenium of corpus callosum, and between MD, AD and RD values with disease duration in right prefrontal WM)
Chang et al. (2014)	22 patients and 22 controls Gender (M/F): 3/19 Age (years): 54.5 ± 14.6 Disease duration (months): 130 ± 107.7 IRLSSGRS score: 25.5 ± 5.8 Data on family history not reported 4 (18.2%) patients on dopaminergic treatment	Voxel-wise analysis of DTI using SPM5 software 3 T scanner	Decreased FA in the genu of the corpus callosum and frontal WM adjacent to the inferior frontal gyrus compared with the control subjects. For areas of decreased FA, both the AD and RD were higher than that in the control subjects (threshold at $P < 0.05$ corrected for multiple comparisons)

Continued

Table 1 Review of the Structural MRI Studies in RLS Patients—cont'd Voxel-Based Morphometry (VBM)

Author	Cohort Studied	Technique	Results
Belke et al. (2015)	12 patients and 12 controls Gender (M/F): 3/9 Age (years): 58.5±8.35 Disease duration (years): 12.1±8.7 IRLSSGRS score not reported (6 pts. with CGI=4 and 6 pts. with CGI=5) Positive family history in 4 (33.3%) 8 (66.6%) patients on dopaminergic treatment	Voxel-wise analysis of DTI using TBSS software 1.5 T scanner	Small areas of altered FA (both increase or decrease) in subcortical white matter bilaterally, mainly in temporal regions as well as in the right internal capsule, the pons, and the right cerebellum (overlapping with changes in AD/RD). $P<0.05$ corrected for multiple comparisons
Lindemann, Muller, Ludolph, Hornyak, and Kassubek (2016)	25 patients and 25 controls Gender (M/F): 8/17 Age (years): 63.2±10.3 Disease duration (years): 11.6±9.5 IRLSSGRS score: 27±9 Data on family history not reported 22 (88%) patients on dopaminergic treatment	Voxel-wise analysis of DTI using TIFT software 1.5 T scanner	Significant FA alterations in two clusters in the midbrain bilaterally ($P<0.05$ corrected for multiple comparisons) Slicewise comparison of the FA maps in the cervical spinal cord showed a trend for lower FA values at the level of the second and third vertebra area in the patient sample ($P<0.05$ uncorrected)
Zhuo et al. (2017)	35 patients and 27 controls Gender (M/F): 13/22 Age (years): 51.83±16.48 Disease duration (years): 15.78±14.61 IRLSSGRS score: 24.53±6.71 Positive family history in 13 (37%) No patient on dopaminergic treatment	Voxel-wise analysis of DTI using TBSS software 3 T scanner	No significant results at $P<0.05$ corrected for multiple comparisons

Volumetric and shape analysis

Rizzo, Manners, et al. (2012)	23 patients and 19 controls Gender (M/F): 9/14 Age (years): 51±9 Disease duration (years): 9±8 IRLSSGRS score: 22±6 Positive family history in 9 (39%) 9 (39%) patients on dopaminergic treatment	Volumetric and shape analysis using FIRST software 1.5 T scanner	Study only focused on the thalamus No significant results at $P<0.05$ corrected for multiple comparisons and at $P<0.001$ uncorrected
Hermesdorf et al. (2018)	39 patients and 117 controls Gender (M/F): 13/26 Age (years): 52.21±8.51 Disease duration (years): 5 (2.5–10; median (IQR) IRLSSGRS score: 22±6 Positive family history in 9 (39%) 9 (39%) patients on dopaminergic treatment	Volumetric and shape analysis using FIRST software 3 T scanner	Study focused on caudate nucleus, hippocampus, globus pallidus, putamen, and thalamus No significant results at $P<0.05$ corrected for multiple comparisons

Cortical thickness analysis

Lee et al. (2018)	28 patients and 51 controls Gender (M/F): 11/17 Age (years): 54.8±14.8 Disease duration (years): 12.6±11.3 IRLSSGRS score: 24.6±6.8 Positive family history in 7 (25%)4 (14%) patients on dopaminergic treatment	Cortical thickness analysis using FreeSurfer software 3 T scanner	Patients with RLS exhibited a 7.5% decrease in average cortical thickness in the bilateral postcentral gyrus and a substantial decrease in the corpus callosum posterior midbody ($P<0.001$ corrected for multiple comparisons)

M, male; *F*, female; *IRLSSGRS*, International Restless Legs Syndrome Study Group Rating Scale; *SPM*, Statistical Parametric Mapping; *TBSS*, Tract-Based Spatial Statistics; *GM*, gray matter; *WM*, white matter; *MD*, mean diffusivity; *AD*, axial diffusivity; *FA*, fractional anisotropy; *RD*, radial diffusivity; *y*, years; *m*, months; *CGI*, Clinical Global Impression ratings; All data are given as mean±SD.

fractional anisotropy and increased radial diffusivity in the right peridentate white matter, a correlation between IRLSSGRS score and mean diffusivity and axial diffusivity in the splenium of corpus callosum and the left sensory-motor white matter, and a correlation between disease duration and mean diffusivity, axial diffusivity and radial diffusivity values in the right prefrontal white matter (Rizzo, Manners, et al., 2012).

A further study aimed to examine the diffusion tensor imaging metrics of the cervical spinal cord and the brainstem up to the midbrain in patients with idiopathic restless legs syndrome, disclosing significant fractional anisotropy alterations in two clusters in the midbrain bilaterally, along with a trend for lower fractional anisotropy values at the level of the second and third vertebra area in the patient sample (not significant after correction for multiple comparisons) (Lindemann et al., 2016). The interpretation of the results was that such fractional anisotropy changes could involve the projections of the hypothalamic A11 neurons, which spread to the midbrain/brainstem and lead to spinal pathways (Lindemann et al., 2016).

Fewer studies used different structural analyses. Volumetric and shape analysis was performed in two studies, one focusing on the thalamus (Rizzo, Tonon, et al., 2012) and another on the caudate nucleus, hippocampus, globus pallidus, putamen, and thalamus (Hermesdorf et al., 2018) without finding differences between restless legs syndrome patients and controls. More recently, some authors used cortical thickness analysis and detected a decrease in average cortical thickness in the bilateral postcentral and in the corpus callosum posterior midbody, wherein the callosal fibers are connected to the postcentral gyrus, suggesting altered white matter properties in the somatosensory pathway (Lee et al., 2018).

Globally, all these structural findings could represent subtle and hard to detect primary changes or variable secondary alterations, the most part belonging to specific brain networks, as sensorimotor and limbic/nociceptive networks. As well as for other imaging studies, the discrepant results could be due to methodological differences in the analysis, the different significance threshold used, heterogeneity in terms of size and clinical features of the samples studied.

2.5 Functional MRI Findings

Functional MRI is an imaging modality based on the blood-oxygen-level-dependent contrast, using hemoglobin as a naturally occurring endogenous contrast agent. It is used to measure brain activity based on the observation that cerebral blood flow and neuronal activation are coupled. We can evaluate the activation of specific brain regions during a task or at rest, or the functional connectivity measuring the level of co-activation of the different brain regions.

An increasing number of studies have used functional MRI in restless legs syndrome patients (Table 2). The first of these investigated the patients under four different conditions: (i) during a symptom free period; (ii) during sensory symptoms of the legs; (iii) during the occurrence of periodic limb movements and sensory symptoms; (iv) while mimicking PLMs. An activation of the thalamus was associated with leg discomfort, activation of red nuclei and brainstem with periodic limb movements, and activation of the cerebellum to both conditions (Bucher et al., 1997). About 10 years after, a second study simultaneously used electromyography and functional MRI in seven restless legs syndrome patients instructed not to move voluntarily, disclosing an association between the tonic electromyography activity of the anterior tibial muscle, negatively correlated with the degree of sensory leg discomfort, and the activation in motor and somatosensory areas such as precentral and postcentral gyri and the cerebellum (Spiegelhalder et al., 2008). Furthermore, electromyography activity was also correlated positively with activation of the posterior cingulate gyrus and negatively with the anterior cingulate gyrus (Spiegelhalder et al., 2008). Two further functional MRI studies used two different motor paradigms. In the first, the motor paradigm consisted of alternating active plantar flexion and dorsiflexion (Astrakas et al., 2008). The result was an activation in primary motor cortex, primary somatosensory cortex, somatosensory association cortex, and the middle cerebellar peduncles in both restless legs syndrome patients and controls, and in the thalamus, putamen, middle frontal gyrus, and anterior cingulate gyrus of the restless legs syndrome patients only (Astrakas et al., 2008). In the second, the paradigm consisted in random active asymmetric self-initiated movements of the feet to relieve their familiar uncomfortable sensations related to the restless legs syndrome and the functional MRI study was performed during the night (Margariti et al., 2012). Patients activated the primary motor and somatosensory cortex, the thalamus, pars opercularis, the ventral anterior cingulum, the striatum, the inferior and superior parietal lobules, the dorsolateral prefrontal cortex, the cerebellum, the midbrain, and the pons interpreted as a striatofrontolimbic activation underlying the repetitive compulsive movements seen in restless legs syndrome (Margariti et al., 2012).

The intrinsic functional architecture in restless legs syndrome has been also investigated by using resting state functional MRI, which reflects spontaneous neuronal activity (Deco, Jirsa, & McIntosh, 2011).

Table 2 Review of the Functional MRI Studies in Restless Legs Syndrome Patients

Author	Cohort Studied	Technique	Results
Bucher, Seelos, Oertel, Reiser, and Trenkwalder (1997)	19 patients and 15 controls Gender (M/F): 5/14 Age (years): 57.9±11.3 Disease duration (years): 2–36 (range) IRLSSGRS score not reported Data on family history not reported 11 (58%) patients on dopaminergic treatment	fMRI with four paradigms: (i) during a symptom free period; (ii) during sensory symptoms of the legs; (iii) during the occurrence of periodic limb movements (PLMs) and sensory symptoms; (iv) while mimicking PLMs 1.5 T scanner	Activation of the thalamus with leg discomfort, activation of red nuclei and brainstem with PLMs, and activation of the cerebellum to both conditions $P < 0.05$ corrected for multiple comparisons
Spiegelhalder et al. (2008)	7 patients Gender (M/F): 4/3 Age (years): 52.5±13.0 (M); 53.7±16.2 (F) Disease duration not reported IRLSSGRS score: 25.0±4 Data on family history not reported 2 patients with minimal dopaminergic treatment exposure	Simultaneous recording of surface EMG and fMRI. Patients instructed not to move voluntarily and asked to estimate their SLD on a scale of 0–10 1.5 T scanner	Association between the tonic EMG activity of the anterior tibial muscle, negatively correlated with the degree of sensory leg discomfort, and the activation in motor and somatosensory areas such as precentral and postcentral gyri and the cerebellum $P < 0.001$ uncorrected
Astrakas et al. (2008)	25 patients and 12 controls Gender (M/F): 11/14 Age (years): 66.5±8.9 Disease duration (years): 6.5±4.5 IRLSSGRS score: not available Positive family history in 12 (48%) No patient on dopaminergic treatment	fMRI with motor paradigm: alternating active plantar flexion and dorsiflexion 1.5 T scanner	Activation in primary motor cortex, primary somatosensory cortex, somatosensory association cortex, and the middle cerebellar peduncles in both RLS patients and controls, and in the thalamus, putamen, middle frontal gyrus, and anterior cingulate gyrus of the RLS patients only $P < 0.001$ uncorrected

Study	Sample characteristics	Method	Findings
Margariti et al. (2012)	11 patients and 11 controls Gender (M/F): 2/9 Age (years): 55.3±8.4 Disease duration not reported IRLSSGRS score 18.4±4.8 Data on family history not reported No patient on dopaminergic treatment	fMRI with motor paradigm: random active asymmetric self-initiated movements of the feet to relieve their uncomfortable sensations related to the RLS 1.5 T scanner	Patients activated the primary motor and somatosensory cortex, the thalamus, pars opercularis, the ventral anterior cingulum, the striatum, the inferior and superior parietal lobules, the dorsolateral prefrontal cortex, the cerebellum, the midbrain $P<0.001$ uncorrected
Ku et al. (2014)	25 patients and 25 controls Gender (M/F): 8/17 Age (years): 55.3±11.7 Disease duration (months): 112=117.8 IRLSSGRS score: 26±7.2 Data on family history not reported No patient on dopaminergic treatment	Resting state fMRI 3 T scanner	Reduced thalamic connectivity with the right parahippocampal gyrus, right precuneus, right precentral gyrus, and bilateral lingual gyri, and enhanced thalamic connectivity with the right superior temporal gyrus, bilateral middle temporal gyrus, and right medial frontal gyrus $P<0.005$ uncorrected
Liu et al. (2015)	15 patients and 14 controls Gender (M/F): 3/12 Age (years): 56.53±9.75 Disease duration (years): 18.87±14.29 IRLSSGRS score 23.4±6.52 Positive family history in 7 (46.6%) 6 (40%) patients on dopaminergic treatment	Resting state fMRI 3 T scanner	Lower ALFF in the sensorimotor and visual processing regions and higher ALFF in the insula, parahippocampal and hippocampal gyri, left posterior parietal areas, and brainstem (partially restored after rTMS treatment) $P<0.05$ corrected for multiple comparisons

Continued

Table 2 Review of the Functional MRI Studies in Restless Legs Syndrome Patients—cont'd

Author	Cohort Studied	Technique	Results
Gorges et al. (2016)	26 patients and 26 controls Gender (M/F): 8/18 Age (years): 63±10 Disease duration (years): 12±10 IRLSSGRS score: 27±6 Data on family history not reported 3 (12%) patients on dopaminergic treatment	Resting state fMRI 1.5 T scanner	Increased connectivity in the sensory thalamic, ventral and dorsal attention, basal ganglia-thalamic, and cingulate networks, positively correlated with increasing symptom severity $P<0.05$ corrected for multiple comparisons
Ku et al. (2016)	16 patients and 16 controls Gender (M/F): 6/10 Age (years): 57.19±10.02 Disease duration (months): 117.81±125.12 IRLSSGRS score: 26.19±6.73 Data on family history not reported No patient on dopaminergic treatment	Resting state fMRI 3 T scanner	Reductions in the DMN connectivity in the left posterior cingulate cortex, the right orbito-frontal gyrus, the left precuneus, and the right subcallosal gyrus, and increased DMN connectivity in sensory-motor-associated circuits, which included the right superior parietal lobule, the right supplementary motor area, and the left thalamus $P<0.05$ corrected for multiple comparisons
Zhuo et al. (2017)	35 patients and 27 controls Gender (M/F): 13/22 Age (years): 51.83±16.48 Disease duration (years): 15.78±14.61 IRLSSGRS score: 24.53±6.71 Positive family history in 13 (37%) No patient on dopaminergic treatment	Resting state fMRI 3 T scanner	Increased regional homogeneity (ReHo) in the bilateral middle frontal gyrus, anterior cingulate cortex, caudate nucleus, insula, thalamus, putamen and left posterior cingulate cortex $P<0.05$ corrected for multiple comparisons

| Liu et al. (2018) | 16 patients and 26 controls
Gender (M/F): 12/4
Age (years): 55±10.35
Disease duration (years): 16.7±12.4
IRLSSGRS score: 23.3±6.5
Positive family history in 7 (43.75%)
No patient on dopaminergic treatment | Resting state fMRI
3 T scanner | Hub analysis: significant decrease in FCS in the right cuneus, fusiform gyrus, paracentral lobe, and left precuneus; significant increase in FCS in the left superior frontal gyrus and bilateral thalamus
Functional connectivity analyses: significantly decreased functional connectivity between the right cuneus and left medial frontal gyrus/paracentral lobule; significantly increased functional connections between the left superior frontal gyrus and right medial prefrontal cortex, and between the bilateral thalamus and left cerebellum posterior lobe, and right middle temporal gyrus
$P<0.05$ corrected for multiple comparisons |
| Ku et al. (2018) | 15 patients and 15 controls
Gender (M/F): 6/9
Age (years): 57.4±10.34
Disease duration (months): 124.07±126.9
IRLSSGRS score: 26.4±6.91
Data on family history not reported
No patient on dopaminergic treatment | Resting state fMRI
3 T scanner | Significant differences in morning and evening DMN: consistent increased connectivity in the parietal lobule in both the morning and evening. In contrast, connectivity in the thalamus increased in the morning and reduced in the evening
$P<0.05$ corrected for multiple comparisons |

M, male; F, female; IRLSSGRS, International Restless Legs Syndrome Study Group Rating Scale; SPM, Statistical Parametric Mapping; TBSS, Tract-Based Spatial Statistics; GM, gray matter; WM, white matter; MD, mean diffusivity; AD, axial diffusivity; RD, radial diffusivity; FA, fractional anisotropy; y, years; m, months; ALFF, amplitude of low-frequency fluctuations; rTMS, repetitive transcranial magnetic stimulation; SLD, sensory leg discomfort; EMG, electromyogram; DMN, default mode network; FCS, functional connectivity strength. All data are given as mean±SD.

The first resting state functional MRI study on restless legs syndrome patients was performed during asymptomatic periods and showed functional connectivity changes, which were associated with their symptoms (Ku et al., 2014). Specifically, reduced thalamic connectivity with the right para-hippocampal gyrus, right precuneus, right precentral gyrus, and bilateral lingual gyri, and enhanced thalamic connectivity with the right superior temporal gyrus, bilateral middle temporal gyrus, and right medial frontal gyrus were found (Ku et al., 2014). Such connectivity changes in brain regions related to the thalamus could indicate that these regions may be associated with sensory monitoring and perception processing in restless legs syndrome patients (Ku et al., 2014) which seem to have abnormalities in network functions dealing with internal stimuli in the resting state even during a symptom-free period (Ku et al., 2014).

A subsequent rest functional MRI study disclosed lower amplitude of low-frequency fluctuations in the sensorimotor and visual processing regions, and higher amplitude of low-frequency fluctuations in the insula, parahippocampal and hippocampal gyri, left posterior parietal areas, and brainstem. Interestingly, after repetitive transcranial magnetic stimulation treatment, amplitude of low-frequency fluctuations in several sensorimotor and visual regions was significantly elevated and IRLSSG Rating Scale scores decreased, indicating improved restless legs syndrome symptoms (Liu et al., 2015).

The same authors applied graph theory methods to analyze rest functional MRI data from restless legs syndrome patients: the hub analysis revealed decreased functional connectivity strength in the cuneus, fusiform gyrus, paracentral lobe, and precuneus, and increased functional connectivity strength in the superior frontal gyrus and thalamus in idiopathic drug-naive restless legs syndrome patients (Liu et al., 2018).

Another study using a resting-state functional MRI in restless legs syndrome patients reported significantly increased connectivity in the sensory thalamic, ventral and dorsal attention, basal ganglia-thalamic, and cingulate networks in restless legs syndrome patients, positively correlated with increasing symptom severity (Gorges et al., 2016). The authors suggested that these altered functional networks might be associated with altered attentional control of sensory inputs (Gorges et al., 2016).

A more recent resting state fMRI study disclosed an increased regional homogeneity in the bilateral middle frontal gyrus, anterior cingulate cortex, caudate nucleus, insula, thalamus, putamen and left posterior cingulate cortex in restless legs syndrome patients compared to the healthy control group

(Zhuo et al., 2017). Furthermore, the restless legs syndrome group showed a smaller cluster size of activation in the default mode network (Zhuo et al., 2017). The default mode network is a network of brain regions that is active when an individual is not focused on the outside world and the brain is at a wakeful rest (Raichle et al., 2001). The default mode network changes are usually observed in people who have brain dysfunctions that cause difficulties in being able to maintain a resting state, such as restless legs syndrome. A default mode network study in restless legs syndrome subjects during the asymptomatic resting state (in the morning) showed disturbances of the default mode network (Ku et al., 2016). In particular, the results showed reductions in the default mode network connectivity in the left posterior cingulate cortex, the right orbito-frontal gyrus, the left precuneus, and the right subcallosal gyrus of the restless legs syndrome subjects, and increased default mode network connectivity in sensory-motor-associated circuits, which included the right superior parietal lobule, the right supplementary motor area, and the left thalamus. In addition, the connectivity between the default mode network and thalamus was negatively correlated with that in the orbito-frontal gyrus and the subcallosal gyrus in the subjects (Ku et al., 2016). The authors commented on these findings suggesting that the extinction processes of the default mode network may allow survival of undesirable sensations generated somewhere in the brain, hypothesizing that thalamic activity could work as a "resting keeper" during the asymptomatic period, necessary to keep the resting state so restless legs syndrome patients can deal with the actual generation of discomfort (Ku et al., 2016). The same authors evaluated diurnal functional default mode network changes in restless legs syndrome patients in both the morning and the evening (Ku et al., 2018). Restless legs syndrome patients showed consistent increased connectivity in the parietal lobule in both the morning and evening, while connectivity in the thalamus was increased in the morning and reduced in the evening (Ku et al., 2018). In addition, there were negative correlations between thalamic connectivity and the Korean version of the international restless legs syndrome scale and the quality-of-life subscore (Ku et al., 2018). They suggested that the circadian expression of restless legs syndrome might relate to changes in arousal cortical-activation thresholds occurring with diurnal changes in the thalamic circuits of the default mode network (Ku et al., 2018).

All these studies indicate that restless legs syndrome patients may have deficits in controlling and managing sensory information, supporting the hypothesis that restless legs syndrome could be a disorder of somatosensory processing.

3. CONCLUDING REMARKS

Neuroimaging findings in restless legs syndrome patients have provided evidences of low brain iron content, dopaminergic dysfunction and involvement of specific brain networks, although there are some discordances probably related to clinical differences in the studied samples and to different methodological approach and technical parameters used.

Most findings from MRI studies using iron sensitive sequences (Allen et al., 2001; Earley et al., 2006; Godau, Klose, et al., 2008; Margariti et al., 2012; Moon et al., 2014, 2015; Rizzo et al., 2013) support the presence of lower brain iron content in restless legs syndrome patients, which appears to be diffuse but regionally variable. Reduced iron level was reported mainly at the level of substantia nigra, but increasing evidence also points to the thalamus. PET and SPECT studies consistently support a dysfunction of dopaminergic pathways. This fits with the hypothesis linking iron and dopamine (Allen & Earley, 2007), as supported by pathological data and animal models (Connor, Patton, Oexle, & Allen, 2017). Overall the D2 receptors and dopamine transporters (mostly the membrane-bound transporters) in the striatum are decreased in most of the molecular imaging studies, compatibly with an increase in synaptic dopamine (Earley et al., 2011, 2013; Lin et al., 2016; Michaud et al., 2002; Oboshi et al., 2012; Ruottinen et al., 2000; Staedt et al., 1993, 1995a,b; Turjanski et al., 1999). This is also supported by the evidence of increased dopaminergic tone in the striatum of patients with lenticulo-striate stroke-related restless legs syndrome (Ruppert et al., 2017). In this scenario, to explain efficacy of dopaminergic drugs in restless legs syndrome patients, it can be speculated that the very low dosage of dopaminergic drugs used in the clinical practice can led to a preferential binding to the dopaminergic autoreceptors, which have higher affinity for dopamine than heteroreceptors, leading to a pre-synaptic inhibition of dopamine release (Ford, 2014). A different interpretation can be that increased dopaminergic stimulation produces a postsynaptic receptor down-regulation that, despite the overall dopamine increase, led to a relative evening and nighttime dopamine deficit, because of circadian dopaminergic activity pattern decreasing in the evening and night and increasing in the morning (Allen, 2015). A more complex dopaminergic change remains possible, with different dopamine levels according to disease duration or clinical severity, or with some brain regions presenting hyperdopaminergic and others presenting hypodopaminergic activity. Regarding the dopaminergic pathways involved, the data supported a dysfunction not only of the nigrostriatal but also of mesolimbic pathway (Cervenka et al., 2006; Earley et al., 2011, 2013; Kim et al., 2012; Lin et al., 2016; Michaud et al., 2002; Oboshi et al., 2012; Ruottinen et al., 2000; Staedt et al., 1993, 1995a,b; Turjanski et al., 1999).

Variable brain structural or microstructural abnormality has been reported in restless legs syndrome patients (Table 1), which could reflect subtle primary changes or, conversely, secondary changes. The results of genome-wide association studies reporting association of restless legs syndrome with variants in genes also involved in developmental processes (Trenkwalder, Hogl, & Winkelmann, 2009), or the fact that myelin synthesis depends on iron, and brain iron deficiency in animals reduces myelin proteins (Ortiz et al., 2004; Yu, Steinkirchner, Rao, & Larkin, 1986) can support a primary structural alteration. In both cases, structural abnormalities seem to involve specific brain networks in restless legs syndrome patients. Indeed, most of the changes reported in the MRI studies were in regions belonging to sensory-motor and limbic/nociceptive networks. Accordingly, the functional MRI studies have demonstrated activation or connectivity changes of cerebral areas belonging to these networks (Table 2). However, the chronic hypo- or hyperactivation of these brain networks could lead to secondary brain modifications. No study investigating the relationship between such structural changes and connectivity modifications is available. Furthermore, no study investigating the relationship between the reduced iron content and the structural and functional alterations is available. Considering the discrepancy in the results reported in the iron-sensitive MRI studies and in the structural MRI studies, it can be supposed that the pathogenic trigger varies in the restless legs syndrome patients, with a subgroup of patients with a primary reduced iron content and a different subgroup with primary structural abnormalities. A pathogenic variability within the whole group of patients with restless legs syndrome can also justify some discrepancies in the molecular imaging studies. Anyway, the resulting dysfunctional networks are the same. Structural and functional data have highlighted the expected involvement of the sensory-motor network. The metabolic alteration in the medial thalami and anterior cingulate cortex disclosed by proton magnetic resonance spectroscopy (Rizzo, Tonon, et al., 2012; Winkelman et al., 2014), by [^{99m}Tc]-HMPAO- SPECT (San Pedro et al., 1998), and by the limbic dopamine and opioid receptor evaluation by PET studies

(Cervenka et al., 2006; von Spiczak et al., 2005) also support an involvement of the limbic structures in RLS patients. The limbic system involvement may underlie restless legs syndrome patients' compulsive urge to move. The overlap of the limbic with the nociceptive system, and its role in the affective-motivational sensory-motor processing of the painful sensory inputs (Price, 2000), fits well with the similarities of the sensory descriptors in restless legs syndrome to those of neuropathic pain (Karroum, Golmard, Leu-Semenescu, & Arnulf, 2012). The limbic dysfunction could also explain the higher prevalence of depressive and anxiety disorders (Celle, Roche, Kerleroux, et al., 2010; Cho et al., 2009; Hornyak, 2010; Winkelmann et al., 2005), mild cognitive deficits (Celle, Roche, Kerleroux, et al., 2010; Fulda, Beitinger, Reppermund, Winkelmann, & Wetter, 2010; Gamaldo, Benbrook, Allen, Oguntimein, & Earley, 2008; Pearson et al., 2006), and sympathetic overactivity (Walters & Rye, 2009) in restless legs syndrome patients.

A single diffusion tensor imaging study (Lindemann et al., 2016) supported the involvement of the A11 diencephalospinal pathway that could be included in a context of a multilevel demodulation of pain stimuli perception. Furthermore, the default mode network circuit seems to be a pathogenic role in the genesis and modulation of the sensory symptoms of restless legs syndrome subjects (Ku et al., 2016, 2018).

In summary, increasing evidences support the notion that restless legs syndrome represents a complex network disorder, with involvement of specific brain networks dynamically interacting with each other, due to a metabolic/functional dysfunction. A crucial node of different networks, including sensory-motor and limbic/nociceptive, is the thalamus, which appears to have lower iron content (Godau, Klose, et al., 2008; Li et al., 2016; Rizzo et al., 2013), metabolic abnormalities (Allen et al., 2013; Rizzo, Tonon, et al., 2012), dopaminergic dysfunction (Cervenka et al., 2006), and changes in activation and functional connectivity (Astrakas et al., 2008; Bucher et al., 1997; Gorges et al., 2016; Ku et al., 2014, 2016, 2018; Liu et al., 2018; Margariti et al., 2012; Zhuo et al., 2017). This could reflect the dysfunction of the networks passing through the thalamus or suggest a specific role in an abnormal sensory-motor integration. The *primum movens* of these network dysfunctions could be the reduction of brain iron content, leading to dysfunction of mesolimbic and nigrostriatal dopaminergic pathways, and in turn to a dysregulation of limbic/nociceptive and sensorimotor networks, although it is possible that some patients have primary structural network abnormalities.

REFERENCES

Allen, R. P. (2015). Restless leg syndrome/Willis-Ekbom disease pathophysiology. *Sleep Medicine Clinics*, *10*(3), 207–214. xi. https://doi.org/10.1016/j.jsmc.2015.05.022.

Allen, R. P., Barker, P. B., Horska, A., & Earley, C. J. (2013). Thalamic glutamate/glutamine in restless legs syndrome: Increased and related to disturbed sleep. *Neurology*, *80*(22), 2028–2034. https://doi.org/10.1212/WNL.0b013e318294b3f6.

Allen, R. P., Barker, P. B., Wehrl, F. W., Song, H. K., & Earley, C. J. (2001). MRI measurement of brain iron in patients with restless legs syndrome. *Neurology*, *56*(2), 263–265.

Allen, R. P., & Earley, C. J. (2007). The role of iron in restless legs syndrome. *Movement Disorders*, *22*(Suppl 18), S440–S448. https://doi.org/10.1002/mds.21607.

Allen, R. P., Picchietti, D. L., Garcia-Borreguero, D., Ondo, W. G., Walters, A. S., Winkelman, J. W., et al. (2014). Restless legs syndrome/Willis-Ekbom disease diagnostic criteria: Updated international restless legs syndrome study group (IRLSSG) consensus criteria—History, rationale, description, and significance. *Sleep Medicine*, *15*(8), 860–873. https://doi.org/10.1016/j.sleep.2014.03.025.

Allen, R. P., Picchietti, D., Hening, W. A., Trenkwalder, C., Walters, A. S., Montplaisi, J., et al. (2003). Restless legs syndrome: Diagnostic criteria, special considerations, and epidemiology. A report from the restless legs syndrome diagnosis and epidemiology workshop at the National Institutes of Health. *Sleep Medicine*, *4*(2), 101–119.

Allen, R. P., Walters, A. S., Montplaisir, J., Hening, W., Myers, A., Bell, T. J., et al. (2005). Restless legs syndrome prevalence and impact: REST general population study. *Archives of Internal Medicine*, *165*(11), 1286–1292. https://doi.org/10.1001/archinte.165.11.1286.

Astrakas, L. G., Konitsiotis, S., Margariti, P., Tsouli, S., Tzarouhi, L., & Argyropoulou, M. I. (2008). T2 relaxometry and fMRI of the brain in late-onset restless legs syndrome. *Neurology*, *71*(12), 911–916. https://doi.org/10.1212/01.wnl.0000325914.50764.a2.

Bassetti, C. L., Mauerhofer, D., Gugger, M., Mathis, J., & Hess, C. W. (2001). Restless legs syndrome: A clinical study of 55 patients. *European Neurology*, *45*(2), 67–74. https://doi.org/10.1159/000052098.

Belke, M., Heverhagen, J. T., Keil, B., Rosenow, F., Oertel, W. H., Stiasny-Kolster, K., et al. (2015). DTI and VBM reveal white matter changes without associated gray matter changes in patients with idiopathic restless legs syndrome. *Brain and Behavior: A Cognitive Neuroscience Perspective*, *5*(9), e00327. https://doi.org/10.1002/brb3.327.

Bucher, S. F., Seelos, K. C., Oertel, W. H., Reiser, M., & Trenkwalder, C. (1997). Cerebral generators involved in the pathogenesis of the restless legs syndrome. *Annals of Neurology*, *41*(5), 639–645. https://doi.org/10.1002/ana.410410513.

Celle, S., Roche, F., Kerleroux, J., Thomas-Anterion, C., Laurent, B., Rouch, I., et al. (2010). Prevalence and clinical correlates of restless legs syndrome in an elderly French population: The synapse study. *The Journals of Gerontology. Series A, Biological Sciences and Medical Sciences*, *65*(2), 167–173. https://doi.org/10.1093/gerona/glp161.

Celle, S., Roche, F., Peyron, R., Faillenot, I., Laurent, B., Pichot, V., et al. (2010). Lack of specific gray matter alterations in restless legs syndrome in elderly subjects. *Journal of Neurology*, *257*(3), 344–348. https://doi.org/10.1007/s00415-009-5320-2.

Cervenka, S., Palhagen, S. E., Comley, R. A., Panagiotidis, G., Cselenyi, Z., Matthews, J. C., et al. (2006). Support for dopaminergic hypoactivity in restless legs syndrome: A PET study on D2-receptor binding. *Brain*, *129*(Pt. 8), 2017–2028. https://doi.org/10.1093/brain/awl163.

Chang, Y., Chang, H. W., Song, H., Ku, J., Earley, C. J., Allen, R. P., et al. (2015). Gray matter alteration in patients with restless legs syndrome: A voxel-based morphometry study. *Clinical Imaging*, *39*(1), 20–25. https://doi.org/10.1016/j.clinimag.2014.07.010.

Chang, Y., Paik, J. S., Lee, H. J., Chang, H. W., Moon, H. J., Allen, R. P., et al. (2014). Altered white matter integrity in primary restless legs syndrome patients: Diffusion tensor imaging study. *Neurological Research*, *36*(8), 769–774. https://doi.org/10.1179/1743132814Y.0000000336.

Cho, S. J., Hong, J. P., Hahm, B. J., Jeon, H. J., Chang, S. M., Cho, M. J., et al. (2009). Restless legs syndrome in a community sample of Korean adults: Prevalence, impact on quality of life, and association with DSM-IV psychiatric disorders. *Sleep*, *32*(8), 1069–1076.

Comley, R. A., Cervenka, S., Palhagen, S. E., Panagiotidis, G., Matthews, J. C., Lai, R. Y., et al. (2012). A comparison of gray matter density in restless legs syndrome patients and matched controls using voxel-based morphometry. *Journal of Neuroimaging*, *22*(1), 28–32. https://doi.org/10.1111/j.1552-6569.2010.00536.x.

Connor, J. R., Patton, S. M., Oexle, K., & Allen, R. P. (2017). Iron and restless legs syndrome: Treatment, genetics and pathophysiology. *Sleep Medicine*, *31*, 61–70. https://doi.org/10.1016/j.sleep.2016.07.028.

Connor, J. R., Ponnuru, P., Lee, B. Y., Podskalny, G. D., Alam, S., Allen, R. P., et al. (2011). Postmortem and imaging based analyses reveal CNS decreased myelination in restless legs syndrome. *Sleep Medicine*, *12*(6), 614–619. https://doi.org/10.1016/j.sleep.2010.10.009.

Deco, G., Jirsa, V. K., & McIntosh, A. R. (2011). Emerging concepts for the dynamical organization of resting-state activity in the brain. *Nature Reviews. Neuroscience*, *12*(1), 43–56. https://doi.org/10.1038/nrn2961.

Earley, C. J., Barker, B. P., Horska, A., & Allen, R. P. (2006). MRI-determined regional brain iron concentrations in early- and late-onset restless legs syndrome. *Sleep Medicine*, *7*(5), 458–461. https://doi.org/10.1016/j.sleep.2005.11.009.

Earley, C. J., Kuwabara, H., Wong, D. F., Gamaldo, C., Salas, R., Brasic, J., et al. (2011). The dopamine transporter is decreased in the striatum of subjects with restless legs syndrome. *Sleep*, *34*(3), 341–347.

Earley, C. J., Kuwabara, H., Wong, D. F., Gamaldo, C., Salas, R. E., Brasic, J. R., et al. (2013). Increased synaptic dopamine in the putamen in restless legs syndrome. *Sleep*, *36*(1), 51–57. https://doi.org/10.5665/sleep.2300.

Eisensehr, I., Wetter, T. C., Linke, R., Noachtar, S., von Lindeiner, H., Gildehaus, F. J., et al. (2001). Normal IPT and IBZM SPECT in drug-naive and levodopa-treated idiopathic restless legs syndrome. *Neurology*, *57*(7), 1307–1309.

Etgen, T., Draganski, B., Ilg, C., Schroder, M., Geisler, P., Hajak, G., et al. (2005). Bilateral thalamic gray matter changes in patients with restless legs syndrome. *NeuroImage*, *24*(4), 1242–1247. https://doi.org/10.1016/j.neuroimage.2004.10.021.

Ford, C. P. (2014). The role of D2-autoreceptors in regulating dopamine neuron activity and transmission. *Neuroscience*, *282*, 13–22. https://doi.org/10.1016/j.neuroscience.2014.01.025.

Fulda, S., Beitinger, M. E., Reppermund, S., Winkelmann, J., & Wetter, T. C. (2010). Short-term attention and verbal fluency is decreased in restless legs syndrome patients. *Movement Disorders*, *25*(15), 2641–2648. https://doi.org/10.1002/mds.23353.

Gamaldo, C. E., Benbrook, A. R., Allen, R. P., Oguntimein, O., & Earley, C. J. (2008). A further evaluation of the cognitive deficits associated with restless legs syndrome (RLS). *Sleep Medicine*, *9*(5), 500–505. https://doi.org/10.1016/j.sleep.2007.07.014.

Godau, J., Klose, U., Di Santo, A., Schweitzer, K., & Berg, D. (2008). Multiregional brain iron deficiency in restless legs syndrome. *Movement Disorders*, *23*(8), 1184–1187. https://doi.org/10.1002/mds.22070.

Godau, J., Schweitzer, K. J., Liepelt, I., Gerloff, C., & Berg, D. (2007). Substantia nigra hypoechogenicity: Definition and findings in restless legs syndrome. *Movement Disorders*, *22*(2), 187–192. https://doi.org/10.1002/mds.21230.

Godau, J., Wevers, A. K., Gaenslen, A., Di Santo, A., Liepelt, I., Gasser, T., et al. (2008). Sonographic abnormalities of brainstem structures in restless legs syndrome. *Sleep Medicine*, *9*(7), 782–789. https://doi.org/10.1016/j.sleep.2007.09.001.

Gorges, M., Rosskopf, J., Muller, H. P., Lindemann, K., Hornyak, M., & Kassubek, J. (2016). Patterns of increased intrinsic functional connectivity in patients with restless legs syndrome are associated with attentional control of sensory inputs. *Neuroscience Letters*, *617*, 264–269. https://doi.org/10.1016/j.neulet.2016.02.043.

Haacke, E. M., Cheng, N. Y., House, M. J., Liu, Q., Neelavalli, J., Ogg, R. J., et al. (2005). Imaging iron stores in the brain using magnetic resonance imaging. *Magnetic Resonance Imaging*, *23*(1), 1–25. https://doi.org/10.1016/j.mri.2004.10.001.

Hermesdorf, M., Sundermann, B., Rawal, R., Szentkiralyi, A., Dannlowski, U., & Berger, K. (2018). Lack of association between shape and volume of subcortical brain structures and restless legs syndrome. *Frontiers in Neurology*, *9*, 355. https://doi.org/10.3389/fneur.2018.00355.

Hornyak, M. (2010). Depressive disorders in restless legs syndrome: Epidemiology, pathophysiology and management. *CNS Drugs*, *24*(2), 89–98. https://doi.org/10.2165/11317500-000000000-00000.

Hornyak, M., Ahrendts, J. C., Spiegelhalder, K., Riemann, D., Voderholzer, U., Feige, B., et al. (2007). Voxel-based morphometry in unmedicated patients with restless legs syndrome. *Sleep Medicine*, *9*(1), 22–26. https://doi.org/10.1016/j.sleep.2006.09.010.

Jhoo, J. H., Yoon, I. Y., Kim, Y. K., Chung, S., Kim, J. M., Lee, S. B., et al. (2010). Availability of brain serotonin transporters in patients with restless legs syndrome. *Neurology*, *74*(6), 513–518. https://doi.org/10.1212/WNL.0b013e3181cef824.

Karroum, E. G., Golmard, J. L., Leu-Semenescu, S., & Arnulf, I. (2012). Sensations in restless legs syndrome. *Sleep Medicine*, *13*(4), 402–408. https://doi.org/10.1016/j.sleep.2011.01.021.

Kim, K. W., Jhoo, J. H., Lee, S. B., Lee, S. D., Kim, T. H., Kim, S. E., et al. (2012). Increased striatal dopamine transporter density in moderately severe old restless legs syndrome patients. *European Journal of Neurology*, *19*(9), 1213–1218. https://doi.org/10.1111/j.1468-1331.2012.03705.x.

Knake, S., Heverhagen, J. T., Menzler, K., Keil, B., Oertel, W. H., & Stiasny-Kolster, K. (2010). Normal regional brain iron concentration in restless legs syndrome measured by MRI. *Nature and Science of Sleep*, *2*, 19–22.

Ku, J., Cho, Y. W., Lee, Y. S., Moon, H. J., Chang, H., Earley, C. J., et al. (2014). Functional connectivity alternation of the thalamus in restless legs syndrome patients during the asymptomatic period: A resting-state connectivity study using functional magnetic resonance imaging. *Sleep Medicine*, *15*(3), 289–294. https://doi.org/10.1016/j.sleep.2013.09.030.

Ku, J., Lee, Y. S., Chang, H., Earley, C. J., Allen, R. P., & Cho, Y. W. (2016). Default mode network disturbances in restless legs syndrome/Willis-Ekbom disease. *Sleep Medicine*, *23*, 6–11. https://doi.org/10.1016/j.sleep.2016.05.007.

Ku, J., Lee, Y. S., Chang, H. W., Earley, C. J., Allen, R. P., & Cho, Y. W. (2018). Diurnal variation of default mode network in patients with restless legs syndrome. *Sleep Medicine*, *41*, 1–8. https://doi.org/10.1016/j.sleep.2017.09.031.

Lee, B. Y., Kim, J., Connor, J. R., Podskalny, G. D., Ryu, Y., & Yang, Q. X. (2018). Involvement of the central somatosensory system in restless legs syndrome: A neuroimaging study. *Neurology*, *90*(21), e1834–e1841. https://doi.org/10.1212/WNL.0000000000005562.

Li, X., Allen, R. P., Earley, C. J., Liu, H., Cruz, T. E., Edden, R. A. E., et al. (2016). Brain iron deficiency in idiopathic restless legs syndrome measured by quantitative magnetic susceptibility at 7 tesla. *Sleep Medicine*, *22*, 75–82. https://doi.org/10.1016/j.sleep.2016.05.001.

Lin, C. C., Fan, Y. M., Lin, G. Y., Yang, F. C., Cheng, C. A., Lu, K. C., et al. (2016). 99mTc-TRODAT-1 SPECT as a potential neuroimaging biomarker in patients with restless legs syndrome. *Clinical Nuclear Medicine, 41*(1), e14–e17. https://doi.org/10.1097/RLU.0000000000000916.

Lindemann, K., Muller, H. P., Ludolph, A. C., Hornyak, M., & Kassubek, J. (2016). Microstructure of the midbrain and cervical spinal cord in idiopathic restless legs syndrome: A diffusion tensor imaging study. *Sleep, 39*(2), 423–428. https://doi.org/10.5665/sleep.5456.

Linke, R., Eisensehr, I., Wetter, T. C., Gildehaus, F. J., Popperl, G., Trenkwalder, C., et al. (2004). Presynaptic dopaminergic function in patients with restless legs syndrome: Are there common features with early Parkinson's disease? *Movement Disorders, 19*(10), 1158–1162. https://doi.org/10.1002/mds.20226.

Liu, C., Dai, Z., Zhang, R., Zhang, M., Hou, Y., Qi, Z., et al. (2015). Mapping intrinsic functional brain changes and repetitive transcranial magnetic stimulation neuromodulation in idiopathic restless legs syndrome: A resting-state functional magnetic resonance imaging study. *Sleep Medicine, 16*(6), 785–791. https://doi.org/10.1016/j.sleep.2014.12.029.

Liu, C., Wang, J., Hou, Y., Qi, Z., Wang, L., Zhan, S., et al. (2018). Mapping the changed hubs and corresponding functional connectivity in idiopathic restless legs syndrome. *Sleep Medicine, 45*, 132–139. https://doi.org/10.1016/j.sleep.2017.12.016.

Margariti, P. N., Astrakas, L. G., Tsouli, S. G., Hadjigeorgiou, G. M., Konitsiotis, S., & Argyropoulou, M. I. (2012). Investigation of unmedicated early onset restless legs syndrome by voxel-based morphometry, T2 relaxometry, and functional MR imaging during the night-time hours. *AJNR. American Journal of Neuroradiology, 33*(4), 667–672. https://doi.org/10.3174/ajnr.A2829.

Michaud, M., Soucy, J. P., Chabli, A., Lavigne, G., & Montplaisir, J. (2002). SPECT imaging of striatal pre- and postsynaptic dopaminergic status in restless legs syndrome with periodic leg movements in sleep. *Journal of Neurology, 249*(2), 164–170.

Moon, H. J., Chang, Y., Lee, Y. S., Song, H. J., Chang, H. W., Ku, J., et al. (2014). T2 relaxometry using 3.0-tesla magnetic resonance imaging of the brain in early- and late-onset restless legs syndrome. *Journal of Clinical Neurology, 10*(3), 197–202. https://doi.org/10.3988/jcn.2014.10.3.197.

Moon, H. J., Chang, Y., Lee, Y. S., Song, H., Chang, H. W., Ku, J., et al. (2015). A comparison of MRI tissue relaxometry and ROI methods used to determine regional brain iron concentrations in restless legs syndrome. *Medical Devices (Auckland, N.Z), 8*, 341–350. https://doi.org/10.2147/MDER.S83629.

Oboshi, Y., Ouchi, Y., Yagi, S., Kono, S., Nakai, N., Yoshikawa, E., et al. (2012). In vivo mesolimbic D2/3 receptor binding predicts posttherapeutic clinical responses in restless legs syndrome: A positron emission tomography study. *Journal of Cerebral Blood Flow and Metabolism, 32*(4), 654–662. https://doi.org/10.1038/jcbfm.2011.201.

Ortiz, E., Pasquini, J. M., Thompson, K., Felt, B., Butkus, G., Beard, J., et al. (2004). Effect of manipulation of iron storage, transport, or availability on myelin composition and brain iron content in three different animal models. *Journal of Neuroscience Research, 77*(5), 681–689. https://doi.org/10.1002/jnr.20207.

Oz, G., Alger, J. R., Barker, P. B., Bartha, R., Bizzi, A., Boesch, C., et al. (2014). Clinical proton MR spectroscopy in central nervous system disorders. *Radiology, 270*(3), 658–679. https://doi.org/10.1148/radiol.13130531.

Pearson, V. E., Allen, R. P., Dean, T., Gamaldo, C. E., Lesage, S. R., & Earley, C. J. (2006). Cognitive deficits associated with restless legs syndrome (RLS). *Sleep Medicine, 7*(1), 25–30. https://doi.org/10.1016/j.sleep.2005.05.006.

Price, D. D. (2000). Psychological and neural mechanisms of the affective dimension of pain. *Science, 288*(5472), 1769–1772.

Raichle, M. E., MacLeod, A. M., Snyder, A. Z., Powers, W. J., Gusnard, D. A., & Shulman, G. L. (2001). A default mode of brain function. *Proceedings of the National Academy of Sciences of the United States of America, 98*(2), 676–682. https://doi.org/10.1073/pnas.98.2.676.

Rizzo, G., Manners, D., Testa, C., Tonon, C., Vetrugno, R., Marconi, S., et al. (2013). Low brain iron content in idiopathic restless legs syndrome patients detected by phase imaging. *Movement Disorders, 28*(13), 1886–1890. https://doi.org/10.1002/mds.25576.

Rizzo, G., Manners, D., Vetrugno, R., Tonon, C., Malucelli, E., Plazzi, G., et al. (2012). Combined brain voxel-based morphometry and diffusion tensor imaging study in idiopathic restless legs syndrome patients. *European Journal of Neurology, 19*(7), 1045–1049. https://doi.org/10.1111/j.1468-1331.2011.03604.x.

Rizzo, G., Tonon, C., Testa, C., Manners, D., Vetrugno, R., Pizza, F., et al. (2012). Abnormal medial thalamic metabolism in patients with idiopathic restless legs syndrome. *Brain, 135*(Pt 12), 3712–3720. https://doi.org/10.1093/brain/aws266.

Ruottinen, H. M., Partinen, M., Hublin, C., Bergman, J., Haaparanta, M., Solin, O., et al. (2000). An FDOPA PET study in patients with periodic limb movement disorder and restless legs syndrome. *Neurology, 54*(2), 502–504.

Ruppert, E., Bataillard, M., Namer, I. J., Tatu, L., Hacquard, A., Hugueny, L., et al. (2017). Hyperdopaminergism in lenticulostriate stroke-related restless legs syndrome: An imaging study. *Sleep Medicine, 30*, 136–138. https://doi.org/10.1016/j.sleep.2016.02.011.

Ryu, J. H., Lee, M. S., & Baik, J. S. (2011). Sonographic abnormalities in idiopathic restless legs syndrome (RLS) and RLS in Parkinson's disease. *Parkinsonism & Related Disorders, 17*(3), 201–203. https://doi.org/10.1016/j.parkreldis.2010.11.014.

San Pedro, E. C., Mountz, J. M., Mountz, J. D., Liu, H. G., Katholi, C. R., & Deutsch, G. (1998). Familial painful restless legs syndrome correlates with pain dependent variation of blood flow to the caudate, thalamus, and anterior cingulate gyrus. *The Journal of Rheumatology, 25*(11), 2270–2275.

Schmidauer, C., Sojer, M., Seppi, K., Stockner, H., Hogl, B., Biedermann, B., et al. (2005). Transcranial ultrasound shows nigral hypoechogenicity in restless legs syndrome. *Annals of Neurology, 58*(4), 630–634. https://doi.org/10.1002/ana.20572.

Spiegelhalder, K., Feige, B., Paul, D., Riemann, D., van Elst, L. T., Seifritz, E., et al. (2008). Cerebral correlates of muscle tone fluctuations in restless legs syndrome: A pilot study with combined functional magnetic resonance imaging and anterior tibial muscle electromyography. *Sleep Medicine, 9*(2), 177–183. https://doi.org/10.1016/j.sleep.2007.03.021.

Staedt, J., Stoppe, G., Kogler, A., Munz, D., Riemann, H., Emrich, D., et al. (1993). Dopamine D2 receptor alteration in patients with periodic movements in sleep (nocturnal myoclonus). *Journal of Neural Transmission. General Section, 93*(1), 71–74.

Staedt, J., Stoppe, G., Kogler, A., Riemann, H., Hajak, G., Munz, D. L., et al. (1995a). Nocturnal myoclonus syndrome (periodic movements in sleep) related to central dopamine D2-receptor alteration. *European Archives of Psychiatry and Clinical Neuroscience, 245*(1), 8–10.

Staedt, J., Stoppe, G., Kogler, A., Riemann, H., Hajak, G., Munz, D. L., et al. (1995b). Single photon emission tomography (SPET) imaging of dopamine D2 receptors in the course of dopamine replacement therapy in patients with nocturnal myoclonus syndrome (NMS). *Journal of Neural Transmission. General Section, 99*(1–3), 187–193.

Trenkwalder, C., Hogl, B., & Winkelmann, J. (2009). Recent advances in the diagnosis, genetics and treatment of restless legs syndrome. *Journal of Neurology, 256*(4), 539–553. https://doi.org/10.1007/s00415-009-0134-9.

Trenkwalder, C., Paulus, W., & Walters, A. S. (2005). The restless legs syndrome. *Lancet Neurology, 4*(8), 465–475. https://doi.org/10.1016/S1474-4422(05)70139-3.

Trenkwalder, C., Walters, A. S., Hening, W. A., Chokroverty, S., Antonini, A., Dhawan, V., et al. (1999). Positron emission tomographic studies in restless legs syndrome. *Movement Disorders, 14*(1), 141–145.

Tribl, G. G., Asenbaum, S., Klosch, G., Mayer, K., Bonelli, R. M., Auff, E., et al. (2002). Normal IPT and IBZM SPECT in drug naive and levodopa-treated idiopathic restless legs syndrome. *Neurology, 59*(4), 649–650.

Turjanski, N., Lees, A. J., & Brooks, D. J. (1999). Striatal dopaminergic function in restless legs syndrome: 18F-dopa and 11C-raclopride PET studies. *Neurology, 52*(5), 932–937.

Unrath, A., Juengling, F. D., Schork, M., & Kassubek, J. (2007). Cortical grey matter alterations in idiopathic restless legs syndrome: An optimized voxel-based morphometry study. *Movement Disorders, 22*(12), 1751–1756. https://doi.org/10.1002/mds.21608.

Unrath, A., Muller, H. P., Ludolph, A. C., Riecker, A., & Kassubek, J. (2008). Cerebral white matter alterations in idiopathic restless legs syndrome, as measured by diffusion tensor imaging. *Movement Disorders, 23*(9), 1250–1255. https://doi.org/10.1002/mds.22074.

von Spiczak, S., Whone, A. L., Hammers, A., Asselin, M. C., Turkheimer, F., Tings, T., et al. (2005). The role of opioids in restless legs syndrome: An [11C]diprenorphine PET study. *Brain, 128*(Pt 4), 906–917. https://doi.org/10.1093/brain/awh441.

Walters, A. S. (1995). Toward a better definition of the restless legs syndrome. The international restless legs syndrome study group. *Movement Disorders, 10*(5), 634–642. https://doi.org/10.1002/mds.870100517.

Walters, A. S., & Rye, D. B. (2009). Review of the relationship of restless legs syndrome and periodic limb movements in sleep to hypertension, heart disease, and stroke. *Sleep, 32*(5), 589–597.

Winkelman, J. W., Schoerning, L., Platt, S., & Jensen, J. E. (2014). Restless legs syndrome and central nervous system gamma-aminobutyric acid: Preliminary associations with periodic limb movements in sleep and restless leg syndrome symptom severity. *Sleep Medicine, 15*(10), 1225–1230. https://doi.org/10.1016/j.sleep.2014.05.019.

Winkelmann, J., Prager, M., Lieb, R., Pfister, H., Spiegel, B., Wittchen, H. U., et al. (2005). "Anxietas tibiarum". Depression and anxiety disorders in patients with restless legs syndrome. *Journal of Neurology, 252*(1), 67–71. https://doi.org/10.1007/s00415-005-0604-7.

Yu, G. S., Steinkirchner, T. M., Rao, G. A., & Larkin, E. C. (1986). Effect of prenatal iron deficiency on myelination in rat pups. *The American Journal of Pathology, 125*(3), 620–624.

Zhuo, Y., Wu, Y., Xu, Y., Lu, L., Li, T., Wang, X., et al. (2017). Combined resting state functional magnetic resonance imaging and diffusion tensor imaging study in patients with idiopathic restless legs syndrome. *Sleep Medicine, 38*, 96–103. https://doi.org/10.1016/j.sleep.2017.06.033.

Zucconi, M., & Ferini-Strambi, L. (2004). Epidemiology and clinical findings of restless legs syndrome. *Sleep Medicine, 5*(3), 293–299. https://doi.org/10.1016/j.sleep.2004.01.004.

FURTHER READING

Connor, J. R., Wang, X. S., Allen, R. P., Beard, J. L., Wiesinger, J. A., Felt, B. T., et al. (2009). Altered dopaminergic profile in the putamen and substantia nigra in restless leg syndrome. *Brain, 132*(Pt. 9), 2403–2412. https://doi.org/10.1093/brain/awp125.

Mrowka, M., Jobges, M., Berding, G., Schimke, N., Shing, M., & Odin, P. (2005). Computerized movement analysis and beta-CIT-SPECT in patients with restless legs syndrome. *Journal of Neural Transmission (Vienna), 112*(5), 693–701. https://doi.org/10.1007/s00702-004-0217-9.

CHAPTER THREE

Neuroimaging Applications in Tourette's Syndrome

Davide Martino*,†,1, Christos Ganos‡, Yulia Worbe§,¶,‖
*Department of Clinical Neurosciences, University of Calgary, Calgary, AB, Canada
†Hotchkiss Brain Institute, Calgary, AB, Canada
‡Department of Neurology, Charité, University Medicine Berlin, Berlin, Germany
§Centre de Référence National Maladie Rare 'Syndrome Gilles de la Tourette, Paris, France
¶Sorbonne Université, UMR S 1127, CNRS UMR 7225, ICM, Paris, France
‖Départment de Physiologie, Hôpital Saint-Antoine, Paris, France
1Corresponding author: e-mail address: davide.martino@ucalgary.ca

Contents

Abstract

Tics are neurodevelopmental hyperkinetic symptoms typically associated with unpleasant sensory experiences called premonitory urges. Tourette syndrome (TS) is the primary chronic tic disorder for which medical surveillance is most frequently required, and is associated with a complex phenotypical spectrum encompassing different types of

65

abnormal behaviors. Animal models of tics support their link to phasic activity changes throughout the sensorimotor loop of the cortico-basal ganglia-thalamo-cortical network. Event-related functional magnetic resonance imaging (fMRI) studies on patients with TS showed that the supplementary motor area relays preparatory signals related to tics to the primary motor area and other cortical regions relevant to action monitoring, following which cortico-basal ganglia-thalamo-cortical activation leads to the manifestation of tics. Despite their methodological heterogeneity, structural MRI studies highlighted the existence of anatomical markers of distinct sub-phenotypes of the TS spectrum. Initial evidence suggests that combining MRI structural methods and functional intrinsic connectivity assessed during resting state could even discriminate between TS patients and control groups. MR-spectroscopy and positron emission tomography studies suggest that TS may be related to a complex interplay between different neurotransmitters (particularly dopamine, GABA and glutamate), but discrepancy across studies prevents firm conclusions. Recent volumetric, cortical thickness and fMRI studies results showed an association between premonitory urges and somatosensory and insular cortical regions, involved in the processing of interoceptive and enteroceptive stimuli and motor output modulation. Finally, both structural and functional MRI studies have provided important support to the subtyping of the TS spectrum with respect to behavioral co-morbidities, in line with a "dimensional approach" to the classification of neuropsychiatric disorders, which is based on the identification of neurocognitive endophenotypes and of their anatomical substrate.

1. PHENOMENOLOGY AND BASIC PATHOLOGICAL MECHANISMS OF TICS

Tics are movements or sounds that resemble voluntary actions, but appear repetitive, in excess and in the absence of appropriate social context (Ganos & Martino, 2015). A perceptual event, labeled as premonitory urge, indicates and precedes the incipient occurrence of a tic. Interestingly, tics can also be inhibited, albeit temporarily. Primary tic disorders are the most common cause of tics. Tourette syndrome (TS) is the primary chronic tic disorder for which medical surveillance is most frequently required, defined by the co-occurrence of motor and phonic tics lasting at least 12 months.

Tics appear during early development, typically in children aged 4–6 years (Knight et al., 2012; Robertson, Eapen, & Cavanna, 2009; Sambrani, Jakubovski, & Muller-Vahl, 2016). However, they may already occur within the first years of life (Sambrani et al., 2016), in parallel with the emergence of voluntary actions. After onset, the clinical course of tics is variable. The worst-ever tic period is typically reported for the age of 12–14 years. From then on, only few studies explored prospectively the persistence of tics over time, suggesting that at least two-thirds of patients will go on to have some tics in their adult life (Bloch & Leckman, 2009). Unfortunately, there is no consensus on robust predictors of tic persistence into adulthood. Poor fine motor and visuospatial skills in childhood could predict subsequent tic severity about 7.5 years later (Bloch et al., 2011); caudate volumes in childhood, as measured by voxel-based morphometry, also seem to correlate with future tic severity (Bloch, Leckman, Zhu, & Peterson, 2005).

Understanding the anatomical substrate of tics has proven to be challenging. Moreover, several associated neuropsychiatric phenomena, including comorbid conditions such as attention-deficit hyperactivity disorder (ADHD) and obsessive-compulsive disorder (OCD), but also self-injurious behaviors, depression and anxiety disorder, are also present in approximately 90% of people with TS (Robertson, 2000). Therefore, the formulation of a precise research question with regard to the underlying pathophysiology of TS, and the definition of the appropriate methodologies to apply, should take into account this complex spectrum of phenotypes. On the other hand, tics and premonitory urges remain the defining core features of TS and chronic tic disorders, and therefore the investigation of the mechanisms that lead to the manifestation of tics is key to the understanding of the pathophysiological mechanisms of TS.

Data from rodents and primate studies have linked models of generation of abnormal tic-like movements to the presence of abnormal motor behaviors, such as stereotypies or overall hyperactivity (Bronfeld, Belelovsky, & Bar-Gad, 2011; McCairn, Bronfeld, Belelovsky, & Bar-Gad, 2009; Worbe et al., 2013; Yael, Vinner, & Bar-Gad, 2015). In the striatal disinhibition model, the local disinhibition of the sensorimotor portion of the striatum by means of γ-aminobutyric acid (GABA)$_A$ receptor antagonists (e.g. bicuculline, picrotoxin) microinfusion causes contralateral movements that have the typical, spatially and temporally discrete pattern of human tics (Bronfeld, Belelovsky, & Bar-Gad, 2011; McCairn et al., 2009; Pogorelov, Xu, Smith, Buchanan, & Pittenger, 2015; Worbe et al., 2013). In addition, this model offers very important insight into some key characteristics of tics, e.g. topographical and temporal variability of tics. Using pharmacological striatal disinhibition, specific tic-related phasic activity changes occur throughout the different regions along the sensorimotor loop of the cortico-basal ganglia-thalamo-cortical network (motor cortex, striatum, globus pallidus external and internal segments, substantia nigra pars reticulate, and thalamus). Second, the tic-inducing effects of striatal disinhibition display a somatotopic organization, whereby disinhibiting the anterior region

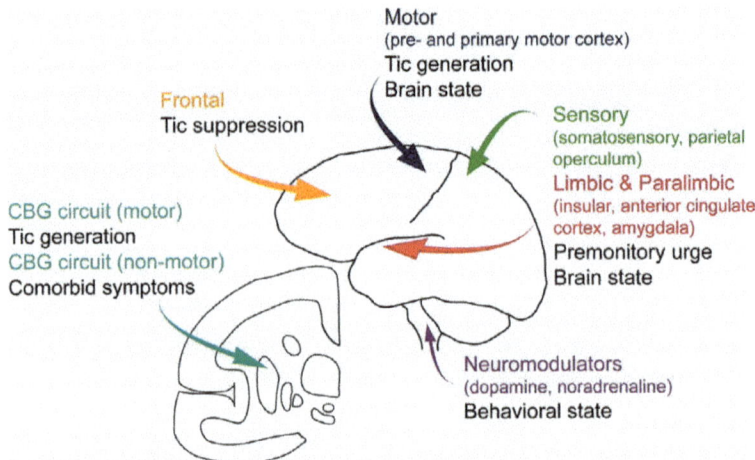

Fig. 1 Schematic diagram of the brain pathways areas associated with the expression and control of tics. *Reproduced with permission from Yael, D., Vinner, E., & Bar-Gad, I. (2015). Pathophysiology of tic disorders. Movement Disorders, 30, 1171–1178.*

of the dorsolateral striatum causes mostly forelimb tics and disinhibiting the posterior region causes primarily hindlimb tics.

Among the different methodologies employed to assess the pathophysiology of tics, neuroimaging has provided unique insights into both the neural mechanisms heralding the onset of tics, and the overall network structure of the brain of patients with tics (see Fig. 1 for a summary diagram). In Section 2, we provide an overview of the neuroimaging studies directly investigating the neural correlates of tic generation, followed by an overview of the studies that investigated the association between structural and functional neuroanatomy and the presence of a diagnosis of tic disorder. In Section 3, we focus on the anatomical and functional correlates of premonitory urges for tics, in the attempt to dissect their pathophysiological substrate from that of tic movements. Finally, in Section 4 we provide an overview of the studies that compared patterns of brain activity during the performance of behavioral tasks in patients with tics with respect to typically developing control subjects.

2. NEURAL CORRELATES OF TICS

2.1 Neural Correlates of Tic Generation

The phenomenological and neurophysiological resemblance of tics with voluntary actions, and the fact that tics undergo volitional inhibitory control but appear patterned and repetitive, suggests that the generation of tics results from the dysfunction of both neural pathways that generate voluntary motor behavior and brain regions associated with the release of involuntary movements. Indeed, two event-related functional neuroimaging studies support this view. First, Bohlhalter et al. (2006) used functional magnetic resonance imaging (fMRI) to investigate the neural antecedents of simple tics in a sample of 10 adult patients by assessing two specific time-points, i.e. 2 s before the onset of a tic and at the exact time of the tic manifestation. MRI-synchronized video/audio monitoring ensured that movements and sounds classified as tics were indeed captured, allowing a time-locked, event-related analysis. In this informative study, researchers also asked patients to imitate their tics in order to compare the functional neuroanatomic signatures of actual tics and imitated tic-like movements. Two seconds before the onset of actual tics, the authors documented an increase of blood oxygen level-dependent (BOLD) signal in the supplementary motor area, alongside a wider cortical and subcortical network, including the parietal operculum, the anterior cingulate and the insular cortex. Interestingly, the pattern of activation changed simultaneously in correspondence to the onset of tics, with increased activity in both the primary motor cortex and the cerebellum. Unfortunately, due to difficulties in methodology (i.e. difficulties in discriminating between imitated and actual tics, motion artifacts resulting from excessive head jerks related to effortful imitation), it was impossible to disentangle the differences in neural activation between actual tics and imitated tic-like behaviors. Addressing this limitation, a subsequent study by Wang et al. (2011) provided evidence that actual tics, when contrasted to imitated tic-like movements, are associated with an excess of activity of sensorimotor pathways within the cortico-basal ganglia circuits. A further exploration into the temporal neural dynamics of tic generation was conducted by Neuner et al. (2014) who investigated three event-related time points relevant to tic occurrence: (i) 2 s prior to tic-occurrence, (ii) 1 s prior to tic occurrence, and (iii) at the exact time point of tic onset. Neuner et al. (2007, 2014) also refined the video-documentation method for capturing tics by specifically developing an MRI-compatible camera system. Their final dataset comprised, however, only 16 TS patients, as an individual tic-to-tic approach necessitated a meticulous characterization of patients where the researchers could unequivocally judge tics as single events. In keeping with the study by Wang et al. (2011), activation patterns between the different time points systematically varied. Two seconds prior to tic occurrence, the authors observed increased BOLD signal at the supplementary motor and primary motor cortices (Fig. 2A), followed by insular,

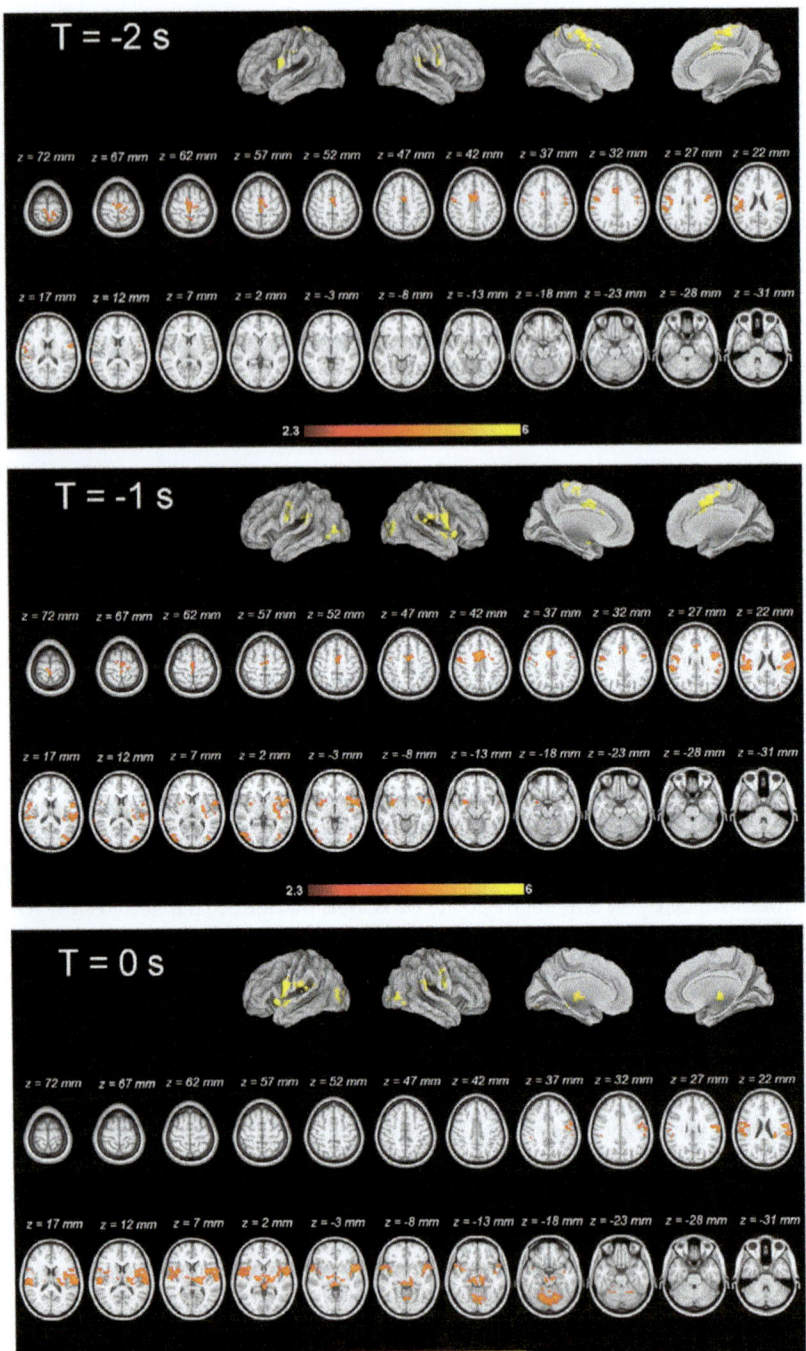

Fig. 2 (A) Tic-related activity 2 s before tic. (B) Tic-related activity 1 s before tic. (C) Tic-related activity. *Reproduced with permission from Neuner, I., Werner, C. J., Arrubla, J., Stocker, T., Ehlen, C., Wegener, H. P., et al. (2014). Imaging the where and when of tic generation and resting state networks in adult Tourette patients. Frontiers in Human Neuroscience, 8, 362.*

anterior cingulate cortex, putaminal and cerebellar hyperactivation at 1 s prior to tic onset (Fig. 2B). In contrast, at the actual time of tic onset, the authors documented hyperactivation of the thalamus, primary motor and somatosensory cortices (Fig. 2C). These results suggest that the supplementary motor area processes the preparatory signals related to motor tic behaviors through an interplay with the primary motor cortex (Neuner et al., 2014). Immediately after, the supplementary and primary motor cortices relay these signals to other cortical regions that are relevant to action monitoring, as for example the anterior cingulate cortex and the insula, following which the activation of output structures within the cortico-basal ganglia loops would ultimately lead to tics. Given the limited temporal resolution of event-related functional neuroimaging, these carefully conducted studies provided unique insights into the genesis of tics, indeed confirming the overlap between the neural structures involved in the generation of tics and voluntary actions. This body of evidence also clearly indicates that the selection of appropriate temporal windows in relation to tic onset represents a key factor in the design of this type of studies.

2.2 Brain Structural Changes Associated With the Diagnosis of TS and Other Chronic Tic Disorders

Several studies analyzed brain morphological changes in patients with primary tic disorders and TS (reviewed in Ganos, Roessner, & Munchau, 2013; Hashemiyoon, Kuhn, & Visser-Vandewalle, 2017; Neuner, Schneider, & Shah, 2013; Worbe, Lehéricy, & Hartmann, 2015). A summary of the findings from the most recent anatomical studies of TS is in Table 1. Despite significant differences among selected clinical samples (e.g. children vs adults, TS without vs TS with clinically relevant neuropsychiatric comorbidities, medication-naïve vs medicated) and applied imaging methodologies, several studies reported decreased prefrontal cortical thickness in patients with TS (Draganski et al., 2010; Fahim et al., 2010; Greene, Williams, Koller, Schlaggar & Black, 2017; Muller-Vahl et al., 2009; Sowell et al., 2008; Worbe et al., 2010). In some of these studies, correlation analyses of this finding with respect to tic severity showed that patients with a more pronounced decrease of prefrontal cortical thickness (or volume) also manifest greater tic severity. This finding led to suggest that the prefrontal cortical alterations could be associated with the emergence of tics, largely due to changes in the inhibition of prepotent motor programs. Although this view has gained consensus, the relationship between the inhibition of motor actions and the presence of tic

Table 1 Main Findings of the Relevant Anatomical MRI Studies ($n=17$) in Patients With Tourette Syndrome (TS) Published in the Last 5 Years

Reference	Age and Number of Participants	Medication-Free Period (or Proportion of Medicated Patients)	Imaging Methods	Statistically Significant Findings in Patients
Liu et al. (2013)	21 TS 7.9 ± 1.95 years 20 HC 8.05 ± 2.3 years	>1 month	T1 images analyzed using DARTEL–VBM; DTI using TBSS	Smaller gray matter volume in L superior temporal gyrus Trend for larger gray matter volume in L paracentral gyrus and R precentral gyrus Smaller white matter volume in R precuneus, R precentral gyrus and trend in L fusiform cortex, R frontal pole (negatively correlated to tic severity), R postcentral gyrus, L lingual gyrus Increased axial and mean diffusivity in anterior thalamic radiation, right cingulum bundle and forceps minor (positively correlated to tic severity and duration)
Cheng et al. (2013)	15 TS 34.5 ± 8.9 years 13 HC 34.6 ± 9.1 years	>3 weeks	DTI; probabilistic fiber tractography	Lower connectivity values in tracts connecting the SMA with basal ganglia and in frontal cortico-cortical circuits
Muller-Vahl et al. (2014)	19 TS 30.4 ± 11 years 20 HC 31.7 ± 10.9 years	>5 months	DTI	Decreased FA index in medial frontal gyrus bilaterally (negatively correlated to tic severity), pars opercularis of L inferior frontal gyrus, L middle occipital gyrus, R cingulate gyrus (negatively correlated to tic severity), R medial premotor cortex Increased ADC maps in L cingulate gyrus, prefrontal areas, L precentral gyrus and L putamen (positively correlated to tic severity)

Reference	Sample		Method	Medication	Findings
Jeppesen et al. (2014)	24 TS and 18 HC (children; no age details provided)	>6 months	DTI	None	
Worbe, Lehéricy, et al. (2015), Worbe, Marrakchi-Kacem, et al. (2015)	49 TS 29.9 ± 10.3 years 28 HC 29.7 ± 11.3 years	≈50% on medication	Diffusion- and T1-weighted images on which streamline probabilistic tractography algorithms were implemented		Enhanced structural connectivity of striatum and thalamus with M1 and S1, paracentral lobule, SMA and parietal cortices (positively correlating with tic severity), and with orbitofrontal and inferior frontal cortices, TPJ, medial temporal cortex and frontal pole
Debes et al. (2015)	22 TS mean age at baseline 14.2 years and mean age at follow-up 18.6 years 21 HC mean time interval 3.8 ± 0.45 years	2/22 patients on medication	VBM; DTI		Decreased parallel and perpendicular diffusivity, most pronounced in patients with persisting tics
Muellner et al. (2015)	52 TS 29.55 ± 8.5 years 52 HC 27.1 ± 6.1 years	52% on medications	Structural analysis of cortical sulci on 3D T1w magnetization-prepared rapid acquisition gradient echo, using BrainVISA		Lower depth and reduced thickness of gray matter in the pre- and post-central as well as superior, inferior, and internal frontal sulci (correlating with tic severity)
Delorme et al. (2016)	17 antipsychotic-medicated TS 29.3 ± 2.6 years 17 unmedicated TS 32.8 ± 3.2 years 17 HC 29.6 ± 2.8 years	Comparison between medicated and unmedicated part of study design	Behavioral habit formation task outside of the scanner and DTI		Engagement in habitual behavior correlated with greater structural connectivity within the right motor cortico-striatal network in medicated patients. Stronger structural connectivity of SMA with the sensorimotor putamen predicted more severe tics in unmedicated patients

Continued

Table 1 Main Findings of the Relevant Anatomical MRI Studies ($n = 17$) in Patients With Tourette Syndrome (TS) Published in the Last 5 Years—cont'd

Reference	Age and Number of Participants	Medication-Free Period (or Proportion of Medicated Patients)	Imaging Methods	Statistically Significant Findings in Patients
Draper, Jackson, Morgan, and Jackson (2016)	29 TS 14 ± 3.1 years 29 HC 14.3 ± 3.1 years	Not specified	Cortical thickness analysis using FreeSurfer	Reduced cortical thickness in bilateral sensorimotor and pre-motor areas, bilateral insular cortex, prefrontal and parietal cortex. Negative correlation with the severity of premonitory urges
Wolff et al. (2016)	26 TS 11.8 ± 1.1 years 24 HC 11.8 ± 1.1 years	All drug-naïve	DTI	Reduced axial diffusivity of all five segments of the corpus callosum (negative correlation with tic severity)
Wen et al. (2016)	27 TS 9 ± 3.4 years 27 HC 10.7 ± 3.3 years	Medication-free for unspecified period	DTI and quantification with TBSS	Decreased FA and increased radial diffusivity in deep white matter tract in cortico-striato–thalamo-cortical circuits and in superficial white matter (all correlating with tic severity)
Greene et al. (2017)	103 TS 11.9 ± 2.1 years 103 HC 11.9 ± 2.1 years	67/103 on medications	VBM	Lower white matter volume bilaterally in orbital and medial prefrontal cortex. Greater gray matter volume in posterior thalamus, hypothalamus and midbrain
Forde et al. (2017)	47 TS 10.5 ± 1.4 years 39 ADHD 10.7 ± 1.3 years 55 HC 11 ± 1 years	Proportion of medicated patients unclear (at least 11/47)	Basal ganglia volumetry and morphology	None

Study	Sample	Medication status	Method	Findings
Wen et al. (2017)	44 TS 9±3.1 years 41 HC 10.3±3.2 years	All drug-naïve	Probabilistic tractography with graph theoretical analysis	Decreased global and local efficiency, increased shortest path length and small worldness. Decreased nodal efficiency, mainly in the default mode, language, visual and sensorimotor systems
Martino, Delorme, et al. (2017)	13 TS 33±10.3 years 13 HC 32.8±6.8 years	5/13 on medications	DTI and probabilistic tractography of interhemispheric corpus callosum	Higher FA in both M1–M1 and SMA–SMA callosal fiber tracts
Schlemm et al. (2017)	13 TS 34.5±8.9 years 13 HC 34.6±9.1 years	>3 weeks	DTI and probabilistic tractography with graph theoretical analysis	Reduced connectivity in right hemispheric networks and reduced local graph parameters (local clustering, efficiency and strength). Increased normalized global efficiency of whole brain and right hemispheric networks (positively correlating with tic severity)
Sigurdsson et al. (2018)	35 TS 14±3.3 years 35 HC 13.9±3.3 years	8/35 on medications	Diffusion MRI analyzed using TBSS and probabilistic tractography	Decreased axial diffusivity and altered white matter connectivity. Positive correlation between M1–caudate connectivity and tic severity. Positive correlation between M1–insula connectivity and premonitory urge severity

To review findings from anatomical studies published prior to 2013, please consult Neuner et al., 2013 (Table 2.1, pp. 43–52). HC, healthy control subjects; VBM, voxel-based morphometry; DTI, diffusion tensor imaging; TBSS, tract-based spatial statistics; SMA, supplementary motor area; FA, fractional anisotropy; ADC, apparent diffusion coefficient; M1, primary motor cortex; S1, primary somatosensory cortex; TPJ, temporoparietal junction; ADHD, attention deficit hyperactivity disorder.

behaviors remains incompletely understood (Ganos, Kuhn, Kahl, Schunke, Feldheim, et al., 2014a; Ganos, Kuhn, Kahl, Schunke, Vrandt, et al., 2014b; Ganos et al., 2018; Morand-Beaulieu et al., 2017).

Beyond the prefrontal cortex, structural alterations have been documented in many other brain areas, involving most brain structures associated with sensorimotor processing (Ganos et al., 2013; Greene, Schlaggar, & Black, 2015; Hashemiyoon et al., 2017; Liu et al., 2013; Neuner et al., 2013; Worbe, Lehéricy, et al., 2015). Studies using larger samples, as the one by Worbe et al. (2010), were able to disentangle distinct structural changes underpinning different clinical phenotypes. Patients with more complex tic behaviors and obsessive-compulsive symptoms had wider patterns of cortical thinning than those with simple motor tics only (Worbe et al., 2010). Importantly, two subsequent studies from the same group further corroborated these results (Muellner et al., 2015; Worbe, Marrakchi-Kacem, et al., 2015). In one of them (Worbe, Marrakchi-Kacem, et al., 2015), the morphology of cortical sulci was investigated in a larger sample of adults with TS and control subjects ($n=52$ per group). Again, TS patients had morphological changes of cortical sulci that were specific for the two distinct phenotypes (patients with tics only and patients with tics + OCD). Importantly, the observed morphological changes correlated with either tic severity (e.g. central and superior temporal sulcus) or with the severity of obsessive-compulsive symptoms, e.g. left intermediate precentral sulcus and posterior calloso-marginal sulcus (Muellner et al., 2015). The other study analyzed cortico-subcortical connectivity in a dataset comprising scans from 49 TS adults and 28 controls (Worbe, Lehéricy, et al., 2015; Worbe, Marrakchi-Kacem, et al., 2015). Enhanced cortico-basal ganglia connectivity within sensorimotor networks (e.g. loops connecting the primary motor and somatosensory cortices with striatum and thalamus) was present in the TS group, which also positively correlated with tic severity. This study also reported distinctive correlations between obsessive-compulsive symptoms in TS patients with OCD and basal ganglia-orbitofrontal cortex connectivity. Taken together, these studies highlight the distinctive structural topographic qualities of TS and direct attention toward neuroimaging markers of sub-phenotyping for the different clinical presentations.

In addition to the above, other studies have assessed further structural connectivity measures in different populations with TS (Cheng et al., 2013; Muller-Vahl et al., 2014; Schlemm et al., 2017; Wen et al., 2017, 2016), generating results of difficult interpretation, often failing to demonstrate a strong correlation with clinical measures. Hence, their clinical relevance remains unclear. For example, two studies from the same group (Cheng et al., 2013; Schlemm et al., 2017) explored white matter connectivity in the same small sample of adults with TS ($n = 15$ in Cheng et al., 2013; $n = 13$ in Schlemm et al., 2017). Both these studies reported structural connectivity changes in fronto-cortico-cortical circuits, supplementary motor area-basal ganglia connectivity, and alterations in local vs global network efficiency in TS patients. However, given their small sample size and the number of examined variables alongside with only weak or partial correlations with clinical measures, it might be difficult to extrapolate the pathophysiological relevance of some of these findings. Interhemispheric connectivity may also occur in TS patients. Axial diffusivity of all five corpus callosum segments was reduced in 26 treatment-naïve boys with pure TS compared to 24 healthy control subjects, and correlated negatively with tic severity (Wolff et al., 2016).

Finally, longitudinal data on the association between the natural course of tics and structural connectivity are very limited. In a recent study following up 22 TS youths and 21 age-matched control subjects, patients with persistent tics at the end of adolescence exhibited a decrease in parallel and perpendicular diffusivity within caudate, thalamus and frontal lobe, at difference with remitting patients who were more similar to control subjects (Debes et al., 2015).

2.3 Brain Functional and Neurochemical Changes Associated With the Diagnosis of TS and Other Chronic Tic Disorders

Several review articles have summarized studies of the functional neuroanatomy of tic disorders (Hashemiyoon et al., 2017; Worbe, Lehéricy, et al., 2015; Zapparoli, Porta, & Paulesu, 2015). These studies have either investigated functional brain networks of an idle "ticcing" state, or assessed the functional neuroanatomical correlates of behavioral performance of patients with TS (Polyanska, Critchley, & Rae, 2017). Here, we focus on the first type of studies, whereas Section 4 will review the second type of studies. A summary of the findings from the most recent functional studies of TS is available in Table 2.

The assessment of an idle "ticcing" state can inform on the brain networks associated with the manifestation of tics, irrespective of specific cognitive (task-related) demands. The analysis of functional connectivity measures provides one such approach. Two seminal fMRI studies have consistently documented an overall reduction in long-range connectivity and an increase in short distance connectivity associated with motor

Table 2 Main Findings of the Relevant Functional (Functional MRI [fMRI], Magnetic Resonance Spectroscopy [MRS] and Positron Emission Tomography [PET]) Studies ($n=23$) in Patients With Tourette Syndrome (TS) Published in the Last 5 years

Reference	Age and Number of Participants	Medication-Free Period (or Proportion of Medicated Patients)	Imaging Methods	Statistically Significant Findings in TS Patients
de Vries et al. (2013)	14 TS 39 ± 13 years 10 HC 38 ± 8 years	Unspecified	H_2 ^{15}O PET associated with provocation of symmetry behaviors	Increased regional cerebral blood flow in the aCC, SMA, and inferior frontal cortex in TS patients during the asymmetry condition. Symmetry ratings during provocation correlated positively with orbitofrontal activation in TS patients
Denys et al. (2013)	12 TS 31 ± 8.1 years 12 OCD 35.8 ± 11.5 years 12 HC 32 ± 12 years	>6 months	^{11}C-raclopride PET at baseline and after administration of D-amphetamine (0.3 mg/kg)	Decreased baseline ^{11}C-raclopride binding potential in bilateral putamen of both patient groups (more pronounced in TS)—Post-amphetamine increase in tic severity correlated with binding potential changes in right ventral striatum
Thomalla et al. (2014)	15 TS 34 ± 9 years 15 HC 35 ± 9 years	>3 weeks	GE-EPI for BOLD signal changes related to Go/NoGo task related to finger movements	Lower task-related activation in primary and secondary motor cortical regions bilaterally Reduced co-activation between the L primary sensory-motor hand area and a network of contralateral sensory-motor areas and ipsilateral cerebellar regions

Cui et al. (2014)	17 TS 10.9 ± 3 years 15 HC 10.8 ± 2.9 years	All drug-naïve	Resting state fMRI computing ALFF and fALFF	Decreased ALFF in the posterior cingulate gyrus/precuneus and bilateral parietal gyrus. Decreased fALFF in anterior cingulate cortex, bilateral middle and superior frontal cortices and superior parietal lobule. Increased fALFF in left putamen and bilateral thalamus (positively correlated to tic severity in R thalamus)
Neuner et al. (2014)	16 TS 32 ± 12.1 years No HC	8/10 patients medicated	Resting state fMRI using two MRI-compatible cameras during data acquisition	2 s before a tic: increased activation of SMA, ventral M1, primary S1 and parietal operculum 1 s before a tic: increased activation of aCC, putamen, insula, amygdala, cerebellum and extrastriatal visual cortex At tic onset: increased activation of thalamus, central operculum, M1 and S1
Ganos, Kuhn, Kahl, Schunke, Feldheim, et al. (2014a)	14 TS 30.6 ± 8.8 years 15 HC 10.8 ± 2.9 years	3/14 medicated	Event-related fMRI during a stop-signal reaction-time task	Weaker dorsal premotor cortex in the StopSuccess than in the Go condition Positive correlation of SMA activation during StopSuccess vs Go with motor tic frequency

Continued

Table 2 Main Findings of the Relevant Functional (Functional MRI [fMRI], Magnetic Resonance Spectroscopy [MRS] and Positron Emission Tomography [PET]) Studies ($n=23$) in Patients With Tourette Syndrome (TS) Published in the Last 5 years—cont'd

Reference	Age and Number of Participants	Medication-Free Period (or Proportion of Medicated Patients)	Imaging Methods	Statistically Significant Findings in TS Patients
Tinaz et al. (2014)	For MRS: 14 TS 31.4±10 years 14 HC 31.5±9 years For fMRI: 13 TS 28±6.7 years 13 HC 31.4±9 years For MEG: 13 TS 32.5±9.6 years 13 HC 32.2±10.1 years	>2 weeks	Focused on sensorimotor cortex: GABA ¹H MRS; MEG; resting state fMRI	Abnormal correlation between the baseline beta band power and GABA+/Cre ratio. Increased functional connectivity of anterior insula to SMA
Kumar, Williams, and Chugani (2015)	12 TS 11±3 years 17 PANDAS 11.4±2.6 years 15 HC 28.7±7.9 years	Unclear	^{11}C-[R]-PK11195 PET	Increased ^{11}C-[R]-PK11195 binding potential in bilateral caudate nuclei
Ganos, Kahl, et al. (2014)	14 TS 30.6±8.8 years 14 HC 31.5±9 years	Not specified	Resting state fMRI during free tic release and voluntary tic inhibition	Increased regional homogeneity in left inferior frontal gyrus (pars orbitalis) during voluntary tic inhibition

Study	Participants	Time	Method	Findings
Deckersbach et al. (2014)	8 TS 26.9 ± 5.4 years 8 HC 25.6 ± 4 years	5/8 on medication and all had CBIT	Visuospatial priming task–based fMRI: assessment before and after CBIT	Decreased putamen activation from pre- to post-treatment Change in visuospatial priming task–related activation from pre- to post-treatment in inferior frontal gyrus that negatively correlated with changes in tic severity. Within a posteriori regions there was a significant negative correlation between initial PUTS score and visuospatial priming task–related activation in the superior temporal gyrus
Abi-Jaoude et al. (2015)	11 TS 32.2 ± 10.1 years 11 HC 34 ± 7.9 years	>3 months	^{11}C-raclopride and ^{11}C-(+)-PHNO PET	None
Tinaz, Malone, Hallett, and Horovitz (2015)	13 TS age range 18–46 years 13 HC age range 22–56 years	>2 weeks	Resting state fMRI	Higher degree of connectivity and node strength in the R thalamus; higher betweenness centrality in the R dorsal anterior insula and R putamen Reduced degree of connectivity in the L mid-cingulate and R supramarginal gyri; reduced betweenness centrality in the L SMA and right mid-insula

Continued

Table 2 Main Findings of the Relevant Functional (Functional MRI [fMRI], Magnetic Resonance Spectroscopy [MRS] and Positron Emission Tomography [PET]) Studies (n = 23) in Patients With Tourette Syndrome (TS) Published in the Last 5 years—cont'd

Reference	Age and Number of Participants	Medication-Free Period (or Proportion of Medicated Patients)	Imaging Methods	Statistically Significant Findings in TS Patients
Puts et al. (2015)	19 TS 10.5 ± 0.9 years 25 HC 10 ± 1.25 years	≈25% on medications	GABA-edited MRS	Reduced SM1 GABA concentration (correlating with motor tic severity)
Zapparoli et al. (2016)	28 TS 28.7 ± 12 years 24 HC 28.8 ± 14 years	17/28 on medications	fMRI related to fully executed and imagined finger opposition movements with either hand; VBM	Increased activation in premotor and prefrontal areas for both motor tasks for both hands, and in R rostral prefrontal and R temporo–parietal regions for motor imagery only. Activation of the premotor cortices during the motor imagery task only correlates with tic severity
Buse, Beste, Herrmann, and Roessner (2016)	22 TS 13.45 ± 1.8 years 22 HC 14.1 ± 1.9 years	9/22 on medications	fMRI related to pre-pulse inhibition of the startle reflex	Decreased pre-pulse inhibition-related activity in the middle frontal gyrus, postcentral gyrus, superior parietal cortex (positive correlation of the latter with pre-pulse inhibition) and caudate

Study	Sample	Medication	Method	Findings
Liao et al. (2017)	24 TS age range 7–15 years 32 HC age distribution not provided	19/24 on medications	Resting state fMRI	Abnormal intrinsic functional connectivity in the bilateral prefronto–striatum–midbrain networks and bilateral sensorimotor and temporal cortices. Negative correlation between functional connectivity of bilateral aCC and tic severity
Eddy, Cavanna, Rickards, and Hansen (2016)	25 TS 31.5±11.5 years 25 HC 29.9±10.1 years	10/25 on medications	Theory of Mind task-related fMRI	Increased activity of the R TPJ, R amygdala and posterior cingulate cortex. Increased TPJ activity correlated with tic, premonitory urges, impulse control problems and echophenomena severity. Amygdala activity correlated with premonitory urges severity and left TPJ activity with ratings of non-obscene socially inappropriate behaviors
Naaijen et al. (2016)	15 TS 10.4±1.2 years 39 ADHD 10.7±1.2 years 29 TS+ADHD 10.7±1.6 years 53 HC 10±1 years	2/15 TS; at least 9/29 TS+ADHD	Point resolved MRS	Obsessive-compulsive symptoms positively correlated with aCC glutamate concentration

Continued

Table 2 Main Findings of the Relevant Functional (Functional MRI [fMRI], Magnetic Resonance Spectroscopy [MRS] and Positron Emission Tomography [PET]) Studies ($n = 23$) in Patients With Tourette Syndrome (TS) Published in the Last 5 years—cont'd

Reference	Age and Number of Participants	Medication-Free Period (or Proportion of Medicated Patients)	Imaging Methods	Statistically Significant Findings in TS Patients
Kanaan et al. (2017)	37 TS at baseline 38.3 ± 11.1 years 15 TS after treatment with aripiprazole 40.1 ± 13.1 years 36 HC 38.4 ± 11.1 years	Aripiprazole medication part of the study design	Point resolved MRS	Decreased glutamate and glutamine concentrations and glutamine/glutamate ratio in L striatum Glutamate concentration in the thalamus was partially normalized by aripiprazole
Fan et al. (2017)	18 TS 35.1 ± 12.6 years 19 OCD 46.1 ± 9.7 years 22 HC 43.3 ± 14.2 years	9/18 TS on medications	MRS	Glutamine concentrations in dorsal aCC correlated negatively with tic severity
Liu et al. (2017)	21 TS 8.7 ± 3 years 29 HC 10.1 ± 3.1 years	Not specified	Resting state fMRI computing ALFF and fALFF	Abnormal ALFF or fALFF in vision-related structures including the calcarine sulcus, cuneus, fusiform gyrus and L insula. Decreased regional homogeneity in R cerebellum (positively correlated with TS duration)

Study	Participants	Medication/Duration	Method	Findings
Zapparoli et al. (2017)	28 TS 28.7 ± 12 years, 24 HC 28.8 ± 14 years	17/28 on medications	fMRI related to finger opposition task analyzed through a dynamic causal modeling	Increased intrinsic connectivity in the premotor network, with stronger connections from the SMA, the dorsolateral premotor cortex and the putamen to the R superior frontal gyrus. Positive correlation between connectivity strength from the R striatum to the R M1 and tic severity. Negative correlation between connectivity strength from the R SMA to the R M1 and tic severity
Mahone, Puts, Edden, Ryan, and Singer (2018)	32 TS 9.9 ± 1.9 years, 43 HC 8.1 ± 1.9 years	>3 months	MRS at 7T	Increased glutamate concentrations in the premotor cortex, associated with selective motor inhibition

To review findings from functional studies published prior to 2013, please consult Neuner et al., 2013 (Table 2.2, pp. 53–60). aCC, anterior cingulate cortex; SMA, supplementary motor area; OCD, obsessive–compulsive disorder; GE-EPI, gradient echo–planar imaging; BOLD, blood oxygenation level–dependent; ALFF, amplitude of low frequency fluctuations; fALFF, fractional amplitude of low frequency fluctuations; M1, primary motor cortex; S1, primary somatosensory cortex; SM1, primary sensorimotor cortex; GABA, γ-aminobutyric acid; Cre, creatine; MEG, magnetoencephalography; PANDAS, pediatric autoimmune neuropsychiatric disorders associated with streptococcal infection; CBIT, comprehensive behavioral intervention for tics; PUTS, Premonitory Urges for Tics Scale; PHNO, propyl-hexahydro-naphtho–oxazin; TPJ, temporo-parietal junction.

processing in TS (Church et al., 2009; Worbe et al., 2012). Resting state fMRI in children with TS showed also abnormally reduced spontaneous neuronal activity in a number of parietal, cingulate, insular and motor cortical regions, as well as the cerebellum (Cui et al., 2014; Liu et al., 2017). Overall, these findings could reflect the aberrant or "immature" brain development of people with TS. However, such changes may not be specific to TS. Indeed, similar connectivity patterns occur in several neurodevelopmental disorders, including autism spectrum disorders and ADHD (Kern et al., 2015). On the other hand, the employment of support vector machine classification on measures of resting state functional neuroimaging accurately discriminated between children with TS and controls ($n = 42$ in each group; Greene et al., 2016). Another multivariate pattern analysis of intrinsic functional connectivity was useful to confirm the presence of abnormal interhemispheric communication and its correlation to tic severity, particularly across homotopic voxels of the anterior cingulate cortices (Liao et al., 2017). It is thus likely that combining structural and functional connectivity methods could even aid clinical decision in patients with tics and TS, and inform counseling on prognosis.

Several neuroimaging protocols assessed the role of different neurotransmitters in the emergence of tics. A hyperdopaminergic state associated with enhanced phasic release of dopamine in the striatum has been postulated for TS. However, recent evidence from PET studies exploring dopamine receptor availability have led to discrepant results (Abi-Jaoude et al., 2015; Denys et al., 2013), and therefore the exact dopaminergic dysfunction associated with tics remains uncertain to date. Beyond dopamine, the inhibitory neurotransmitter GABA has been recently investigated, given the neuropathological evidence of altered distribution of GABAergic striatal interneurons (Kalanithi et al., 2005; Kataoka et al., 2010) and the striatal GABAergic disinhibition animal models of tic-like movements (McCairn et al., 2009). One study using positron emission tomography (PET) with [^{11}C]flumazenil, a GABA$_A$ receptor radioligand, in 11 adults with TS reported widespread decreased GABA$_A$ receptor binding potential in patients (Lerner et al., 2012). A subsequent magnetic resonance spectroscopy (MRS) study focusing on the primary somatosensory cortex of children with TS and control subjects documented reduced GABA spectral peak in TS (Puts et al., 2015). However, a recent study that investigated 32 children with TS and 43 controls using 7-T ^{1}H MRS (Mahone et al., 2018) failed to replicate these findings in a priori regions of interest (dorsolateral and ventromedial prefrontal cortex, premotor cortex, and striatum). Interestingly, this study also investigated concentrations of glutamate in the same areas, which were significantly increased in the premotor cortex, correlating with behavioral measures of motor inhibition but not with tic severity. A further study also documented MRS changes in glutamatergic neurotransmission in drug-free adults with TS (Kanaan et al., 2017), showing this time that patients with more tics at the time of study had lower striatal glutamate concentrations. However, another study on children with TS, TS + ADHD and ADHD and healthy controls failed to find group differences in glutamate concentration in the anterior cingulate cortex and striatum, whereas obsessive-compulsive symptoms severity was positively correlated to glutamate levels in the anterior cingulate cortex (Naaijen et al., 2016). This is in line with another study failing to show major differences in metabolite spectra between TS adults and control subjects (Fan et al., 2017). In summary, imaging data using radioactive tracers or spectroscopy provided some evidence in favor of neurochemical changes associated with the diagnosis of TS and/or with tic severity, but a number of reports also failed to confirm these findings, probably reflecting patient group heterogeneity. Despite these existing discrepancies, the classic view of aberrant dopaminergic neurotransmission dominating the pathophysiology of TS seems no longer tenable. Instead, recent evidence highlights that a more complex interplay between different neurotransmitters, including GABA and glutamate, in widespread neural locations may underpin the neurochemistry of tic generation in TS. Other monoaminergic neurotransmitters like norepinephrine and histamine may also contribute to these basic mechanisms through their down-regulation of striatal dopamine levels, but these still need to be investigated with imaging techniques in TS.

In the past few years, a few studies highlighted the potential involvement of neuroinflammatory mechanisms in the neurodevelopmental abnormalities underlying the genesis of tics. Lennington et al. (2016) have reported an increased number of CD45 + microglial cells and enriched expression of inflammatory genes in the striatum of post mortem brains from adults with pharmacologically refractory TS. PET imaging with [^{11}C]-[R]-PK11195, which binds to the translocator protein (TSPO) also known as the peripheral benzodiazepine receptor, is used to examine in vivo the presence of activated microglia. Children with TS manifested an increased binding of this ligand to the caudate nucleus bilaterally, suggestive of local increase of activated microglia, although these patients were compared to healthy adult subjects (Kumar et al., 2015).

3. NEURAL CORRELATES OF PU FOR TICS

3.1 Premonitory Urges for Tics: Basic Phenomenology and Assessment

Sensory phenomena commonly labeled premonitory urges typically precede tics, forming a phenomenological construct in which urges are considered the driving force behind the manifestation of tics (Martino, Madhusadan, Zis, & Cavanna, 2013). The construct of premonitory urges for tics displays strong similarities to the more general construct of urge-for-action. A general definition of an urge-for-action is a physical need to respond to a sensory stimulus with a motor action (Davenport, Sapienza, & Bolser, 2002). Therefore, the main difference between any type of bodily sensation and an urge-for-action is that the latter represents the combination of a bodily sensation charged with the potential to induce a motor response. Hence, a bodily sensation that does not generate a need to produce a motor response is not an urge-for-action, no matter how intense this sensation may be. In fact, the conscious awareness of the urge may vary considerably. This variability may depend on the intensity of the physiological afferent fueling the urge, and such intensity may be strongly influenced by the duration of the time during which the individual is actively suppressing the motor response to the urge (Jackson, Parkinson, Kim, Schuermann, & Eickhoff, 2011). For example, the urge to yawn may progressively increase in intensity as the yawning action is voluntarily inhibited. Moreover, the motor response to the urge temporarily abolishes the urge, which may reappear later in time under appropriate conditions. Similar to urges-for-action, premonitory urges for tics are perceived as sensory experiences generating the motor discharge that produces tics, are temporarily abolished by tic execution, and increase in intensity if the related tics are voluntarily inhibited. There is evidence showing a degree of overlap in body distribution for tics and premonitory urges. However, some tics may not be preceded by a premonitory urge, potentially indicating the existence of different types of tic behaviors in patients (Banaschewski, Woerner, & Rothenberger, 2003).

The majority of patients with tics and TS describe premonitory urges, although it is less clear whether their self-report is sensitive and reliable in children younger than 10 years of age. Despite the clinical belief that patients start manifesting the ability to report and describe premonitory urges about 3 years after the onset of tics, more recent clinimetric studies do not confirm this assumption (Gulisano, Calì, Palermo, Robertson, & Rizzo, 2015;

Raines et al., 2018), showing similar premonitory urge intensity in younger and older youth, and attributing this age difference to developmental changes in the ability to verbalize interoceptive experiences.

The phenomenological construct of premonitory urges does not explain the origin of the sensory experience from which the premonitory urge develops. The clinical description of the premonitory urges may also elude its basic mechanisms, as the verbalization of the urge differs substantially across individuals, particularly with respect to the quality of the bodily sensation. Nevertheless, its basic characteristic of being unpleasant remains ineludible, constantly producing negative aversion. Even differentiating whether a premonitory urge is an exteroceptive or an interoceptive sensation may be difficult. Hence, the subjective nature of premonitory urges and the lack of uniformity of patients' descriptions of premonitory urges represent an obstacle to the investigation of their neuroanatomical correlates. Also, there are very few clinimetric instruments to assess and rate premonitory urges. The Premonitory Urge for Tics Scale is recommended for clinical use in the assessment of premonitory urges, but it has also been repeatedly used to measure premonitory urge intensity in brain imaging and neurobehavioral studies (Woods, Piacentini, Himle, & Chang, 2005). However, this tool is currently under revision (McGuire et al., 2016), and it does not measure premonitory urge intensity in real time. More recently, a psychophysical method to monitor premonitory urges in real time has been proposed (Brandt et al., 2016), but not yet applied to neuroimaging paradigms.

3.2 Structural Correlates of Premonitory Urges for Tics

The brain structural correlates of the intensity of premonitory urges for tics have been investigated directly only by two studies. Draganski et al. (2010) performed voxel-based morphometry, voxel-based cortical thickness mapping and diffusion tensor imaging in 40 adults with TS, 14 of whom with co-morbid OCD and 19 of whom with co-morbid ADHD. These authors detected a statistically significant positive correlation between urge intensity and both gray matter volume and cortical thickness in the left somatosensory and prefrontal cortices. At the same time, fractional anisotropy values on diffusion tensor imaging indicated worse structural organization in the parietal portion of the superior longitudinal fascicle in patients with higher urge intensity. Overall, these findings suggest that, in adult patients with TS, an expansion of projections to the somatosensory cortex may be associated with greater premonitory urge intensity. These initial findings were replicated and expanded by Draper et al. (2016), who performed a cortical thickness analysis using FreeSurfer on 29 older children/adolescents with TS, exhibiting a lower degree of co-morbidities compared to the adult cohort in the study by Draganski et al. (2010). This latter investigation demonstrated a negative correlation between urge intensity and gray matter thickness in clusters located within sensorimotor cortical areas, as well as within the left insula. Conversely, there were no significant clusters of positive correlations between urge intensity and cortical gray matter thickness. Apart from confirming the association between premonitory urges and somatosensory cortical regions, the findings from Draper et al. (2016) show for the first time also a relationship between premonitory urges and the structure of the insular lobe. A subsequent larger study from the same group complemented this finding demonstrating an association between the frequency of premonitory urges and increased connectivity between primary motor cortex and the insula (Sigurdsson et al., 2018). These important finding are in keeping with the physiological role postulated for the insula in the generation of urges-for-action (see below).

3.3 What Is the Functional Correlate of Premonitory Urges for Tics?

Only until recently, our knowledge of the neural correlates of premonitory urges for tics was entirely based on functional changes in tic-related circuits that occur immediately prior to the release of tics, i.e. when premonitory urges are most likely to be experienced. As summarized above, Bohlhalter et al. (2006) recorded the BOLD signal from adults with TS during a resting condition (while relaxing and letting tics emerge without any inhibitory effort) and during active imitation of tics in random time intervals of 10–20s during which they did not feel the urge. Before tic onset, they observed bilaterally a significant, relative over-activation of the supplementary motor area, anterior cingulate cortex, parietal operculum, insula, putamen, thalamus and cerebellum. This finding suggested a role of paralimbic areas and parietal operculum as a potential substrate for premonitory urges for tics, although the authors did not perform a direct measurement of premonitory urges in this study. A very similar study design was proposed by Hampson, Tokoglu, King, Constable, and Leckman (2009), who demonstrated that patients reporting premonitory urges exhibited over-activation of the supplementary motor area prior and during tics compared to when intentionally imitating their tics. As previously mentioned, Wang et al. (2011)

used task-based fMRI obtaining two types of run from adult patients with TS, one allowing spontaneous tics to occur and the other involving emulation of a right-sided facial tic at a rate that eliminated the urge preceding that tic. They also obtained two run types from age-matched healthy volunteers, one with self-paced mimicking tics and the other with auditory cue-paced mimicked tics. A stronger activity and interregional causality were observed throughout the motor pathway, within primary somatosensory areas, the posterior parietal cortex, amygdala and hippocampus while releasing spontaneous tics; at the same time they reported weaker activity in the caudate and the anterior cingulate cortex, regions that regulate top–down control over motor pathways. Wang et al. (2011) also observed stronger connectivity between primary somatosensory and primary motor cortices during spontaneous tics compared to self-paced mimicked tics. These authors comment their findings as indirect evidence of an involvement of the somatosensory cortical areas and their connection to the primary motor cortex in the modulation of premonitory urges, as well as of amygdala and hippocampus in the genesis of the emotional discomfort associated with the urges.

Ganos, Kahl, et al. (2014) were the first to correlate brain activity to a direct measure of premonitory urges, performing resting state fMRI in 14 young adults with TS without comorbid diagnoses ADHD and OCD (3 of whom treated with dopamine receptor blockers). Their design included two conditions, one in which subjects were instructed to simply keep their eyes closed and one in which subjects were instructed to maximally inhibit their tics with their eyes closed. These authors found a correlation between the regional homogeneity in the left inferior frontal gyrus (pars orbitalis) and capacity to inhibit their tics, but not with number of motor tics or with urge intensity. Although not indicating a direct functional correlate of premonitory urges for tics, these results suggest a functional dissociation between premonitory urges and tic inhibition (voluntary tic control), as these two behavioral states did not share neural activation patterns. Subsequently, Tinaz et al. (2015) compared the whole brain functional connectivity of 13 adults with TS (5 with comorbid ADHD and 9 with comorbid OCD) to that in 13 age-matched control subjects while keeping their eyes closed at rest in the scanner. Although tics were not directly captured during the fMRI session, these authors documented a significant, positive correlation of urge intensity to the unthresholded connectivity between right dorsal anterior insula and right or left supplementary motor area. Interestingly, they did not find any significant correlation between connectivity parameters and tic severity score in any brain region.

Overall, this evidence shows consistently a link between premonitory urges for tics and the degree of connectivity between cortical regions involved in the processing of interoceptive and enteroceptive stimuli, including their emotional characterization, and motor output modulation. These regions comprise primary and secondary somatosensory areas as well as the supplementary motor area, which are strongly connected to both the primary motor cortex and the insula. This latter region has a crucial role for the convergence and integration of somatosensory and visceral sensations, and their emotional characterization. Using a quantitative activation likelihood estimation method of meta-analysis of neuroimaging data, Jackson et al. (2011) demonstrated overlap in the brain activation of the right insula and the dorsal anterior cingulate cortex in association with prototypical examples of urge-for-action, i.e. swallowing, yawning and micturition. This supports the conceptualization of a neural circuit that links bodily sensations to urges-for-action, subsequently contributing to the selection of motor actions that have an advantageous outcome, e.g. the resolution of the aversive sensory experience that generated the same urge (Fig. 3). The absence of a reliable physiological marker

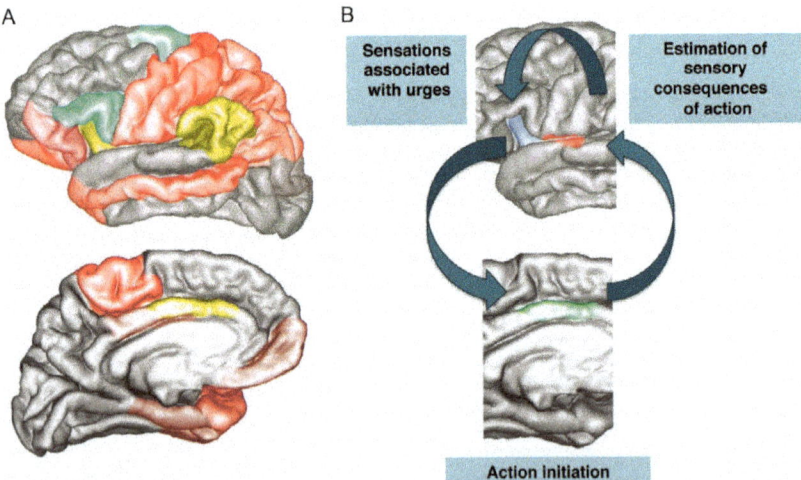

Fig. 3 (A) Schematic representation of the structural and functional abnormalities in patients with Tourette syndrome: in red—cortical areas manifesting abnormal structural connectivity with basal ganglia in adult Tourette syndrome patients compared to controls; in green—main cortical areas involved in mechanisms of cognitive control over tics (supplementary motor area and inferior frontal gyrus); in yellow—cortical areas involved with premonitory urges. (B) Schematic representation of the premonitory urges neuroanatomical model, where motor area of anterior cingulate cortex is represented in green, anterior insula in blue and middle part of the insula in red. *Reproduced with permission from Worbe, Y., Lehericy, S., & Hartmann, A. (2015). Neuroimaging of tic genesis: Present status and future perspectives.* Movement Disorders, 30, 1179–1183.

(e.g. an electrophysiological correlate) of premonitory urges for tics poses serious limitations to the possibility of studying the simultaneous functional neural correlate of this sensory experience using fMRI. Real-time monitoring approaches to measure urges are limited by the use of a motor feedback that would likely confound the interpretation of neural activation patterns. Indirect approaches, e.g. assessing the correlation between amelioration of urge intensity and connectivity of putative urge-related networks after a therapeutic intervention, might represent valuable alternatives.

3.4 What Is the Neurochemical Correlate of Premonitory Urges for Tics?

Only one study has directly analyzed neurotransmitter changes in vivo in relation to premonitory urge severity. Kanaan et al. (2017) compared 37 adults with TS, 12 of whom with ADHD and 11 with OCD, to 36 age-matched control subjects using MR Point-RESolved Spectroscopy. Applying an accurate, automated voxel relocalization method and absolute metabolite quantitation considering within voxel compartmentation, these authors observed reduced concentrations of glutamine and glutamate, as well as decreased glutamine/glutamate ratio, in the left striatum, and decreased concentrations of glutamate only in the thalamus of TS patients. These changes were partially reversed by treatment with a partial dopamine receptor agonist like aripiprazole. On multiple regression analyses, there were strong, negative correlations between tic severity measured using the Rush Video-based Tic Rating Scale and striatal glutamine levels, and between urge intensity and thalamic glutamate levels. These findings point to a potential alteration in the GABA-glutamate-glutamine cycling and metabolism within subcortical structures of the cortico-basal ganglia loop, which rely on balanced astrocyte-neuron coupling.

3.5 Can We Model the Urge-Tic Complex? Computational Approaches and Therapeutic Implications

Overall, neuroimaging data are contributing to elucidate the main neural substrates of premonitory urges and their role in tic execution and consolidation. All the pathophysiological models for the premonitory urge-tic complex propose the sensorimotor loop of the cortico-basal ganglia circuitry as the key network, but structural and functional MRI studies have indicated the involvement of other cortical regions that may communicate with this loop or with other parallel, segregated cortico-basal ganglia loops to subdue premonitory urge and tic generation. In the case of premonitory urges specifically, key cortical regions involved are the primary and secondary somatosensory areas of the parietal lobe, as well as the supplementary motor area, the anterior lobe of the insula (dorsal aspect), and the cingulate motor area. The connections between pure sensory areas and the insula could be relevant to the integration between sensory and emotional information required to define the primary aversive character of the premonitory urges. This would then be key in sustaining tic execution through the strong interconnectivity between insula and supplementary motor area and cingulate motor areas, ultimately modulating primary motor cortex output for tic execution. The perpetuation of this loop would promote the learning and consolidation of the premonitory urge-tic behavioral complex. The substrate of this consolidation mechanism corresponds to the basal ganglia modulation of these cortical connections, which depends on the levels of striatal dopamine, in particular on phasic dopamine release (Conceicao, Dias, Farinha, & Maia, 2017; Maia & Conceicao, 2017). Whether cortical regions such as the insula, which defines the aversive nature of the premonitory urges that drive tics, directly modulate striatal dopamine levels remains to be determined. Likewise, the direct relationship between premonitory urge intensity and striatal dopamine levels remains to be elucidated. MR spectroscopy data like those provided by Kanaan et al. (2017), however, indirectly show that chronic perturbations in the flux of metabolites in the GABA-glutamate-glutamine cycle could alter the balance between excitatory and inhibitory neural transmission in specific foci of the striatum and thalamus, possibly leading to alterations in the dynamics (tonic/phasic) of basal ganglia dopamine signaling. These authors have indeed documented that such changes in neurotransmitter/metabolites strikingly correlate to both premonitory urge and tic severity.

Finally, some of these findings from neuroimaging studies might have translational therapeutic implications. For instance, current views of behavioral interventions for tics postulate, as a primary therapeutic goal, the gradual abolition of premonitory urges with a progressive decrease of their intensity during the course of treatment (McGuire et al., 2015). If the generation of premonitory urges and their aversive characterization are necessary to the execution and subsequent learning and consolidation of tics, then pre-treatment activation levels of sensory and insular cortical regions would be helpful to predict response to behavioral therapies. On a similar vein, these cortical regions could be targeted directly by non-invasive brain stimulation techniques (e.g. repetitive transcranial magnetic stimulation and transcranial current stimulation techniques), either as a standalone treatment or in association with behavioral treatment, in order to facilitate premonitory urge extinction.

4. NEURONAL SIGNATURE OF BEHAVIORAL DISORDERS AND COGNITIVE FUNCTIONS ASSOCIATED WITH TICS

Behavioral co-morbidities are frequently associated with tics in the context of primary TS. In particular, OCD, depression and ADHD are the most frequent comorbidities associated with TS in both children and adults. Clinically, these associations have considerable impact on the tic severity and long-term outcome of the disorder. For instance, one recent study has highlighted the burden of anxiety and depression on disability in youngsters with TS (Rizzo, Gulisano, Martino, & Robertson, 2017). The authors found that depression was significantly associated with factors such as tic severity. Another study that addressed the question of prognostic factors in TS showed that tic severity was strongly associated, among other factors, also with the overall number of psychiatric co-morbidities (Sambrani et al., 2016).

At a pathophysiological level, the association of tics with psychiatric co-morbidities can result from the disruption of several cortico-basal ganglia loops, typically considered as a main pathophysiological substrate in this disorder. This disruption can operate through abnormalities of cognitive processes common to both TS and specific co-morbid disorder, according to a "dimensional approach" to neuropsychiatric disorders, a pathophysiological concept proposed during the past decade (Robbins, Gillan, Smith, de Wit, & Ersche, 2012). According to this conceptual framework, a neuropsychiatric disorder could be characterized by a specific "neurocognitive endophenotype," based on changes in behavioral or cognitive processes, which are associated with deficits in defined neural systems. For instance, reliance on habitual, at the expense of goal-directed, behavioral control has been proposed as a potential cognitive endophenotype of compulsive disorders (Voon et al., 2015) that encompass conditions such as OCD, food or substance addictions.

In TS, several cognitive functions have been probed as potential predictors of symptom severity, remission and response to treatment. For instance, sensitivity to reward (Palminteri et al., 2009; Worbe et al., 2011) and habit formation mechanisms (Delorme et al., 2016) have been highlighted as promising cognitive markers in TS. In children with TS, the introduction of rewards within behavioral therapy protocols can improve the outcome of these therapeutic approaches (Greene et al., 2015).

In this section, we will focus on neuroimaging research aiming at the identification of the neuronal substrates for potential "cognitive endophenotypes" of TS and associated co-morbid disorders.

4.1 Neural Correlates of Common Behavioral Co-morbidities Associated With TS

4.1.1 ADHD in the Context of TS

The frequent comorbidity of TS and ADHD likely reflects a common underlying neurobiological substrate. Several behavioral and neuroimaging studies suggested the involvement of partly overlapping, albeit still separate, fronto-striatal circuits in both TS and ADHD. From these studies, poor inhibitory control mediated by volumetric reductions in fronto-striatal circuits appears to be a core feature of ADHD, whereas the abnormalities within the motor cortico-basal networks seem to be core features of TS (Draganski et al., 2010; Plessen, Royal, & Peterson, 2007; Worbe et al., 2010).

Despite this evidence, neuroimaging studies comparing patients with TS-alone, TS + ADHD, and ADHD without tics remain limited in number. A recent study of basal ganglia morphology in children with TS and/or ADHD within an 8–12 years age range could not find any between group difference in basal ganglia volume and shape (Forde et al., 2017). A number of studies that explored the morphology of other deep brain structures highlighted abnormalities in the amygdala and hippocampal complex in both children and adults with TS + ADHD and suggested that the left inferior frontal gyrus might represent a common underlying structural correlate of TS with co-morbid ADHD (Ludolph et al., 2008; Peterson et al., 2007; Wittfoth et al., 2012).

Interestingly, one study tackled the issue of diagnostic accuracy of various neuropsychiatric disorders using imaging-based measures and semi-supervised machine learning algorithms (Bansal et al., 2012). The results from this study showed that patterns of variations across the surface morphology of brain structures derived from the scans were highly predictive of the diagnosis of specific chronic neuropsychiatric disorders. In particular, the morphology of distinct parts of the globus pallidus, caudate nucleus, putamen, thalamus and amygdalo-hippocampal complex could discriminate between patients with a diagnosis of ADHD and patients with a diagnosis of TS. These results suggest that basal ganglia morphology differs substantially between ADHD and TS, even if these differences are not captured by conventional imaging modalities.

4.1.2 OCD in the Context of TS

OCD is a frequent co-morbidity in patients with TS (Martino, Ganos, & Pringsheim, 2017), frequently manifesting as a "tic-related" phenomenological subtype typically associated with "just right" phenomena. However, it is not clear if this particular subtype differs from "classical" OCD at the network connectivity level, since imaging studies comparing "classical" OCD to "tic-related" OCD have not been published to date.

Similar to "pure" OCD without tics, the presence of OCD was found to be associated with volume reduction in the caudate nucleus (Bloch et al., 2005), as well as with a reduction of cortical thickness in the ventromedial prefrontal cortex and hippocampus (Worbe et al., 2010). Interestingly, in an fMRI study in which patients were asked to perform a simple probabilistic learning task, patients with TS and associated OCD did not exhibit efficient learning within this task, due to impaired sensitivity to reward reflected by underactivation of the ventromedial prefrontal cortex and ventral striatum. The activity in the ventromedial prefrontal cortex was also inversely correlated to the severity of compulsive behaviors (Worbe et al., 2012). Such aberrant activity in incentive motivation neuronal network was found also in OCD patients without tics (Jung et al., 2011; Kaufmann et al., 2013), suggesting that this could be a common cognitive mechanism between these two potentially distinct OCD subtypes. At a specific symptom level, symmetry and ordering behaviors related to "just-right" phenomena were associated with increased activity in a limbic network encompassing anterior cingulate, supplementary motor and inferior frontal cortices (de Vries et al., 2013).

At a large-scale brain network level, the "tic-related" OCD subtype was characterized by abnormal functional and structural connectivity within associative and limbic cortico-basal ganglia loops. In particular, applying the small-world theoretical approach to resting state functional neuroimaging data and diffusion tensor imaging, it was shown that the strength of connections of the medial orbito-frontal and dorsolateral-prefrontal cortices with other cortical areas and with the basal ganglia correlated to OCD severity in TS patients (Worbe, Lehéricy, et al., 2015; Worbe et al., 2012; Worbe, Marrakchi-Kacem, et al., 2015). This is in line with the previous structural (Chamberlain et al., 2008) and functional connectivity (Sakai et al., 2011) results obtained from patients with "pure" OCD without tics. In addition, a dysfunction of the dorsolateral prefrontal cortex in OCD patients without tics has been shown to lead to a lack of a cognitive and attention flexibility, which maintained the repetitive character of OCD (Menzies et al., 2008). More recent results on the "pure" OCD population also highlighted that reduced connectivity between putamen and dorsolateral prefrontal cortex is associated with greater obsessive-compulsive symptom severity (Vaghi et al., 2017). Interestingly, one study also highlighted an abnormal connectivity of cerebellar regions in these patients, in particular a stronger connection of cerebellar regions with the basal ganglia compared to control subjects. In TS, co-morbid OCD was associated with a relative enlargement of several cerebellar regions, including crus and lobules VI, VIIB, and VIIIA, in proportion to the increasing severity of obsessive-compulsive symptoms, whereas tic severity was correlated to a decrease in volumes of these cerebellar regions (Tobe et al., 2010). Overall, it is possible to hypothesize, based on these findings, that both "tic-related" and "pure" OCD rely on at least partially overlapping neural circuits and share underlying cognitive mechanisms.

4.2 Neural Correlates of Cognitive Control in TS: Insights From Social Cognition, Habit Formation and Cognitive Control of Motor Behavior

4.2.1 Neural Correlates of Social Cognition in TS

Social cognition is defined as the ability to understand mental states, such as beliefs and emotions of other people, which are crucial for a successful interaction within the society. Alterations of social cognition and decision-making can be seen in many neurological and psychiatric disorders including schizophrenia, substance addiction, anxiety and personality disorders (Hinterbuchinger, Kaltenboeck, Baumgartner, Mossaheb, & Friedrich, 2018). In experimental settings, social cognition domains are usually assessed using a battery of tests tapping on its main domains that comprise social perception abilities (e.g. facial recognition), Theory of Mind, empathy, and social behavior (including imitation abilities).

In TS, 25–30% of patients report experiencing behavioral symptoms consisting of urges to act in a socially inappropriate way or make socially inappropriate remarks, including insults, a group of behaviors labeled as non–obscene socially inappropriate behaviors (Eddy & Cavanna, 2013).

Recently, a dysfunction of neuronal networks that underlies social decision-making has been proposed as one important pathophysiological mechanism operating in TS (Albin, 2018; Vicario & Martino, 2018). Behavioral and neuroimaging studies investigating directly social cognition in TS are limited. In one of these studies, TS patients were invited to recognize emotionally salient expressions of human eyes during the acquisition of functional neuroimaging scans. The results of this work showed that, during this task, TS patients exhibit greater activity than control subjects within several brain areas including orbitofrontal cortex, posterior cingulate, right amygdala and right temporo-parietal junction, as well as decreased activity in the left inferior parietal cortex. In addition, the level of activity in the temporo-parietal junction also correlated with global ratings of the urge to tic (Eddy, Cavanna, & Hansen, 2017). Overall, these results highlighted important modifications in neuronal activity in response to emotional and social stimuli in TS. Another study from the same group also investigated the neuronal activity in TS during the performance of a task measuring Theory of Mind, in which subjects were invited to reason about an incorrect belief (false belief condition) and to reason about an out-dated physical representation (false photo condition) (Eddy et al., 2016). Compared to healthy volunteers, TS patients showed a lower activity in temporo-parietal junction, right amygdala and posterior cingulate cortex on this task. Further analyses revealed that the metabolic activity across several brain regions correlated with the severity of different symptom domains, including echophenomena, impulsivity and premonitory urges. In particular, notable findings were a positive correlation between the activity of the amygdala and the severity of premonitory urges, as well as a positive correlation between the activity of the left temporo-parietal junction and the severity of non-obscene socially inappropriate behaviors.

4.2.2 Habitual Control in TS and Its Neural Correlates

In everyday life, optimal behavioral performance results from a balance between adaptive flexible behavioral choices and more rigid, repetitive choices, which are supported respectively by brain networks known as goal-directed and habitual brain systems. In the case of TS, tics are sometimes perceived as intentional actions performed in response to sensory stimuli and thus could be described as voluntary movements performed in an automatic or habitual way (Leckman & Riddle, 2000). Habit formation in TS has been the object of a number of studies that adopted behavioral paradigms, only one of which evaluated also the anatomical correlates of the goal-directed and habitual systems in this condition. Compared to healthy volunteers, TS patients show a propensity to rely on habitual control at expense of goal-directed behavioral control. This behavioral pattern is more prominent in un-medicated TS compared to both healthy volunteers and medicated TS patients. At an anatomical level, the tendency to rely on habitual behavioral control was predicted by structural connectivity of white matter within the premotor cortex and putamen (Delorme et al., 2016). Interestingly, behavioral interventions used to treat tics, in particular habit reversal training, focus on training patients to become more aware of premonitory urges and to initiate competing movements that counteract the occurrence of tics. Only one imaging study so far has directly addressed the question of the neuroanatomical correlates of the mechanisms by which habit reversal training improves tics. In this study, Deckersbach et al. (2014) observed a significant decrease in putamen activation comparing the pre-treatment to the post-treatment phases in a group of TS patients undergoing this intervention.

4.2.3 Cognitive Control of Motor Behavior in TS and Its Neural Correlates

The ability to suppress competing attentional and behavioral responses is known as cognitive or "self-regulatory" control. This ability develops throughout childhood and is underpinned by the maturation of the functional activity of a relevant portion of the frontal cortex and the basal ganglia. One model of cognitive control suggests that different regions of the frontal cortex are involved in the programming and execution of context-appropriate motor behaviors, whereas the basal ganglia are engaged in the inhibition of competing, undesired motor behaviors (Casey, Tottenham, & Fossella, 2002). In the case of neurodevelopmental disorders such as TS, a disruption in cognitive control of motor behaviors is hypothesized to occur at the level of both these brain structures.

Cognitive control has been initially investigated in TS using a version of the Stroop task, in which a subject is required to inhibit an automatized response (usually reading) in favor of another response (usually color naming). Neuroimaging studies in healthy volunteers showed that the performance on this task improves with age (from childhood to adulthood), and that this improvement is accompanied, on one hand, by an increased activation of fronto-striatal networks and, on the other hand, by greater deactivation of the ventral and posterior cingulate cortices (Marsh, Maia, & Peterson, 2009).

A study on a large group of patients with TS, including both children and adults performing the Stroop task, showed that TS patients deactivate the ventral prefrontal and posterior cingulate cortices to a lesser extent than healthy subjects with the advancing age. Moreover, in order to maintain a level of performance on the task that is comparable to that of healthy volunteers, patients with TS manifest increased activation of the prefrontal dorsolateral cortex, striatum, globus pallidus, and thalamus (Marsh, Zhu, Wang, Skudlarski, & Peterson, 2007). In another study employing this task in children with TS without major behavioral comorbidities, those with more severe tics exhibited a slower performance. On imaging, children with more severe tics showed greater activity in the substantia nigra/ventral tegmental area, striatum, thalami, and subthalamic regions; also, TS patients displayed stronger activation of the left prefrontal cortex than healthy volunteers during the performance of this task (Baym, Corbett, Wright, & Bunge, 2008). On a task assessing action inhibition like the Go/NoGo reaction time task, TS patients showed slower performance and lower brain activation in primary and secondary motor cortices only on the Go trials but not during response inhibition (Thomalla et al., 2014).

Overall, these results suggest delayed development of cognitive control systems in TS, which is associated with a greater prefrontal cortex activity associated with the preservation of a normal performance on the cognitive task, and with greater activity of cortico-basal ganglia networks, possibly reflecting a compensation for the relative inability to inhibit the undesired motor actions, e.g. tics and socially inappropriate behaviors.

5. CONCLUSIONS

During the past decade, brain imaging methodologies have contributed to advance our understanding of the basic mechanisms of tics, premonitory urges and associated cognitive and behavioral abnormalities. These approaches are also paving the way to the exploration of other functional characteristics of tic disorders, in particular to electrophysiological research on oscillatory dysrhythmias within the cortico-basal ganglia networks associated with the generation and control of tics (Hashemiyoon et al., 2017). Event-related fMRI has helped to unfold the brain activation patterns related to the preparation and execution of tics. Beyond the understanding of the core phenomenon of tic disorders, the urge-tic complex, new analytic methods exploring structural and intrinsic functional connectivity have a great potential to identify specific network-based signatures of sub-phenotypes of the complex TS spectrum. More research is also necessary to understand how these neural changes modify over time in association with clinical evolution. Although molecular imaging methods (e.g. PET and MR spectroscopy) could provide invaluable information with strong translational impact, the discrepancies across the studies available so far suggest that sample size, inclusion criteria, adjustment for confounders such as treatments, and selection of regions of interest are key components to address very rigorously in designing new studies. Finally, combined structural and functional imaging methods have the potential to provide disease markers that could be exploited to monitor response and optimize patient selection for novel treatment approaches in TS like behavioral and neuromodulatory strategies.

REFERENCES

Abi-Jaoude, E., Segura, B., Obeso, I., Cho, S. S., Houle, A. E., Lang, A. E., et al. (2015). Similar striatal D2/D3 dopamine receptor availability in adults with Tourette syndrome compared with healthy controls: A [^{11}C]-(+)-PHNO and [^{11}C]raclopride positron emission tomography imaging study. *Human Brain Mapping, 36*, 2592–2601.

Albin, R. L. (2018). Tourette syndrome: A disorder of the social decision-making network. *Brain, 141*, 332–347.

Banaschewski, T., Woerner, W., & Rothenberger, A. (2003). Premonitory sensory phenomena and suppressibility of tics in Tourette syndrome: Developmental aspects in children and adolescents. *Developmental Medicine and Child Neurology, 45*, 700–703.

Bansal, R., Staib, L. H., Laine, A. F., Hao, X., Xu, D., Liu, J., et al. (2012). Anatomical brain images alone can accurately diagnose chronic neuropsychiatric illnesses. *PLoS One, 7*, e50698.

Baym, C. L., Corbett, B. A., Wright, S. B., & Bunge, S. A. (2008). Neural correlates of tic severity and cognitive control in children with Tourette syndrome. *Brain, 131*, 165–179.

Bloch, M. H., & Leckman, J. F. (2009). Clinical course of Tourette syndrome. *Journal of Psychosomatic Research, 67*, 497–501.

Bloch, M. H., Leckman, J. F., Zhu, H., & Peterson, B. S. (2005). Caudate volumes in childhood predict symptom severity in adults with Tourette syndrome. *Neurology, 65*, 1253–1258.

Bloch, M. H., Sukhodolsky, D. G., Dombrowski, P. A., Panza, K. E., Craiglow, B. G., Landeros-Weisenberger, A., et al. (2011). Poor fine-motor and visuospatial skills predict persistence of pediatric-onset obsessive-compulsive disorder into adulthood. *Journal of Child Psychology and Psychiatry, 52*, 974–983.

Bohlhalter, S., Goldfine, A., Matteson, S., Garraux, G., Hanakawa, T., Kansaku, K., et al. (2006). Neural correlates of tic generation in Tourette syndrome: An event-related functional MRI study. *Brain, 129*, 2029–2037.

Brandt, V. C., Beck, C., Sajin, V., Baaske, M. K., Baumer, T., Beste, C., et al. (2016). Temporal relationship between premonitory urges and tics in Gilles de la Tourette syndrome. *Cortex, 77*, 24–37.

Bronfeld, M., Belelovsky, K., & Bar-Gad, I. (2011). Spatial and temporal properties of tic-related neuronal activity in the cortico-basal ganglia loop. *Journal of Neuroscience, 31*, 8713–8721.

Buse, J., Beste, C., Herrmann, E., & Roessner, V. (2016). Neural correlates of altered sensorimotor gating in boys with Tourette syndrome: A combined EMG/fMRI study. *World Journal of Biological Psychiatry, 17*, 187–197.

Casey, B. J., Tottenham, N., & Fossella, J. (2002). Clinical, imaging, lesion, and genetic approaches toward a model of cognitive control. *Developmental Psychobiology, 40*, 237–254.

Chamberlain, S. R., Menzies, L., Hampshire, A., Suckling, J., Fineberg, N. A., del Campo, N., et al. (2008). Orbitofrontal dysfunction in patients with obsessive-compulsive disorder and their unaffected relatives. *Science, 321*, 421–422.

Cheng, B., Braass, H., Ganos, C., Treszl, A., Biermann-Ruben, K., Hummel, F. C., et al. (2013). Altered intrahemispheric structural connectivity in Gilles de la Tourette syndrome. *NeuroImage. Clinical, 3*, 174–181.

Church, J. A., Fair, D. A., Dosenbach, N. U., Cohen, A. L., Miezin, F. M., Petersen, S. E., et al. (2009). Control networks in paediatric Tourette syndrome show immature and anomalous patterns of functional connectivity. *Brain, 132*, 225–238.

Conceicao, V. A., Dias, A., Farinha, A. C., & Maia, T. V. (2017). Premonitory urges and tics in Tourette syndrome: Computational mechanisms and neural correlates. *Current Opinion in Neurobiology, 46*, 187–199.

Cui, Y., Jin, Z., Chen, X., He, Y., Liang, X., & Zheng, Y. (2014). Abnormal baseline brain activity in drug-naïve patients with Tourette syndrome: A resting-state fMRI study. *Frontiers in Human Neuroscience*, 7, 913.

Davenport, P. W., Sapienza, C. M., & Bolser, D. C. (2002). Psychophysical assessment of the urge-to-cough. *European Respiratory Review*, 12, 249–253.

Debes, N. M., Jeppesen, S., Raghava, J. M., Groth, C., Rostrup, E., & Skov, L. (2015). Longitudinal magnetic resonance imaging (MRI) analysis of the developmental changes of Tourette syndrome reveal reduced diffusion in the cortico-striato-thalamo-cortical pathways. *Journal of Child Neurology*, 30, 1315–1326.

Deckersbach, T., Chou, T., Britton, J. C., Carlson, L. E., Reese, H. E., Siev, J., et al. (2014). Neural correlates of behavior therapy for Tourette's disorder. *Psychiatry Research*, 224, 269–274.

Delorme, C., Salvador, A., Valabrègue, R., Roze, E., Palminteri, S., Vidailhet, M., et al. (2016). Enhanced habit formation in Gilles de la Tourette syndrome. *Brain*, 139, 605–615.

Denys, D., de Vries, F. E., Cath, D., Figee, M., Vulink, N., Veltman, D. J., et al. (2013). Dopaminergic activity in Tourette syndrome and obsessive-compulsive disorder. *European Neuropsychopharmacology*, 23, 1423–1431.

de Vries, F. E., van den Heuvel, O. A., Cath, D. C., Groenewegen, H. J., van Balkom, A. J., Boellaard, R., et al. (2013). Limbic and motor circuits involved in symmetry behavior in Tourette's syndrome. *CNS Spectrums*, 18, 34–42.

Draganski, B., Martino, D., Cavanna, A. E., Hutton, C., Orth, M., Robertson, M. M., et al. (2010). Multispectral brain morphometry in Tourette syndrome persisting into adulthood. *Brain*, 133, 3661–3675.

Draper, A., Jackson, G. M., Morgan, P. S., & Jackson, S. R. (2016). Premonitory urges are associated with decreased grey matter thickness within the insula and sensorimotor cortex in young people with Tourette syndrome. *Journal of Neuropsychology*, 10, 143–153.

Eddy, C. M., & Cavanna, A. E. (2013). Altered social cognition in Tourette syndrome: Nature and implications. *Behavioural Neurology*, 27, 15–22.

Eddy, C. M., Cavanna, A. E., & Hansen, P. C. (2017). Empathy and aversion: The neural signature of mentalizing in Tourette syndrome. *Psychological Medicine*, 47, 507–517.

Eddy, C. M., Cavanna, A. E., Rickards, H. E., & Hansen, P. C. (2016). Temporo-parietal dysfunction in Tourette syndrome: Insights from an fMRI study of Theory of Mind. *Journal of Psychiatric Research*, 81, 102–111.

Fahim, C., Yoon, U., Das, S., Lyttelton, O., Chen, J., Amaoutelis, R., et al. (2010). Somatosensory-motor bodily representation cortical thinning in Tourette: Effects of tic severity, age and gender. *Cortex*, 46, 750–760.

Fan, S., Cath, D. C., van den Heuvel, O. A., van der Werf, Y. D., Schols, C., Veltman, D. J., et al. (2017). Abnormalities in metabolite concentrations in Tourette's disorder and obsessive-compulsive disorder—A proton magnetic resonance spectroscopy study. *Psychoneuroendocrinology*, 77, 211–217.

Forde, N. J., Zwiers, M. P., Naaijen, J., Akkermans, S. E. A., Openneer, T. J. C., Visscher, F., et al. (2017). Basal ganglia structure in Tourette's disorder and/or attention-deficit/hyperactivity disorder. *Movement Disorders*, 32, 601–604.

Ganos, C., Kahl, U., Brandt, V., Schunke, O., Baumer, T., Thomalla, G., et al. (2014). The neural correlates of tic inhibition in Gilles de la Tourette syndrome. *Neuropsychologia*, 65, 297–301.

Ganos, C., Kuhn, S., Kahl, U., Schunke, O., Feldheim, J., Gerloff, C., et al. (2014a). Action inhibition in Tourette syndrome. *Movement Disorders*, 29, 1532–1538.

Ganos, C., Kuhn, S., Kahl, U., Schunke, O., Vrandt, V., Baumer, T., et al. (2014b). Prefrontal cortex volume reductions and tic inhibition are unrelated in uncomplicated GTS adults. *Journal of Psychosomatic Research*, 76, 84–87.

Ganos, C., & Martino, D. (2015). Tics and Tourette syndrome. *Neurologic Clinics, 33,* 115–136.

Ganos, C., Roessner, V., & Munchau, A. (2013). The functional anatomy of Gilles de la Tourette syndrome. *Neuroscience and Biobehavioral Reviews, 37,* 1050–1062.

Ganos, C., Rothwell, J., & Haggard, P. (2018). Voluntary inhibitory motor control over involuntary tic movements. *Movement Disorders, 33,* 937–946.

Greene, D. J., Church, J. A., Dosenbach, N. U., Nielsen, A. N., Adeyemo, B., Nardos, B., et al. (2016). Multivariate pattern classification of pediatric Tourette syndrome using functional connectivity MRI. *Developmental Science, 19,* 581–598.

Greene, D. J., Schlaggar, B. L., & Black, K. J. (2015). Neuroimaging in Tourette syndrome: Research highlights from 2014-2015. *Current Developmental Disorders Reports, 2,* 300–308.

Greene, D. J., Williams, A. C., III, Koller, J. M., Schlaggar, B. L., & Black, K. J. (2017). Brain structure in pediatric Tourette syndrome. *Molecular Psychiatry, 22,* 972–980.

Gulisano, M., Calì, P., Palermo, F., Robertson, M. M., & Rizzo, R. (2015). Premonitory urges in patients with Gilles de la Tourette syndrome: An Italian translation and a 7-year follow-up. *Journal of Child and Adolescent Psychopharmacology, 25,* 810–816.

Hampson, M., Tokoglu, F., King, R. A., Constable, R. T., & Leckman, J. F. (2009). Brain areas coactivating with motor cortex during chronic motor tics and intentional movements. *Biological Psychiatry, 65,* 594–599.

Hashemiyoon, R., Kuhn, J., & Visser-Vandewalle, V. (2017). Putting the pieces together in Gilles de la Tourette syndrome: Exploring the link between clinical observations and the biological basis of dysfunction. *Brain Topography, 30,* 3–29.

Hinterbuchinger, B., Kaltenboeck, A., Baumgartner, J. S., Mossaheb, N., & Friedrich, F. (2018). Do patients with different psychiatric disorders show altered social decision-making? A systematic review of ultimatum game experiments in clinical populations. *Cognitive Neuropsychiatry, 23,* 117–141.

Jackson, S. R., Parkinson, A., Kim, S. Y., Schuermann, M., & Eickhoff, S. B. (2011). On the functional anatomy of the urge-for-action. *Cognitive Neuroscience, 2,* 227–257.

Jeppesen, S. S., Debes, N. M., Simonsen, H. J., Rostrup, E., Larsson, H. B. W., & Skov, L. (2014). Study of medication-free children with Tourette syndrome do not show imaging abnormalities. *Movement Disorders, 29,* 1212–1216.

Jung, W. H., Kang, D. H., Han, J. Y., Jang, J. H., Gu, B. M., Choi, J. S., et al. (2011). Aberrant ventral striatal responses during incentive processing in unmedicated patients with obsessive-compulsive disorder. *Acta Psychiatrica Scandinavica, 123,* 376–386.

Kalanithi, P. S., Zheng, W., Kataoka, Y., DiFiglia, M., Grantz, H., Saper, C. B., et al. (2005). Altered parvalbumin-positive neuron distribution in basal ganglia of individuals with Tourette syndrome. *Proceedings of the National Academy of Sciences of the United States of America, 102,* 13307–13312.

Kanaan, A. S., Gerasch, S., Garcia-Garcia, I., Lampe, L., Pampel, A., Anwander, A., et al. (2017). Pathological glutamatergic neurotransmission in Gilles de la Tourette syndrome. *Brain, 140,* 218–234.

Kataoka, Y., Kalanithi, P. S., Grantz, H., Schwartz, M. L., Saper, C., Leckman, J. F., et al. (2010). Decreased number of parvalbumin and cholinergic interneurons in the striatum of individuals with Tourette syndrome. *Journal of Comparative Neurology, 518,* 277–291.

Kaufmann, C., Beucke, J. C., Preuße, F., Endrass, T., Schlagenhauf, F., Heinz, A., et al. (2013). Medial prefrontal brain activation to anticipated reward and loss in obsessive-compulsive disorder. *NeuroImage. Clinical, 2,* 212–220.

Kern, J. K., Geier, D. A., King, P. G., Sykes, L. K., Mehta, J. A., & Geier, M. R. (2015). Shared brain connectivity issues, symptoms, and comorbidities in autism spectrum disorder, attention deficit/hyperactivity disorder, and Tourette syndrome. *Brain Connectivity, 5,* 321–335.

Knight, T., Steeves, T., Day, L., Lowerison, M., Jette, N., & Pringsheim, T. (2012). Prevalence of tic disorders: A systematic review and meta-analysis. *Pediatric Neurology, 47*, 77–90.

Kumar, A., Williams, M. T., & Chugani, H. T. (2015). Evaluation of basal ganglia and thalamic inflammation in children with pediatric autoimmune neuropsychiatric disorders associated with streptococcal infection and Tourette syndrome: A positron emission tomographic (PET) study using ^{11}C-[R]-PK11195. *Journal of Child Neurology, 30*, 749–756.

Leckman, J. F., & Riddle, M. A. (2000). Tourette's syndrome: When habit-forming systems form habits of their own? *Neuron, 28*, 349–354.

Lennington, J. B., Coppola, G., Kataoka-Sasaki, Y., Fernandez, T. V., Palejev, D., Li, Y., et al. (2016). Transcriptome analysis of the human striatum in Tourette syndrome. *Biological Psychiatry, 79*, 372–382.

Lerner, A., Bagic, A., Simmons, J. M., Mari, Z., Bonne, O., Xu, B., et al. (2012). Widespread abnormality of the gamma-aminobutyric acid-ergic system in Tourette syndrome. *Brain, 135*, 1926–1936.

Liao, W., Yu, Y., Miao, H. H., Feng, Y. X., Ji, G. J., & Feng, J. H. (2017). Inter-hemispheric intrinsic connectivity as a neuromarker for the diagnosis of boys with Tourette syndrome. *Molecular Neurobiology, 54*, 2781–2789.

Liu, Y., Miao, W., Wang, J., Gao, P., Yin, G., Zhang, L., et al. (2013). Structural abnormalities in early Tourette syndrome children: A combined voxel-based morphometry and tract-based spatial statistics study. *PLoS One, 8*, e76105.

Liu, Y., Wang, J., Zhang, J., Wen, H., Zhang, Y., Kang, H., et al. (2017). Altered spontaneous brain activity in children with early Tourette syndrome: A resting-state fMRI study. *Scientific Reports, 7*, 4808.

Ludolph, A. G., Pinkhardt, E. H., Tebartz van Elst, L., Libal, G., Ludolph, A. C., Fegert, J. M., et al. (2008). Are amygdalar volume alterations in children with Tourette syndrome due to ADHD comorbidity? *Developmental Medicine and Child Neurology, 50*, 524–529.

Mahone, E. M., Puts, N. A., Edden, R. A. E., Ryan, M., & Singer, H. S. (2018). GABA and glutamate in children with Tourette syndrome: A (1)H MR spectroscopy study at 7T. *Psychiatry Research. Neuroimaging, 273*, 46–53.

Maia, T. V., & Conceicao, V. A. (2017). The roles of phasic and tonic dopamine in tic learning and expression. *Biological Psychiatry, 82*, 401–412.

Marsh, R., Maia, T. V., & Peterson, B. S. (2009). Functional disturbances within frontostriatal circuits across multiple childhood psychopathologies. *American Journal of Psychiatry, 166*, 664–674.

Marsh, R., Zhu, H., Wang, Z., Skudlarski, P., & Peterson, B. S. (2007). A developmental fMRI study of self-regulatory control in Tourette's syndrome. *American Journal of Psychiatry, 164*, 955–966.

Martino, D., Delorme, C., Pelosin, E., Hartmann, A., Worbe, Y., & Avanzino, L. (2017). Abnormal lateralization of fine motor actions in Tourette syndrome persists into adulthood. *PLoS One, 12*, e0180812.

Martino, D., Ganos, C., & Pringsheim, T. M. (2017). Tourette syndrome and chronic tic disorders: The clinical spectrum beyond tics. *International Review of Neurobiology, 134*, 1461–1490.

Martino, D., Madhusadan, N., Zis, P., & Cavanna, A. E. (2013). An introduction to the clinical phenomenology of Tourette syndrome. *International Review of Neurobiology, 112*, 1–33.

McCairn, K. W., Bronfeld, M., Belelovsky, K., & Bar-Gad, I. (2009). The neurophysiological correlates of motor tics following focal striatal disinhibition. *Brain, 132*, 2125–2138.

McGuire, J. F., McBride, N., Piacentini, J., Johnco, C., Lewin, A. B., Murphy, T. K., et al. (2016). The premonitory urge revisited: An individualized premonitory urge for tics scale. *Journal of Psychiatric Research*, *83*, 176–183.

McGuire, J. F., Piacentini, J., Scahill, L., Woods, D. W., Villarreal, R., Wilhelm, S., et al. (2015). Bothersome tics in patients with chronic tic disorders: Characteristics and individualized treatment response to behavior therapy. *Behaviour Research and Therapy*, *70*, 56–63.

Menzies, L., Chamberlain, S. R., Laird, A. R., Thelen, S. M., Sahakian, B. J., & Bullmore, E. T. (2008). Integrating evidence from neuroimaging and neuropsychological studies of obsessive-compulsive disorder: The orbitofronto-striatal model revisited. *Neuroscience and Biobehavioral Reviews*, *32*, 525–549.

Morand-Beaulieu, S., Grot, S., Lavoie, J., Leclerc, J. B., Luck, D., & Lavoie, M. E. (2017). The puzzling question of inhibitory control in Tourette syndrome: A meta-analysis. *Neuroscience and Biobehavioral Reviews*, *80*, 240–262.

Muellner, J., Delmaire, C., Valabrégue, R., Schupbach, M., Mangin, J. F., Viailhet, M., et al. (2015). Altered structure of cortical sulci in gilles de la Tourette syndrome: Further support for abnormal brain development. *Movement Disorders*, *30*, 655–661.

Muller-Vahl, K. R., Grosskreutz, J., Prell, T., Kaufmann, J., Bodammer, N., & Peschel, T. (2014). Tics are caused by alterations in prefrontal areas, thalamus and putamen, while changes in the cingulate gyrus reflect secondary compensatory mechanisms. *BMC Neuroscience*, *15*, 6.

Muller-Vahl, K. R., Kaufmann, J., Grosskreutz, J., Dengler, R., Emrich, H. M., & Peschel, T. (2009). Prefrontal and anterior cingulate cortex abnormalities in Tourette syndrome: Evidence from voxel-based morphometry and magnetization transfer imaging. *BMC Neuroscience*, *10*, 47.

Naaijen, J., Forde, N. J., Lythgoe, D. J., Akkermans, S. E., Openneer, T. J., Dietrich, A., et al. (2016). Fronto-striatal glutamate in children with Tourette's disorder and attention-deficit/hyperactivity disorder. *NeuroImage. Clinical*, *13*, 16–23.

Neuner, I., Schneider, F., & Shah, N. J. (2013). Functional neuroanatomy of tics. *International Review of Neurobiology*, *112*, 35–71.

Neuner, I., Wegener, P., Stoecker, T., Kircher, T., Schneider, F., & Shah, N. J. (2007). Development and implementation of an MR-compatible whole body video system. *Neuroscience Letters*, *420*, 122–127.

Neuner, I., Werner, C. J., Arrubla, J., Stocker, T., Ehlen, C., Wegener, H. P., et al. (2014). Imaging the where and when of tic generation and resting state networks in adult Tourette patients. *Frontiers in Human Neuroscience*, *8*, 362.

Palminteri, S., Lebreton, M., Worbe, Y., Grabli, D., Hartmann, A., & Pessiglione, M. (2009). Pharmacological modulation of subliminal learning in Parkinson's and Tourette's syndromes. *Proceedings of the National Academy of Sciences of the United States of America*, *106*, 19179–19184.

Peterson, B. S., Choi, H. A., Hao, X., Amat, J. A., Zhu, H., Whiteman, R., et al. (2007). Morphologic features of the amygdala and hippocampus in children and adults with Tourette syndrome. *Archives of General Psychiatry*, *64*, 1281–1291.

Plessen, K. J., Royal, J. M., & Peterson, B. S. (2007). Neuroimaging of tic disorders with co-existing attention-deficit/hyperactivity disorder. *European Child & Adolescent Psychiatry*, *16*(Suppl. 1), 60–70.

Pogorelov, V., Xu, M., Smith, H. R., Buchanan, G. F., & Pittenger, C. (2015). Corticostriatal interactions in the generation of tic-like behaviors after local striatal disinhibition. *Experimental Neurology*, *265*, 122–128.

Polyanska, L., Critchley, H. D., & Rae, C. L. (2017). Centrality of prefrontal and motor preparation cortices to Tourette syndrome revealed by meta-analysis of task-based neuroimaging studies. *NeuroImage. Clinical*, *16*, 257–267.

Puts, N. A., Harris, A. D., Crocetti, D., Nettles, C., Singer, H. S., Tommerdahl, M., et al. (2015). Reduced GABAergic inhibition and abnormal sensory symptoms in children with Tourette syndrome. *Journal of Neurophysiology, 114*, 808–817.

Raines, J. M., Edwards, K. R., Sherman, M. F., Higginson, C. I., Winnick, J. B., Navin, K., et al. (2018). Premonitory urge for tics scale (PUTS): Replication and extension of psychometric properties in youth with chronic tic disorders (CTDs). *Journal of Neural Transmission (Vienna), 125*, 727–734.

Rizzo, R., Gulisano, M., Martino, D., & Robertson, M. M. (2017). Gilles de la Tourette syndrome, depression, depressive illness, and correlates in a child and adolescent population. *Journal of Child and Adolescent Psychopharmacology, 27*, 243–249.

Robbins, T. W., Gillan, C. M., Smith, D. G., de Wit, S., & Ersche, K. D. (2012). Neurocognitive endophenotypes of impulsivity and compulsivity: Towards dimensional psychiatry. *Trends in Cognitive Sciences, 16*, 81–91.

Robertson, M. M. (2000). Tourette syndrome, associated conditions and the complexities of treatment. *Brain, 123*, 425–462.

Robertson, M. M., Eapen, V., & Cavanna, A. E. (2009). The international prevalence, epidemiology, and clinical phenomenology of Tourette syndrome: A cross-cultural perspective. *Journal of Psychosomatic Research, 67*, 475–483.

Sakai, Y., Narumoto, J., Nishida, S., Nakamae, T., Yamada, K., Nishimura, T., et al. (2011). Corticostriatal functional connectivity in non-medicated patients with obsessive-compulsive disorder. *European Psychiatry, 26*, 463–469.

Sambrani, T., Jakubovski, E., & Muller-Vahl, K. R. (2016). New insights into clinical characteristics of Gilles de la Tourette syndrome: Findings in 1032 patients from a single German center. *Frontiers in Neuroscience, 10*, 415.

Schlemm, E., Cheng, B., Fischer, F., Hilgetag, C., Gerloff, C., & Thomalla, G. (2017). Altered topology of structural brain networks in patients with Gilles de la Tourette syndrome. *Scientific Reports, 7*, 10606.

Sigurdsson, H. P., Pépés, S. E., Jackson, G. M., Draper, A., Morgan, P. S., & Jackson, S. R. (2018). Alterations in the microstructure of white matter in children and adolescents with Tourette syndrome measured using tract-based spatial statistics and probabilistic tractography. *Cortex, 104*, 75–89.

Sowell, E. R., Kan, E., Yoshii, J., Thompson, P. M., Bansal, R., Xu, D., et al. (2008). Thinning of sensorimotor cortices in children with Tourette syndrome. *Nature Neuroscience, 11*, 637–639.

Thomalla, G., Jonas, M., Baumer, T., Siebner, H. R., Biermann-Ruben, K., Ganos, C., et al. (2014). Costs of control: Decreased motor cortex engagement during a Go/NoGo task in Tourette's syndrome. *Brain, 137*, 122–136.

Tinaz, S., Belluscio, B. A., Malone, P., van der Veen, J. W., Hallett, M., & Horovitz, S. G. (2014). Role of the sensorimotor cortex in Tourette syndrome using multimodal imaging. *Human Brain Mapping, 35*, 5834–5846.

Tinaz, S., Malone, P., Hallett, M., & Horovitz, S. G. (2015). Role of the right dorsal anterior insula in the urge to tic in Tourette syndrome. *Movement Disorders, 30*, 1190–1197.

Tobe, R. H., Bansal, R., Xu, D., Hao, X., Liu, J., Sanchez, J., et al. (2010). Cerebellar morphology in Tourette syndrome and obsessive-compulsive disorder. *Annals of Neurology, 67*, 479–487.

Vaghi, M. M., Hampshire, A., Fineberg, N. A., Kaser, M., Bruhl, A. B., Sahakian, B. J., et al. (2017). Hypoactivation and dysconnectivity of a frontostriatal circuit during goal-directed planning as an endophenotype for obsessive-compulsive disorder. *Biological Psychiatry. Cognitive Neuroscience and Neuroimaging, 2*, 655 663.

Vicario, C. M., & Martino, D. (2018). Social communication in Tourette syndrome: A glimpse at the contribution of the insula and the prefrontal cortex. *Brain*. https://doi.org/10.1093/brain/awy140 [Epub ahead of print].

Voon, V., Derbyshire, K., Ruck, C., Irvine, M. A., Worbe, Y., Enander, J., et al. (2015). Disorders of compulsivity: A common bias towards learning habits. *Molecular Psychiatry*, *20*, 345–352.

Wang, Z., Maia, T. V., Marsh, R., Colibazzi, T., Gerber, A., & Peterson, B. S. (2011). The neural circuits that generate tics in Tourette's syndrome. *American Journal of Psychiatry*, *168*, 1326–1337.

Wen, H., Liu, Y., Rekik, I., Wang, S., Zhang, J., Zhang, Y., et al. (2017). Disrupted topological organization of structural networks revealed by probabilistic diffusion tractography in Tourette syndrome children. *Human Brain Mapping*, *38*, 3988–4008.

Wen, H., Liu, Y., Wang, J., Rekik, I., Zhang, J., Zhang, Y., et al. (2016). Combining tract- and atlas-based analysis reveals microstructural abnormalities in early Tourette syndrome children. *Human Brain Mapping*, *37*, 1903–1919.

Wittfoth, M., Bornmann, S., Peschel, T., Grosskreutz, J., Glahn, A., Buddensiek, N., et al. (2012). Lateral frontal cortex volume reduction in Tourette syndrome revealed by VBM. *BMC Neuroscience*, *13*, 17.

Wolff, N., Luehr, I., Sender, J., Ehrlich, S., Schmidt-Samoa, C., Dechent, P., et al. (2016). A DTI study on the corpus callosum of treatment-naïve boys with 'pure' Tourette syndrome. *Psychiatry Research. Neuroimaging*, *247*, 1–8.

Woods, D. W., Piacentini, J., Himle, M. B., & Chang, S. (2005). Premonitory urge for tics scale (PUTS): Initial psychometric results and examination of the premonitory urge phenomenon in youths with tic disorders. *Journal of Developmental and Behavioral Pediatrics*, *26*, 397–403.

Worbe, Y., Gerardin, E., Hartmann, A., Valabréque, R., Chupin, M., Tremblay, L., et al. (2010). Distinct structural changes underpin clinical phenotypes in patients with Gilles de la Tourette syndrome. *Brain*, *133*, 3649–3660.

Worbe, Y., Lehéricy, S., & Hartmann, A. (2015). Neuroimaging of tic genesis: Present status and future perspectives. *Movement Disorders*, *30*, 1179–1183.

Worbe, Y., Malherbe, C., Hartmann, A., Pélégrini-Issac, M., Messé, A., Vidailhet, M., et al. (2012). Functional immaturity of cortico-basal ganglia networks in Gilles de la Tourette syndrome. *Brain*, *135*, 1937–1946.

Worbe, Y., Marrakchi-Kacem, L., Lecomte, S., Valabréque, R., Poupon, F., Guevara, P., et al. (2015). Altered structural connectivity of cortico-striato-pallido-thalamic networks in Gilles de la Tourette syndrome. *Brain*, *138*, 472–482.

Worbe, Y., Palminteri, S., Hartmann, A., Vidailhet, M., Lehéricy, S., & Pessiglione, M. (2011). Reinforcement learning and Gilles de la Tourette syndrome: Dissociation of clinical phenotypes and pharmacological treatments. *Archives of General Psychiatry*, *68*, 1257–1266.

Worbe, Y., Sgambato-Faure, V., Epinat, J., Chiagneau, M., Tandé, D., Francois, C., et al. (2013). Towards a primate model of Gilles de la Tourette syndrome: Anatomo-behavioural correlation of disorders induced by striatal dysfunction. *Cortex*, *49*, 1126–1140.

Yael, D., Vinner, E., & Bar-Gad, I. (2015). Pathophysiology of tic disorders. *Movement Disorders*, *30*, 1171–1178.

Zapparoli, L., Porta, M., Gandola, M., Invernizzi, P., Colajanni, V., Servello, D., et al. (2016). A functional magnetic resonance imaging investigation of motor control in Gilles de la Tourette syndrome during imagined and executed movements. *European Journal of Neuroscience*, *43*, 494–508.

Zapparoli, L., Porta, M., & Paulesu, E. (2015). The anarchic brain in action: The contribution of task-based fMRI studies to the understanding of Gilles de la Tourette syndrome. *Current Opinion in Neurology*, *28*, 604–611.

Zapparoli, L., Tettamanti, M., Porta, M., Zerbi, A., Servello, D., Banfi, G., et al. (2017). A tug of war: Antagonistic effective connectivity patterns over the motor cortex and the severity of motor symptoms in Gilles de la Tourette syndrome. *European Journal of Neuroscience*, *46*, 2203–2213.

CHAPTER FOUR

Neuroimaging Applications in Chronic Ataxias

Mario Mascalchi*,†,1, Alessandra Vella‡

*Meyer Children Hospital, Florence, Italy
†Department of Experimental and Clinical Biomedical Sciences "Mario Serio", University of Florence, Florence, Italy
‡Nuclear Medicine, "Le Scotte" University Hospital, Siena, Italy
1Corresponding author: e-mail address: mario.mascalchi@unifi.it

Contents

Abstract

Magnetic resonance imaging (MRI), single photon emission computed tomography (SPECT) and positron emission tomography (PET) are the main instruments for neuroimaging investigation of patients with chronic ataxia. MRI has a predominant diagnostic role in the single patient, based on the visual detection of three patterns of atrophy, namely, spinal atrophy, cortical cerebellar atrophy and olivopontocerebellar atrophy, which correlate with the aetiologies of inherited or sporadic ataxia. In fact spinal atrophy is observed in Friedreich ataxia, cortical cerebellar atrophy in Ataxia Telangectasia, gluten ataxia and Sporadic Adult Onset Ataxia and olivopontocerebellar atrophy in Multiple System Atrophy cerebellar type. The 39 types of dominantly inherited spinocerebellar ataxias show either cortical cerebellar atrophy or olivopontocerebellar atrophy. T2 or T2* weighted MR images can contribute to the diagnosis by revealing abnormally increased or decreased signal with a characteristic distribution. These include symmetric T2 hyperintensity of the posterior and lateral columns of the cervical spinal cord in Friedreich ataxia, diffuse and symmetric hyperintensity of the cerebellar cortex in Infantile Neuro-Axonal Dystrophy, symmetric hyperintensity of the peridentate white matter in Cerebrotendineous Xanthomatosis, and symmetric hyperintensity of the middle cerebellar peduncles and peridentate white matter, cerebral white matter and corpus callosum in Fragile X Tremor Ataxia Syndrome.

International Review of Neurobiology, Volume 143
ISSN 0074–7742
https://doi.org/10.1016/bs.irn.2018.09.011

Abnormally decreased T2 or T2* signal can be observed with a multifocal distribution in Ataxia Telangectasia and with a symmetric distribution in the basal ganglia in Multiple System Atrophy. T2 signal hypointensity lining diffusely the outer surfaces of the brainstem, cerebellum and cerebrum enables diagnosis of superficial siderosis of the central nervous system. The diagnostic role of nuclear medicine techniques is smaller. SPECT and PET show decreased uptake of radiotracers investigating the nigrostriatal system in Multiple System Atrophy and in patients with Fragile X Tremor Ataxia Syndrome. Semiquantitative or quantitative MRI, SPECT and PET data describing structural, microstructural and functional changes of the cerebellum, brainstem, and spinal cord have been widely applied to investigate physiopathological changes in patients with chronic ataxias. Moreover they can track diseases progression with a greater sensitivity than clinical scales. So far, a few small-size and single center studies employed neuroimaging techniques as surrogate markers of treatment effects in chronic ataxias.

1. INTRODUCTION

Magnetic resonance imaging (MRI), single photon emission computed tomography (SPECT) and positron emission tomography (PET), both using a variety of radiotracers, are the main neuroimaging techniques available for evaluation of the cerebellum and cerebellar diseases. Technical features of MRI and SPECT and PET were reviewed elsewhere (Mascalchi & Vella, 2012; Vella & Mascalchi, 2018).

2. CLASSIFICATION OF CHRONIC ATAXIAS

Classification of cerebellar diseases associated with chronic ataxia is difficult and can be organized according to different criteria including frequency, age of onset (children, adolescents and adults), type of presentation (acute, subacute or chronic), course (stable or progressive), etiology and pattern of inheritance if any (Table 1), and type of molecular genetic abnormality.

From the neuroimaging point of view we feel that it is still reasonable to follow the classification of chronic ataxias based on classical neuropathological descriptions made between 1887 and 1922 by eminent European neurologists in autoptic cases identifying three archetypal patterns of atrophy, namely, spinal atrophy (Friedreich, 1877), cortical cerebellar atrophy (Marie, Foix, & Alajouanine, 1922) and olivopontocerebellar atrophy (Déjérine & Thomas, 1900). Schematically, the spinal atrophy pattern

Table 1 Classification of Chronic Ataxias

Inherited

  Recessive

    Friedreich Ataxia
    Ataxia Telangectasia
    Ataxia with Vitamin E Deficiency
    Infantile Neuronal Axonal Dystrophy
    Ataxia with Oculomotor Apraxia type 1 and 2
    Autosomal Recessive Spastic Ataxia of Charlevoix-Saguenay
    Autosomal Recessive Ataxia Cerebellar type 1 and 2
    CerebroTendineous Xanthomatosis
    Niemann-Pick type C

  Autosomal dominant

    Dentato-Rubro-Pallido-Luysian Atrophy
    Spinocerebellar ataxias (SCAs) type 1-8, 10-23, 25-32, 34-39

  Sex-linked

    Fragile X Tremor Ataxia Syndrome

Sporadic

  Toxic

    Alcohol
    Phenytoin and other antiepileptic drugs
    Solvents
    Heavy metals

  Immuno-mediated

    Gluten ataxia
    Anti-glutamic Acid Decarboxylase Antibody Associated Ataxia
    Paraneoplastic Cerebellar Degeneration
    Primary Autoimmune Cerebellar Ataxia

  Sporadic Adult Onset Ataxia

  Multi-System Atrophy cerebellar type

  Siderosis of the Central Nervous System

reflects the predominant damage of the neurons of the sensory ganglion of the spinal nerves and of the Clarke dorsal nucleus in the spinal gray matter, the cortical cerebellar atrophy pattern reflects the predominant damage of the Purkinje cells of the cerebellar cortex, and the olivopontocerebellar atrophy pattern reflects the predominant damage to the pontine gray nuclei, deep cerebellar nuclei and cerebellar cortex.

This choice is based on three arguments. First, this distinction has been confirmed by visual assessment of MRI *in vivo* (Fig. 1). Second, the MRI pattern may be useful for diagnostic purposes, especially in case of recessively inherited and sporadic chronic ataxias (Table 2). Third, it enables to appreciate similarities from the physiopathological point of view among diverse entities which may have therapeutic implications.

However, the following arguments weakening the absolute value of the distinction among three types of atrophy pattern must be taken into account. First, in presymtpomatic and early phases of disease, the atrophy may not be apparent, while more sensitive techniques as PET, diffusion weighted imaging or diffusion tensor imaging or proton MR spectroscopy may detect some abnormalities. Second, the cerebellum is part of several networks connecting it with brainstem, striata, cerebral cortex and spinal cord. Hence functional and subtle volumetric changes, especially in the advanced phases of a disease, can be observed in other central nervous system structures besides those qualifying the atrophy pattern. Third, unfortunately, genetic classification of autosomal dominant ataxias in spinocerebellar ataxias (SCAs) does not take into account the neuropathological and MRI atrophy patterns, with most of the 39 types of SCAs discovered to date showing the cortical cerebellar atrophy pattern and a minority (SCA1, SCA2, SCA3, SCA7, SCA34 and SCA36) the olivopontocerebellar atrophy pattern. Fourth, some inherited or sporadic ataxias, for instance Autosomal Recessive Spastic Ataxia of Charlevoix-Saguenay, CerebroTendineous Xanthomatosis and Fragile X Tremor Ataxia Syndrome, do not match any of the three patterns.

3. APPLICATIONS OF NEUROIMAGING TECHNIQUES TO CHRONIC ATAXIAS

Neuroimaging techniques in patients with chronic ataxias can schematically have four applications. They can serve as diagnostic markers, physiopathological markers, markers of disease progression and surrogate markers to assess efficacy of new treatments (Baldarçara et al., 2015). The different distribution frequency of the types of chronic ataxias and the variable availability of the modern neuroimaging instruments over the world justify the fact that so far physiopathological markers, progression markers and surrogate markers of new treatments, all applications that require between or within groups analyses, have been explored only in the most frequent ataxias in developed countries.

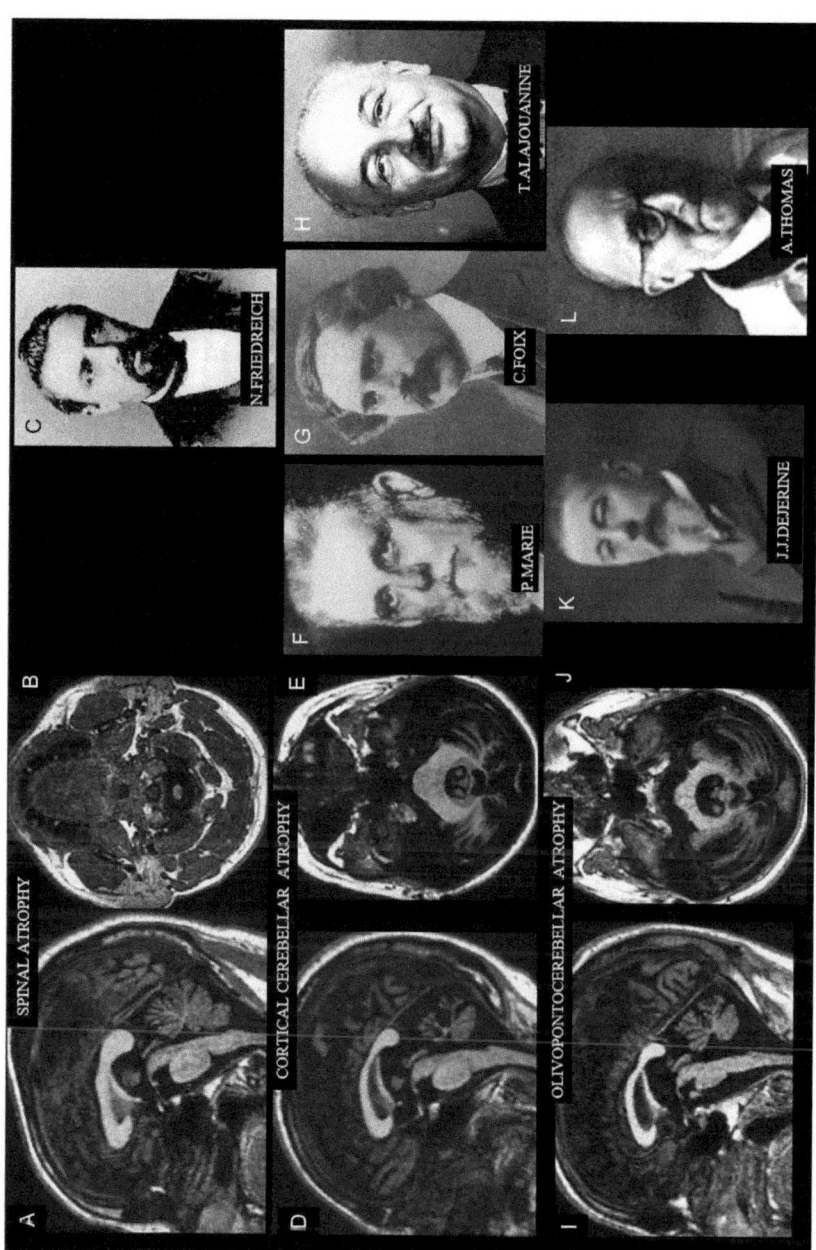

Fig. 1 See figure legend on next page.

Table 2 MRI-Aetiological Correlation in Chronic Ataxias

Spinal Atrophy	Cortical Cerebellar Atrophy	Olivo-Ponto-Cerebellar Atrophy
Friedreich ataxia	Ataxia Telangectasia	Dentato-Rubro-Pallido-Luysian Atrophy
Vitamin E Deficiency Ataxia	Infantile Neuro-Axonal Dystrophy	SCA1
	SCA4	SCA2
	SCA5	SCA3
	SCA6	SCA7
	SCA8	SCA34
	SCA10	SCA36
	SCA11	Early Onset Cerebellar Ataxia
	SCA12	Multiple System Atrophy cerebellar type
	SCA13	
	SCA14	
	SCA15/16	
	SCA17	
	SCA18	

Fig. 1 (A–L) MRI evidence of the patterns of atrophy in chronic ataxias and pictures of the people who originally described the latter in autoptic cases. (A and B) MR sagittal and axial T1 weighted images show the typical features of spinal atrophy, namely, thinned medulla and cervical spinal cord with normal volume of the pons and cerebellar vermis, in a patient with a genetically proven disease that was originally described by Nikolaus Friedreich (C) in 1887. (D and E) MR sagittal and axial T1 weighted images show the typical features of cortical cerebellar atrophy with loss of bulk of the vermis and cerebellar hemispheres with normal volume of the pons, middle cerebellar peduncles and of the cervical spinal cord in a patient with sporadic adult onset ataxia matching the pattern originally described by Pierre Marie, (F) Charles Foix (G) and Théophile Alajouanine (H) in 1922. (I and J) MR sagittal and axial T1 weighted images show the typical features of olivopontocerebellar atrophy in a patients with genetically proven SCA2 including atrophy of the brainstem, more pronounced in the inferior portion of the basis pontis, of the vermis, of the middle cerebellar peduncles and of the cerebellar hemispheres matching the pattern originally described by Joseph Jules Déjerine (K) and André Thomas (L) in 1900.

Table 2 MRI-Aetiological Correlation in Chronic Ataxias—cont'd

Spinal Atrophy	Cortical Cerebellar Atrophy	Olivo-Ponto-Cerebellar Atrophy
	SCA19	
	SCA20	
	SCA21	
	SCA22	
	SCA23	
	SCA25	
	SCA26	
	SCA27	
	SCA28	
	SCA29	
	SCA31	
	SCA32	
	SCA35	
	SCA37	
	SCA38	
	SCA39	
	Gluten and immuno-mediated Ataxias	
	Alcoholic Cerebellar Degeneration	
	Phenytoin and Anti-epileptics	
	Heavy metals	
	Early onset Cerebellar Ataxia	
	Sporadic Adult Onset Ataxia	

SCA, spinocerebellar ataxia.

Table 3 summarizes the neuroimaging features which can serve as diagnostic markers, physiopathology markers, progression markers and and surrogate markers in the most common chronic ataxias.

3.1 Diagnostic Markers

In the single patient with chronic ataxia MRI is the pivotal neuroimaging technique for the diagnosis. Its contribute is however variable and is usually based on the visual evaluation. In some instances, especially in the early phase of the disease, semiquantitative measurements might be of support in identifying the pattern of atrophy (Mascalchi & Vella, 2012; Wüllner et al., 1993).

The wide availability of molecular genetic diagnosis for most of the inherited causes of chronic ataxias has decreased the diagnostic role of MRI, especially for dominantly inherited diseases. However, this role remains valuable in cases of recessive or X-linked conditions and, especially, in sporadic ataxias.

In patients with chronic ataxia visual assessment of MRI demostrates three fundamental patterns of atrophy, namely, spinal atrophy, cortical cerebellar atrophy, olivopontocerebellar atrophy (Fig. 1), with a substantial correlation between the MRI pattern and the different causes of inherited or acquired chronic ataxias (Table 2). In fact spinal atrophy is observed in Friedreich ataxia, cortical cerebellar atrophy in Ataxia Telangectasia, gluten ataxia, and Sporadic Adult Onset Ataxia and olivopontocerebellar atrophy in Multiple System Atrophy cerebellar type. The 39 types of dominantly inherited spinocerebellar ataxias show either cortical cerebellar atrophy or olivopontocerebellar atrophy.

In some diseases the patterns of volume changes are associated with signal changes in T2, T2* or susceptibility weighted images with a characteristic distribution that may serve as "red flags" and suggest or support the diagnosis (Baldarçara et al., 2015; Mascalchi & Vella, 2012) (Fig. 2). These include symmetric T2 hyperintensity of the posterior and lateral columns of the cervical spinal cord in Friedreich ataxia, diffuse and symmetric T2 hyperintensity of the cerebellar cortex in Infantile Neuro-Axonal Dystrophy, symmetric T2 hyperintensity of the peridentate white matter in Cerebrotendineous Xanthomatosis, and symmetric T2 hyperintensity of the middle cerebellar peduncles and peridentate white matter, cerebral white matter and corpus callosum in Fragile X Tremor Ataxia Syndrome. Finally, increased signal in proton density and T2 weighted images of the

Table 3 Neuroimaging Applications in the Most Common Chronic Ataxias

Disease	Diagnostic Markers	Physiopathology Markers	Progression Markers	Surrogate Markers	References
Friedreich ataxia	Spinal atrophy; Symmetric T2 hyperintensity posterior and lateral columns of the cervical spinal cord	Decreased size and T2 signal of the dentate; Diffusion changes in the brainstem, inferior and superior cerebellar peduncle, corticospinal tracts and optic pathways; Decreased NAA/Cr and Cho/Cr in pons and deep cerebellum; Decreased activation of the cerebellar and cerebral cortex and dentate during motor and cognitive functional MRI with compensatory hyperactivated areas in cerebellum and cerebrum; Reduced coupling between cerebellar and frontal cerebral cortex and increased connectivity between areas within the cerebral cortex on resting state functional MRI; Cerebellar increased FDG uptake in early phase and decreased uptake thereafter	Diffusion tensor imaging indexes of the cerebral white matter; T2* signal of the dentate	Volume and diffusion tensor imaging indexes after therapy with recombinant human erythropoietin	Gilman, Junck, Markel, Koeppe, and Kluin (1990), Wüllner, Klockgether, Petersen, Naegele, and Dichgans (1993), Mascalchi, Salvi, Piacentini, and Bartolozzi (1994), Schöls et al. (1997), Mascalchi et al. (2002), Viau and Boulanger (2004), Della Nave, Ginestroni, Giannelli, et al. (2008), Akhlaghi et al. (2011), Ginestroni et al. (2012), Akhlaghi et al. (2012), Chevis et al. (2013), Egger et al. (2014), Solbach et al. (2014), Bonilha da Silva et al. (2014), Santner, Schocke, Boesch, Nachbauer, and Egger (2014), Stefanescu et al. (2015), Stefanescu et al. (2015), Rezende et al. (2016), Rezende et al. (2016), Mascalchi et al. (2016), Dogan et al. (2016), Harding et al. (2016), Gramegna et al. (2017), Dogan et al. (2018), Rezende et al. (2018), and Cocozza et al. (2018)

Continued

Table 3 Neuroimaging Applications in the Most Common Chronic Ataxias—cont'd

Disease	Diagnostic Markers	Physiopathology Markers	Progression Markers	Surrogate Markers	References
Ataxia Telangectasia	Cortical cerebellar atrophy; Multifocal T2 o T2* hypointense areas	Increased apparent diffusion coefficient in the cerebellar cortex and white matter; Decreased NAA and Cr and increased Cho in the vermis and cerebellar hemispheres; Decreased FDG uptake in cerebellum, occipital and temporal lobe and increased FDG uptake in the pallidus	—	Activation of the motor cortex on a simple motor task at functional MRI after oral betamethasone	Fırat, Karakaş, Fırat, and Yakinci (2005), Lin, Crawford, Lederman, and Barker (2006), Wallis et al. (2007), Quarantelli et al. (2013), and Volkow et al. (2014)
SCA1	Olivopontocerebellar atrophy; ponto-medullary volume loss before clinical onset; Diffuse T2 high signal in pons, middle cerebellar peduncles and cerebellar white matter with "hot cross bun" sign in advanced phases	Diffusion changes in the brainstem, cerebellar peduncles and cerebellum; Decreased NAA, Glu and NAA/Cr, Cho/Cr and increased mI and Cr in the pons, cerebellar white matter and vermis; Decreased activation of the cerebellum and compensatory activations in the cerebral cortex and thalamus on motor task functional MRI; altered intrinsic functional connectivity in lateral cerebellum and thalamus and patchy appearance of cerebellar functional clusters on resting state functional MRI; Decreased FDG uptake in brainstem, cerebellum, cerebral cortex and subcortical gray matter	Volume of the pons, brainstem, cerebellum and basal ganglia	—	Gilman et al. (1996), Mascalchi et al. (1998), Guerrini et al. (2004), Wüllner et al. (2005), Mandelli et al. (2007), Jayakumar et al. (2008), Della Nave, Ginestroni, Tessa, Salvatore, De Grandis, et al. (2008), Prakash et al. (2009), Oz et al. (2010), Solodkin, Peri, Chen, Ben-Jacob, and Gomez (2011), Reetz et al. (2013), Jacobi et al. (2013), Adanyeguh et al. (2015), Adanyeguh et al. (2018), and Joers et al. (2018)

| SCA2 | Olivopontocerebellar atrophy; brainstem volume loss before clinical onset; Diffuse T2 high signal in pons, middle cerebellar peduncles and cerebellar white matter with "hot cross bun" sign; Thalamus and parietal cortical atrophy in advanced phases | Diffusion changes in the brainstem, cerebellar peduncles, cerebellum, cerebral peduncles, corticospinal tracts, thalamus and frontal and temporal cortical and subcortical regions; Decreased NAA/Cr, Cho/Cr NAA and Glu and increased mI and Cr in pons and cerebellum also in presymptomatic subjects; Altered connectivity between pons and cerebellum and sensorimotor, frontal parietal and temporal cortices on resting state functional MRI; Decreased FDG uptake in brainstem, cerebellum, also in presymptomatic subjects, and in frontal, parietal and limbic cortex; Decreased dopamine transporter in the striata also in presymptomatic subjects | — | Volume of the brainstem and cerebellum; Diffusion tensor imaging indexes of the brainstem and cerebellum | Klockgether et al. (1998), Boesch et al. (2001), Shan et al. (2001), Furtado et al. (2002), Brenneis, Bösch, Schocke, Wenning, and Poewe (2003), Varrone et al. (2004), Boesch et al. (2004), Guerrini et al. (2004), Wüllner et al. (2005), Viau and Boulanger (2004), Inagaki, Iida, Matsubara, and Inagaki (2005), Wang, Liu, Yang, and Soong (2007), Wüllner et al. (2005), Mandelli et al. (2007), Boesch et al. (2007), Kim et al. (2007), Della Nave, Ginestroni, Tessa, Cosottini, et al. (2008), Della Nave, Ginestroni, Tessa, Salvatore, et al. (2008), Wang et al. (2009), D'Agata et al. (2011), Goel et al. (2011), Lirng et al. (2012), Jung, Choi, Du, Cuzzocreo, Ying, et al. (2012b), Jakobi et al. (2013), Wu et al. (2013), Mascalchi et al. (2014), Salvatore et al. (2014), Cocozza et al. (2015), Hernandez-Castillo et al. (2015), Adanyeguh et al. (2015), Olivito, Lupo, et al. (2017), Oh, Kim, Oh, Lee, and Chung (2017), Adanyeguh et al. (2018), Joers et al. (2018), Olivito, Cercignani, et al. (2017), Reetz et al. (2018), and Mascalchi et al. (2018) |

Continued

Table 3 Neuroimaging Applications in the Most Common Chronic Ataxias—cont'd

Disease	Diagnostic Markers	Physiopathology Markers	Progression Markers	Surrogate Markers	References
SCA3	Olivopontocerebellar atrophy; Diffuse T2 high signal in pons, middle cerebellar peduncles and cerebellar white matter with "hot cross bun" sign; Atrophy of frontal and temporal cortex, basal ganglia and spinal cord	Diffusion changes in the cerebellum brainstem, thalami and cerebral hemispheres; Decreased NAA/Cr, NAA/mI, Glu and increased mI and Cr in pons, cerebellum and cerebral white matter also in presymptomatic subjects; Decreased uptake of FDG in brainstem, cerebellum, putamen, thalamus and limbic and occipital cortex also in presymptomatic subjects; Decreased F-dopa and dopamine transporter in the striata also in presymptomatic subject	Volume of the brainstem, cerebellum and basal ganglia, especially putamen	—	Soong, Cheng, Liu, and Shan (1997), Shinotoh et al. (1997), Soong et al. (1997), Shinotoh et al. (1997), Klockgether et al. (1998), Soong and Liu (1998), Murata et al. (1998), Yen et al. (2000), Etchebehere et al. (2001), Yen et al. (2002), Tokumaru et al. (2003) Wüllner et al. (2005), Lukas et al. (2006), Wang et al. (2007), Lukas et al. (2008), Wang et al. (2009), D'Abreu, França, Appenzeller, Lopes-Cendes, and Cendes (2009), Schulz et al. (2010), Goel et al. (2011), Eichler et al. (2011), Lirng et al. (2012), Guimarães et al. (2013), Reetz et al. (2013), Kang et al. (2014), Stefanescu et al. (2015), Adanyeguh et al. (2015), Huang et al. (2017), Joers et al. (2018), and Adanyeguh et al. (2018)

| SCA6 | Cortical cerebellar atrophy | Decreased size and diffusion changes in superior and middle cerebellar peduncles and cerebral peduncles also in presymptomatic subjects; Decreased NAA/Cr, NAA/Cho, NAA/mI in the cerebellum also in presymptomatic subjects; Decreased activation of cerebellar cortex and dentate and cerebral cortex on motor functional MRI; shift of activation during visual functional MRI; Decreased FDG uptake in cerebellum, frontal cortex, putamen and brainstem; Decreased density of striatal dopamine transporter | — | Volume of the cerebellum, and basal ganglia | Boesch et al. (2001), Soong, Liu, Wu, Lu, and Lee (2001), Wüllner et al. (2005), Lukas et al. (2006), Wang et al. (2007), Schulz et al. (2010), Kim et al. (2010), Eichler et al. (2011), Lirng et al. (2012), Reetz et al. (2013), Stefanescu et al. (2015), Falcon, Gomez, Chen, Shereen, and Solodkin (2016), Oh et al. (2017), Kang et al. (2017), and Joers et al. (2018) |

Continued

Table 3 Neuroimaging Applications in the Most Common Chronic Ataxias—cont'd

Disease	Diagnostic Markers	Physiopathology Markers	Progression Markers	Surrogate Markers	References
SCA17	Cortical cerebellar atrophy and volume loss of putamen, caudate, thalamus and of limbic, frontal, parietal, and occipital lobes	Decreased NAA/Cr and Cho/Cr in cerebellar hemispheres and vermis; Decreased FDG uptake in cerebellum, caudate, putamen and parietal cortex; Decreased uptake of dopamine transporter and of ^{11}C-raclopride uptake in the striata also in presymptomatic subjects	Volume of the cerebellum, limbic system and parietal precuneus	—	Minnerop et al. (2005), Lasek et al. (2006), Kim et al. (2009), Reetz et al. (2010), Brockmann et al. (2012), and Lirng et al. (2012)
Gluten Ataxia	Cortical cerebellar atrophy	Decreased NAA, NAA/Cr and increased Cho/Cr in deep hemispheric cerebellum	—	NAA/Cr ratio in the cerebellar vermis after gluten free diet	Bürk et al. (2001), Wilkinson et al. (2005), Hadjivassiliou et al. (2012), and Hadjivassiliou, Grünewald, Sanders, Shanmugarajah, and Hoggard (2017)
Sporadic Adult Onset Ataxia	Cortical cerebellar atrophy	Diffusion changes in the cerebellum and brainstem; Decreased NAA/Cr in cerebellum, pons and frontal cortex; Decreased perfusion in cerebellum and frontal cortex	—	Brain perfusion after therapy with thyrotropin releasing hormone	Tachibana, Kawabata, Tomino, and Sugita (1999), Terakawa et al. (1999); Mascalchi et al. (2002), Della Nave et al. (2004), Bürk et al. (2004), Waragai, Yamada, and Matsuda (2007), and Kimura et al. (2009b)

| Multi-System Atrophy cerebellar type | Olivopontocerebellar atrophy; Diffuse T2 high signal in pons, middle and cerebellar white matter with "hot cross bun" sign; Symmetric T2 and T2* low signal in the striatum; putaminal and thalamic atrophy | Diffusion changes in cerebellum, cerebellar peduncles, pons, middle cerebellar peduncles, putamen, caudate and cerebral white matter and cortex; Increased R2 relaxation rate in the putamen; Decreased NAA/Cr and Cho/Cr in cerebellum, pons, putamen and frontal cortex; Altered regional homogeneity in cerebral cortical regions in resting state functional MRI; Decreased FDG uptake in pons, cerebellum, striata and frontal-parietal cortex; Decreased presynaptic and postsynaptic striatal uptake of tracers of the dopaminergic system; Cholinergic pathways dysfunction | Volume of the middle cerebellar peduncles, cerebellum, putamen and corpus callosum; Diffusion indexes of the putamen, pons, cerebellar white matter, thalamus and frontal white matter | Cerebellar perfusion after therapy with thyrotropin releasing hormone | Terakawa et al. (1999), Mascalchi et al. (2002), Specht, Minnerop, Müller–Hübenthal, and Klockgether (2005), Brenneis et al. (2006), Hauser et al. (2006), Taoka et al. (2007), Boesch et al. (2007), Lyoo et al. (2008), Hirano et al. (2008), Lee, An, Yong, and Yoon (2008), Prakash et al. (2009), Pellecchia et al. (2011), Minnerop et al. (2010), Tha et al. (2010), Kimura et al. (2011), You et al. (2011), Pellecchia et al. (2011), Lirng et al. (2012), Mascalchi and Vella (2012), Mazere et al. (2013), and Oh et al. (2017) |

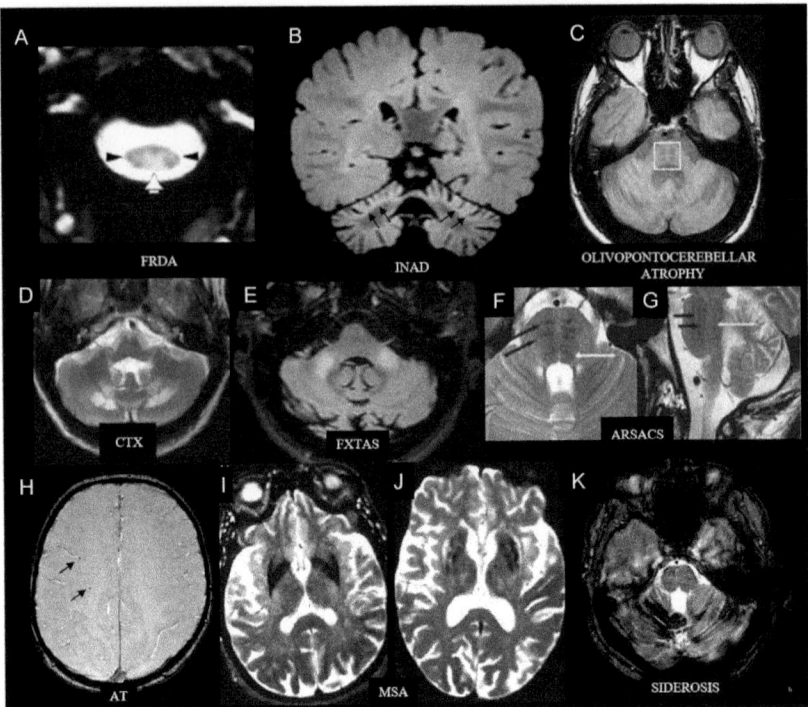

Fig. 2 (A–K) T2 or T2* signal changes that might support the diagnosis in chronic inherited and sporadic ataxias. (A) Symmetric increased T2 signal in the posterior (white arrow) and lateral (black arrowheads) columns of the cervical spinal cord in Freidreich ataxia (FRDA). (B) Increased T2 signal of the superior cerebellar cortex (black arrows) associated with thinning of the folia in Infantile Neuro-Axonal Dystrophy (INAD). (C) Increased signal in a proton density image of the white matter of the pons, middle cerebellar peduncels and cerebellar white matter with sparing of the corticospinal tracts in the central portion of basis pontis featuring the "hot cross bun" sign (white box) in a case of olivopontocerebellar atrophy due to SCA2. (D) Symmetric T2 hyperintensity of the peridentate cerebellar WM in cerebrotendineous xanthomatosis (CTX). (E) Symmetric T2 hyperintensity of the middle cerebellar peduncles in Fragile X Tremor Ataxia Syndrome (FXTAS). (F and G) T2 low signal intensity stripes corresponding to the corticospinal tracts (black arrows) and medial lemniscus (white arrows) in a bulky basis pontis in Autosomal Recessive Spastic Ataxia of Charlevoix-Saguenay (ARSACS). (H) Two small low signal dots (black arrows) corresponding to cerebral telangectasias in a susceptibility weighted image in Ataxia Telangectasia (AT). (I and J) Symmetric abnormally low T2 signal intensity of the putamen in two different patients with multiple system atrophy (MSA), which is surrounded by a regular hyperintense rim in (J). (K) Extensive areas of T2* low signal in the cerebellar superior cortex in siderosis of the CNS. *Modified from Baldarçara, L., Currie, S., Hadjivassiliou, M., Hoggard, N., Jack, A., Jackowski, A.P., et al. (2015). Consensus paper: Radiological biomarkers of cerebellar diseases.* Cerebellum, 14 (2), 175–196, https://doi.org/10.1007/s12311-014-0610-3.

white matter of the pons, middle cerebellar peduncels and cerebellar white matter with sparing of the corticospinal tracts in the central portion of basis pontis featuring the "hot cross bun" sign are characteristics of advanced olivopontocerebellar atrophy. Abnormally decreased T2 or T2* signal can be observed with a multifocal distribution in Ataxia Telangectasia and with a symmetric distribution in the basal ganglia in Multiple System Atrophy. T2 low signal intensity stripes corresponding to the corticospinal tracts and medial lemniscus in a bulky basis pontis are characteristic of Autosomal Recessive Spastic Ataxia of Charlevoix-Saguenay. Finally, a diffuse T2 or T2* hypointense lining of the outer surfaces of the brainstem, cerebellum and cerebrum enables diagnosis of superficial siderosis of the central nervous system.

MRI is also fundamental for differential diagnosis in children with ataxia, which is often non-progressive, and has a number of acquired and rare genetic causes (Alves, Fragoso, Gonçalves, Marussi, & Amaral, 2018).

The contribute of SPECT and PET to the diagnosis of the single patient with chronic ataxia is limited (Mascalchi & Vella, 2012). Brain SPECT and PET with radiotracers targeted to the nigrostriatal system are useful in patients with adult-onset sporadic ataxia for the differential diagnosis between multi system atrophy, in which overt striatal abnormalities are observed in terms of decreased uptake of dopamine, dopamine trasporters and D2 receptors and of 2-18F-fluoro-2-deoxy-D-glucose (FDG), and Sporadic Adult Onset Ataxia, a condition previously termed Idiopathic Late Onset Cerebellar Ataxia, in which no such striatal abnormalities are detected. In patients with Fragile X Tremor Ataxia Syndrome, SPECT shows decreased uptake of the dopamine transporter and post-synaptic D2 receptors in the striata that account for the extrapyramidal symptoms associated with ataxia and can assist in the differential diagnosis with essential tremor.

Table 3 summarizes the neuroimaging features which are diagnostic markers of the most common chronic ataxias.

3.2 Physiopathological Markers

Cross-sectional case vs control studies have provided information about the type and distribution of neuroimaging abnormalities characterizing each ataxia disease.

3.2.1 Spinal Atrophy (Friedreich Ataxia)

Friedreich ataxia is the most common inherited ataxia in the world.

In the Friedreich's original description of the disease emphasis was given to the atrophy of the spinal nerve ganglions and of the spinal cord (1877). The relatively few studies that addressed the evaluation with MRI of the spinal cord in patients with Friedreich ataxia (Chevis et al., 2013; Dogan et al., 2018; Mascalchi et al., 1994; Rezende et al., 2018; Schöls et al., 1997; Wüllner et al., 1993) have consistently demonstrated a diffusely thinned spinal cord already present in pediatric age (Rezende et al., 2018). Many MRI studies evaluated the brain in Friedreich ataxia patients using several quantitative techniques. The volumes of the medulla and of the inferior and superior cerebellar peduncles were decreased (Akhlaghi et al., 2011; Della Nave, Ginestroni, Giannelli, Tessa, et al., 2008). The volume of the cerebellum was decreased in some studies (Della Nave, Ginestroni, Giannelli, Tessa, et al., 2008; Dogan et al., 2018; França et al., 2009; Selvadurai et al., 2016) but normal in others (Stefanescu et al., 2015). Several morphometry studies have revealed gray matter atrophy in the supratentorial brain including posterior cyngulate gyrus, paracentral lobule and middle frontal gyrus, premotor and supplementary motor areas, the cuneus and precuneus, posterior aspects of the medial and lateral prefrontal cortices, insula, temporal poles and diencephalon (Dogan et al., 2018; França et al., 2009; Selvadurai et al., 2016).

The sensitivity of T2* weighted and susceptibility weighted images to the iron content of the dentate nuclei has permitted an indirect evaluation of the size of this nucleus that is crucially affected in Friedreich ataxia but whose complex morphology hindered its evaluation in T1 weighted images. Two studies have demonstrated that the size of the dentate is decreased in Friedreich ataxia (Solbach et al., 2014; Stefanescu et al., 2015). Moreover one study has shown that T2 relaxation time of the dentate nuclei was significantly shorter in patients with Friedreich ataxia compared to controls (Bonilha da Silva et al., 2014).

Using texture analysis Santos et al. (2015) documented altered tissue organization in the medulla but not in the pons of patients with Friedreich ataxia.

Several diffusion weighted imaging and diffusion tensor imaging studies using regions of interest or voxel wise analyses have demonstrated abnormalities of the water diffusion in terms of increased apparent diffusion coefficient, mean, radial and axial diffusivity and decreased fractional anisotropy, in patients with Friedreich ataxia. The affected structures include inferior and superior cerebellar peduncels, the medulla, the cerebellar white matter, the cortico-spinal tracts, thalamic radiations, optic chiasm and radiations, intrahemispheric associative white matter tracts and corpus callosum

(Clemm von Hohenberg et al., 2013; Della Nave et al., 2011; Della Nave, Ginestroni, Tessa, Salvatore, Bartolomei, et al., 2008; Fortuna et al., 2009; Pagani et al., 2010; Rezende et al., 2016; Rizzo et al., 2011; Vieira Karuta et al., 2015) (Fig. 3). Microstructural damage has also been reported in the dentate (Gramegna et al., 2017), dentate-rubral, dentate-thalamic (both contained in the superior cerebellar peduncles) and thalamocortical white matter fibers (Akhlaghi et al., 2014) and this has provided support to the hypotheisis that a cerebello-cerebral connectivity deficit may contribute with the cerebral cortical GM atrophy to the cognitive deficit that can be observed in patients with Friedreich ataxia (Selvadurai, Harding, Corben, & Georgiou-Karistianis, 2018; Zalesky et al., 2014).

Finally, using diffusion tensor imaging a more widespread white matter damage and more severe microstructural damage of the corticospinal tracts were observed in patients with late-onset in comparison with classical Friedreich ataxia (Rezende et al., 2017).

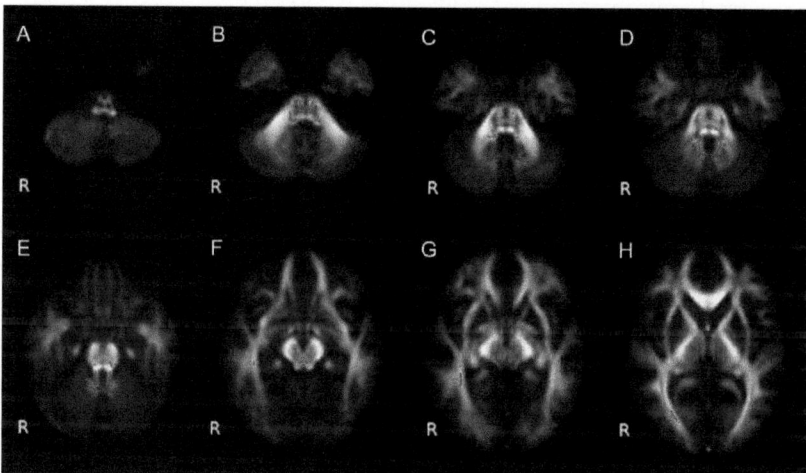

Fig. 3 (A–H) Tract Based Spatial Statistics analysis of diffusion tensor imaging data showing white matter tracts microstructural damage associated with decreased fractional anisotropy (in red) in patients with Friedreich ataxia compared to controls. They include corticospinal tracts at the medullary pyramids (A); inferior cerebellar peduncles and right cerebellar hemisphere white matter (B); superior cerebellar peduncles (C–E); decussation of the superior cerebellar peduncles in the midbrain (F and G) and right inferior fronto-occipital and inferior longitudinal fasciculus (F–H). *Reproduced with permission from Della Nave, R., Ginestroni, A., Tessa, C., Salvatore, E., Bartolomei, I., Salvi, F., et al. (2008). Brain white matter tracts degeneration in Friedreich ataxia. An in vivo MRI study using tract-based spatial statistics and voxel-based morphometry.* NeuroImage, *40(1), 19–25. https://doi.org/10.1016/j.neuroimage.2007.11.050.*

Results of proton MR spectroscopy in patients with Friedreich ataxia are fragmentary. These include a mild decrease of the ratio between the neuronal marker N-acetyl aspartate (NAA) and the putative glial marker creatine (Cr) (NAA/Cr) ratio in the pons and deep cerebellar hemisphere (Gramegna et al., 2017; Mascalchi et al., 2002), a decrease of the concentration of Choline (Cho) a marker of membrane turnover, in the vermis (Viau & Boulanger, 2004) and a decrease of the Cho/Cr in the dentate (Gramegna et al., 2017). However, another study found no abnormalities in the vermis (Hadjivassiliou et al., 2012). A decrease of the NAA/Cr ratio but not of the Cho/Cr ratio was also reported in the cerebral white matter (França et al., 2009).

Few functional MRI studies investigated patients Friedreich ataxia using motor and cognitive tasks. Simple motor tasks demonstrated decreased activation of the cerebellar cortex and dentate (Akhlaghi et al., 2012; Ginestroni et al., 2012; Stefanescu et al., 2015) but also of cerebral cortical regions including the primary motor cortex, supplementary motor area and dorsolateral prefrontal cortex (Akhlaghi et al., 2012; Ginestroni et al., 2012). More complex motor tasks with the right dominant hand revealed additional areas of increased activation of possible compensatory significance in the bilateral parietal and frontal cortices, right globus pallidus and putamen and left cerebellum (Akhlaghi et al., 2012; Ginestroni et al., 2012).

Overall the above data confirm that the motor impairment in Friedreich ataxia is associated with dysfunction of the spinal cord and cerebellum but also of additional cerebral cortical and subcortical structures.

Two functional MRI studies investigated patients with Friedreich ataxia and controls using cognitive tasks.

During a working memory task patients with Friedreich ataxia showed deficits in brain activations both in the lateral cerebellar hemispheres, principally encompassing lobule VI, and the prefrontal cortex, including regions of the anterior insular and rostrolateral prefrontal cortices, and a decreased functional connectivity between cerebellum and prefrontal areas (Harding et al., 2016). These findings were felt to support a diaschisis model of brain dysfunction, implying primary cerebellar damage and secondary functional changes in downstream fronto-cerebellar networks.

As compared to controls, patients with Friedreich ataxia showed enhanced activity in the cerebellum (VI, Crus I) fronto-insular, premotor and temporo-occipital regions in a phonemic fluency task, a reduced activity in the anterior cerebellum during a semantic task and increased activation in the motor cortex for overt speech (Dogan et al., 2016). Analysis of functional connectivity revealed impaired interregional coupling between the cerebellum and fronto-insular cortex for phonemic processing, which was related to poorer task performance. The pattern of increased cerebellar and cerebral cortical neural response underlying executive functioning was interpreted as indicating functional reorganization driven by disease-related structural damage in Friedreich ataxia.

The only resting-state functional MRI study published to date showed reduced coupling between cerebellar and frontal cerebral cortex and increased connectivity between areas within the cerebral cortex in patients with Friedreich ataxia as compared to controls (Cocozza et al., 2018), consistent with a cerebello–cerebral diaschisis model of brain dysfunction.

While perfusion SPECT studies have shown non-specific reduced blood flow in the cerebellum and cerebral hemispheres of patients with Friedreich ataxia (see Mascalchi & Vella, 2012), the FDG-PET pattern is distinctive. In fact FDG-PET demonstrated a diffuse increase of glucose metabolism in the brain of patients with Friedreich ataxia including the cerebellum when the patients had mild to moderate clinical impairment. A normal pattern or a regionally specific decrease of glucose metabolism, including the cerebral cortex, the cerebellum, and the brainstem, were present in patients with more severe deficit (Gilman et al., 1990). The diffuse hypermetabolism of early Friedreich ataxia has been assumed may reflect the abnormalities in the mitochondrial function associated with deficiency of functional frataxin. A PET study has shown that brain benzodiazepine receptors are preserved in patients with Friedreich ataxia (Chavoix et al., 1990).

The number of abnormal Friedreich ataxia triplets correlated with volume loss in the lateral cerebellar hemispheres (Selvadurai et al., 2016), the NAA/Cr ratio in the dentate (Gramegna et al., 2017) and cerebellar dysfunction in a functional MRI study with a working memory task (Harding et al., 2016).

3.2.2 Cortical Cerebellar Atrophy (Ataxia Telangectasia, SCA6, SCA17, Gluten Ataxia, Sporadic Adult Onset Ataxia)

3.2.2.1 Ataxia Telangectasia

In a diffusion weighted imaging study, patients with ataxia telangectasia showed increased mean apparent diffusion coefficient in the cerebellar white matter and cortex as compared to healthy controls, whereas the mean apparent diffusion coefficient in the cerebral hemispheres including basal ganglia was not significantly different (Firat et al., 2005).

One spectroscopic proton MR study reported a decrease of the normalized NAA, Cho, and Cr signal in the vermis and cerebellar hemisphere in patients with ataxia telangectasia as compared to controls without

abnormalities in the basal ganglia (Lin et al., 2006), while a single-voxel proton MR spectroscopy study reported a significantly lower NAA/Cho ratio
and higher Cho/Cr ratio with a normal NAA/Cr ratio, suggesting an
increased Cho content, in the cerebellar dentate of patients with ataxia
telangectasia as compared to controls (Wallis et al., 2007).

Nuclear Medicine studies have contribute to explain the extrapyramidal
dysfunction sometimes observed in ataxia telangectasia. In particular, a
SPECT study showed a bilaterally decreased dopamine D2 receptor binding
in the striatum of a child with ataxia telangectasia presenting with progressive dystonia (Koepp, Schelosky, Cordes, Cordes, & Poewe, 1994) and a
FDG-PET study of patients with ataxia telangectasia and siblings heterozy-
gotes for the *ATM* gene (Volkow et al., 2014) showed that the former had
lower metabolism in the cerebellum, fusiform gyrus and hippocampus but
higher metabolism in globus pallidus compared with controls. Asymptomatic relatives had lower metabolism in anterior vermis and hippocampus than
controls suggesting that heterozygocity influences the function of these brain
regions.

3.2.2.2 SCA6

Cerebellar atrophy, especially of the cortical gray matter, has been demonstrated by voxel based morphometry or volumetry studies in SCA6 (Eichler
et al., 2011; Falcon et al., 2016; Jung, Choi, Du, Cuzzocreo, Geng, et al.,
2012a; Lukas et al., 2006; Schulz et al., 2010; Stefanescu et al., 2015).
Atrophy of the dentate was also reported (Stefanescu et al., 2015). Volumetry has also shown a minor but significant atrophy of the brainstem
(Eichler et al., 2011; Schulz et al., 2010). The spinal cord has a normal size
in SCA6 patients (Lukas et al., 2008).

The white matter tracts in the middle and superior cerebellar peduncles
and the cerebral peduncle tracts linking indirectly the cerebellum with the
cerebral cortex showed decreased size and altered diffusivity, in terms of
decreased fractional anisotropy and increased radial and mean diffusivity,
in SCA6 patients with moderate or advanced disease as compared to controls
(Falcon et al., 2016). Presymptomatic SCA6 subjects showed increased fractional anisotropy and decreased radial diffusivity in the same white matter
tracts.

The NAA/Cr ratio and NAA/Cho ratios in the cerebellum were
decreased in SCA6 patients also in early disease phase (Boesch et al.,
2001; Lirng et al., 2012). In a recent study of pre- and early symptomatic
SCA6 gene carriers (Joers et al., 2018) the ratio of total NAA/myo-inositol

(mI), a glial marker, (tNAA/mI) was significantly lower in SCA6 as compared to controls in vermis and cerebellar white matter but not in the pons.

In a functional MRI study (Stefanescu et al., 2015), activation of the cerebellar cortex and nuclei during simple hand movements was significantly lower in patients with SCA6 as compared to controls.

During a handgrip motor task SCA6 patients showed decreased activation in the sensorimotor cortex, supplementary motor area and cerebellar vermis and hemispheres, that was accompanied by a reduced task-based functional connectivity between cerebral cortical motor regions and the cerebellar regions (Kang et al., 2017).

Functional MRI during a visual (passive smooth-pursuit) task demonstrated a shift in activation from vermis in SCA6 presymptomatic individuals to lateral cerebellum in moderate-to-severe cases. Effective connectivity between regions of cerebral cortex and cerebellum was increased in moderate cases, but disappeared in severe cases (Falcon et al., 2016).

Perfusion SPECT studies have shown decreased regional blood flow in the cerebellum but also in the prefrontal regions (Honjo et al., 2004; Kawai et al., 2008).

FDG-PET studies have demonstrated that the decreased metabolism in the cerebellum in SCA6 is diffuse, with preferential involvement of the anterior lobe (Oh et al., 2017), and is accompanied by hypometabolism in cerebral frontal and prefrontal cortex, putamen and brainstem (Soong et al., 2001; Wang et al., 2007; Wüllner et al., 2005) (Fig. 4).

One PET study reported preserved density of striatal dopamine transporter in SCA6 (Wüllner et al., 2005), while a SPECT study reported a decreased density of striatal dopamine transporter in five of six SCA6 patients (Kim et al., 2010).

A PET study revealed a decreased uptake of ^{11}C-flumazenil in the cerebellum consistent with an altered function of the gamma-aminobutyric acid type A/benzodiazepine receptors in two SCA6 patients (Kono, Terada, Ouchi, & Miyajima, 2014).

No significant correlations were observed between trinucleotide repeat lengths and degree of atrophy of the cerebellar cortex and nuclei (Stefanescu et al., 2015).

3.2.2.3 SCA17
Voxel based morphometry studies demonstrated atrophy in the cerebellum and in cortical and subcortical gray matter structures including the limbic system and frontal, occipital and parietal lobes, the putamen and thalamus

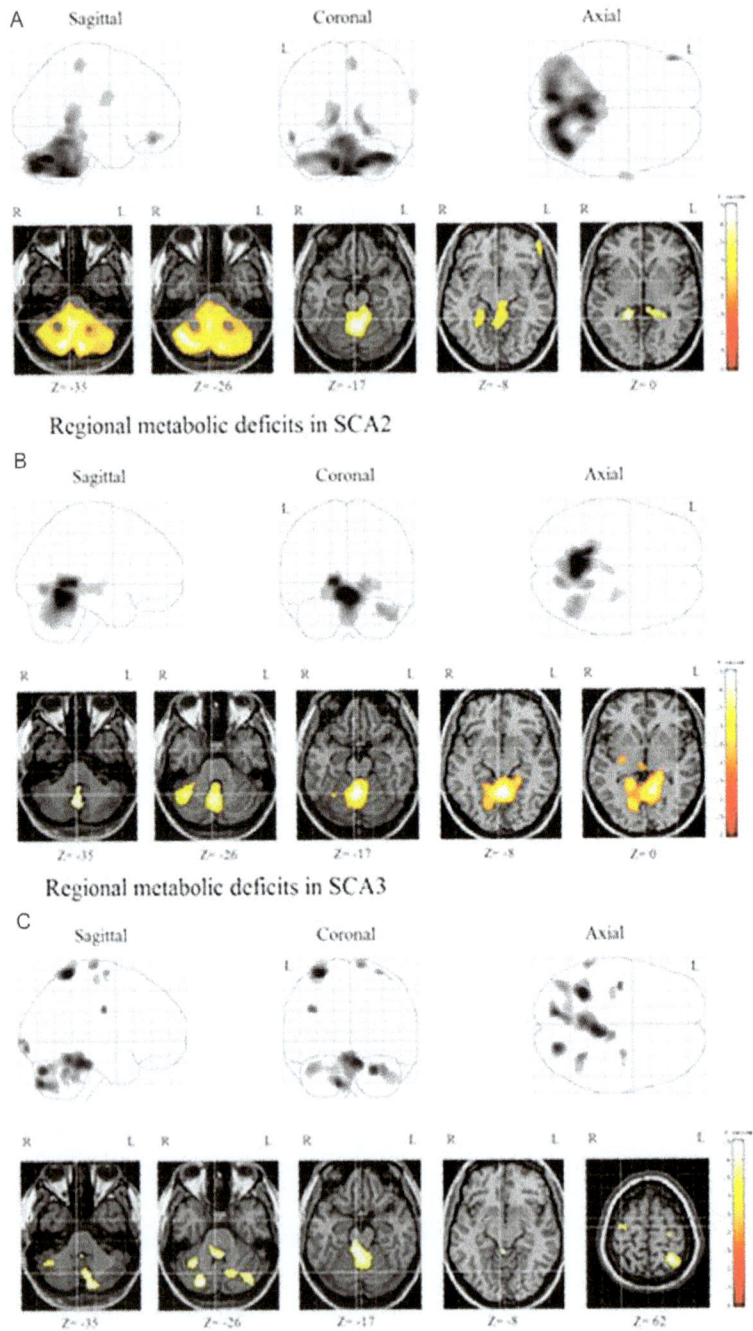

Regional metabolic deficits in SCA2

Regional metabolic deficits in SCA3

Regional metabolic deficits in SCA6

Fig. 4 See figure legend on next page.

in SCA17 patients compared controls (Lasek et al., 2006; Reetz et al., 2010). Volumetry revealed atrophy of the cerebellum and caudate already in presymptomatic patients (Brockmann et al., 2012). The NAA/Cr and NAA/Cho ratios in the cerebellar hemipheres and vermis were lower in SCA17 patients compared to controls (Lirng et al., 2012).

FDG-PET studies have shown hypometabolism in the cerebellum, putamen and caudate and parietal cortex (Brockmann et al., 2012; Minnerop et al., 2005). In addition, a decrease of dopamine transporter has been constantly reported in SPECT studies in SCA17 patients (Kim et al., 2009; Minnerop et al., 2005). In a PET study also decreased uptake of ^{11}C-raclopride in the putamen and caudate consistent with impairment of the postsynaptic dopaminergic compartment was observed (Brockmann et al., 2012). All the above abnormalities of the tracers uptake were generally observed already in presymptomatic SCA17 gene carries (Brockmann et al., 2012; Kim et al., 2009).

Overall the neuroimaging findings, while confirming that cerebellar involvement exhibits a cortical cerebellar atrophy pattern, clearly indicate that SCA17 genetic abnormality is associated with widespread central nervous system involvement.

3.2.2.4 Gluten Ataxia

MRI demonstrates a pattern of cerebellar cortical atrophy in patients with gluten ataxia (Bürk et al., 2001; Hadjivassiliou et al., 2012).

Proton MR spectroscopy reveals significantly lower NAA, NAA/Cr and NAA/Cho ratios and increased Cho/Cr ratio in the hemispheric deep cerebellum in patients with gluten ataxia as compared to controls (Hadjivassiliou et al., 2012; Wilkinson et al., 2005).

Fig. 4 (A–C) Sagittal, coronal, and transverse glass brain images and Statistical Parametric Mapping projection images on normalized MRI slices illustrating regions of decreased 18F-FDG metabolism in SCA2, 3 and 6 subtypes compared to controls. They include the cerebellum, pons, parahippocampal gyrus and frontal cortex in SCA2 (A); the cerebellum, parahippocampal gyrus of the limbic system, and lentiform nucleus in SCA3 (B) and the cerebellum and frontal cortex in SCA6 (C). *Reproduced with permission from Wang, P.-S., Liu, R.-S., Yang, B.-H., & Soong, B.-W. (2007). Regional patterns of cerebral glucose metabolism in spinocerebellar ataxia type 2, 3 and 6: A voxel-based FDG-positron emission tomography analysis. Journal of Neurology, 254(7), 838–845. https://doi.org/ 10.1007/s00415-006-0383-9.*

3.2.2.5 Sporadic Adult Onset Ataxia/Idiopathic Late Onset Cerebellar Ataxia

Volumetry has revealed overt cortical cerebellar atrophy and a mild but significant decrease of the brainstem in patients with idiopathic cerebelalr ataxia with pure cerebellar syndrome as compared to controls (Bürk et al., 2004).

In a diffusion weighted imaging study (Della Nave et al., 2004), the apparent diffusion coefficient in the cerebellum and brainstem of patients with Idiopathic Late Onset Cerebellar Ataxia was significantly increased compared to controls, possibly revealing a mild diffuse structural damage of the cerebellar cortical and subcortical gray matter and white matter, whereas the cerebral apparent diffusion coefficient was not significantly different.

In patients with sporadic cortical cerebellar atrophy, a decrease of the NAA/Cr ratio was observed in the cerebellum, pons, and frontal cortex but not in the putamen (Mascalchi et al., 2002; Terakawa et al., 1999).

SPECT studies in patients with Idiopathic Late Onset Cerebellar Ataxia showed hypoperfusion in the cerebellum and frontal cortex with significant correlation between the perfusion of the frontal lobe and cerebellum (Tachibana et al., 1999; Waragai et al., 2007). Disruption of the neural circuits between the cerebellum and frontal lobe could explain the cognitive impairment observed in patients with cerebellar degeneration.

3.2.3 Olivopontocerebellar Atrophy (SCA1, SCA2, SCA3, Multi-System Atrophy)

3.2.3.1 SCA1

Volumetry and voxel-based morphometry demonstrated lower size of the brainstem, especially the pons, cerebellar gray and white matter, putamen and caudate wih normal cerebral cortex in patients with SCA1 compared to controls (Ginestroni et al., 2008; Goel et al., 2011; Guerrini et al., 2004; Klockgether et al., 1998; Schulz et al., 2010). Presymptomatic subjects already show decreased volume of the pons and medulla oblongata (Jacobi et al., 2013). Also the cervical spinal cord showed decreased cross-sectional area and increased eccentricity in symptomatic SCA1 patients compared to controls (Martins et al., 2017).

A diffusion weighted imaging study revealed diffuse increase of the brainstem and cerebellum apparent diffusion coefficient with normal value of the cerebral hemispheres (Guerrini et al., 2004), whereas diffusion tensor imaging studies (Della Nave, Ginestroni, Tessa, Salvatore, De Grandis, et al., 2008; Mandelli et al., 2007; Prakash et al., 2009; Solodkin et al., 2011) have demonstrated a distributed pattern of microstructural damage of the white matter tracts in the brainstem, cerebellar peduncles and cerebellar white matter in patients with SCA1.

A decrease of neuronal markers including NAA, total NAA and Glutamine (Glu) concentrations and NAA/Cr ratio was reported in the pons, deep cerebellar hemisphere and vermis of asymptomatic and symptomatic carriers of the SCA1 mutation compared to healthy controls (Adanyeguh et al., 2015; Guerrini et al., 2004; Joers et al., 2018; Mascalchi et al., 1998; Oz et al., 2010). An increase of mI and total Cr and a decrease of Cho/Cr ratios consistent with gliosis in the the same regions were also detected (Adanyeguh et al., 2015; Joers et al., 2018; Mascalchi et al., 1998; Oz et al., 2010).

Functional MRI during a simple motor task revealed paucity or absence of cerebellar activation with additional activations in the contralateral cerebral cortex and thalamus of possible compensatory significance in SCA1 patients compared to controls (Jayakumar et al., 2008). A preliminary resting state functional fMRI study showed alteration of the intrinsic functional connectivity in lateral cerebellum and thalamus in SCA1 compared to controls with increase of the absolute value of the correlation coefficients and a patchy appearance of cerebellar functional clusters (Solodkin et al., 2011).

Hypometabolism in brainstem and cerebellum was reported in FDG-PET studies of SCA1 patients (Gilman et al., 1996; Wüllner et al., 2005) which was combined in one study (Gilman et al., 1996) with reduced FDG uptake in the cerebral cortex and subcortical gray matter nuclei.

PET studies have demonstrated no abnormalities in the dopamine transporter in the striatum (Wüllner et al., 2005) and of the benzodiazepine receptor distribution volume (Gilman et al., 1996).

Length of the SCA1 abnormal triplets correlated with pontine and spinal cord atrophy (Martins et al., 2017; Schulz et al., 2010).

3.2.3.2 SCA2
SCA2 is the second most frequent autosomal dominant inherited ataxia in the world.

Volumetry and voxel based morphometry studies in patients with SCA2 demonstrate a marked decrease of the volume of the brainstem, especially the pons, and of cerebellum (Brenneis et al., 2003; D'Agata et al., 2011; Della Nave, Ginestroni, Tessa, Cosottini, et al., 2008; Goel et al., 2011; Guerrini et al., 2004; Jung, Choi, Du, Cuzzocreo, Geng, et al., 2012a; Jung, Choi, Du, Cuzzocreo, Ying, et al., 2012b; Klockgether et al., 1998; Reetz et al., 2018) that was combined with cerebral cortical atrophy in patients with advanced disease (Brenneis et al., 2003; D'Agata et al., 2011). The basal ganglia are usually of normal size (Della Nave, Ginestroni, Tessa, Cosottini, et al., 2008; Klockgether et al., 1998).

Notably the volume loss in the brainstem and volume is already present in asymptomatic SCA2 carriers (Reetz et al., 2018) supporting the possibility that a developmental component takes part in SCA2 pathogenesis (Guerrini et al., 2004; Jung, Choi, Du, Cuzzocreo, Ying, et al., 2012b).

Fractal dimension, a measure of tissue organization, was reduced in the cerebellum and cerebral cortical gray matter in SCA2 patients (Marzi et al., 2018).

A diffuse increase of the apparent diffusion coefficent in the brainstem and cerebellum was reported in SCA2 (Guerrini et al., 2004).

Region of interest and voxel-wise analyses demonstrated additional areas of cerebral increased diffusivity including thalamus and frontal and temporal regions (Salvatore et al., 2014). Two diffusion tensor imaging studies (Della Nave, Ginestroni, Tessa, Salvatore, De Grandis, et al., 2008; Mandelli et al., 2007) in patients with SCA2 have documented a distributed pattern of microstructural damage of white matter tracts in the brainstem, cerebellar peduncles and within the cerebellum, as well as in the corticospinal tracts at the level of the internal capsule and of the cerebral peduncles.

Tractography has demonstrated severe alterations of the cerebellar white matter and of the middle and superior cerebellar peduncles in SCA2 (Olivito, Lupo, et al., 2017) supporting the view that the genetic alteration might determine a cerebello-cerebral dysregulation underlying the cognitive symptoms observed in patients.

Proton MR spectroscopy demonstrated a decrease of the NAA/Cr, NAA/Cho and Cho/Cr ratios and of the NAA, total NAA and Glu concentration in the pons and cerebellum (Adanyeguh et al., 2015; Boesch et al., 2001, 2007; Guerrini et al., 2004; Lirng et al., 2012; Viau & Boulanger, 2004). This was accompanied by an increase of the putative glial markers mI and total Cr in the pons and in the cerebellum (Adanyeguh et al., 2015; Viau & Boulanger, 2004).

Two studies documented an increase of lactate in the cerebellum (Boesch et al., 2001, 2007).

Importantly, a recent study of presymptomatic and early symptomatic SCA2 patients revealed a significant decrease of the total NAA/mI in the pons, cerebellar white matter and vermis (Joers et al., 2018).

Four groups have investigated brain functional connectivity in SCA2 patients using resting state functional MRI and a variety of analytical approaches reporting fragmentary results. These include: a decreased connectivity between pons and putamen and between rostral sensorimotor cortex and both pons and cerebellum in symptomatic SCA2 patients presenting with parkinsonism, but an increased connectivity in asymptomatic gene carriers of the same families (Wu et al., 2013); a decrease of the functional connectivity of the cerebellar components of the default mode, executive and right fronto-parietal networks in symptomatic SCA2 patients (Cocozza et al., 2015); reduced functional connectivity within the cerebellum and between the cerebellum and frontal/parietal cortices that was combined with increase of the cerebellar functional connectivity with parietal, frontal, and temporal areas in symptomatic SCA2 patients (Hernandez-Castillo et al., 2015). Finally, more posterior regions in the cerebellum and regions in the cerebral cortex related to cognition and emotion showed altered inter-nodal connectivity as well more anterior cerebellar lobules and motor and somatosensory cerebral regions (Olivito, Cercignani, et al., 2017) (Fig. 5).

FDG-PET studies demonstrated reduced glucose metabolism in the brainstem, cerebellum, and frontal, parietal and limbic cortex of patients with SCA2 (Oh et al., 2017; Wang et al., 2007; Wüllner et al., 2005) (Fig. 4). Hypometabolism in the pons and cerebellum was also observed in asymptomatic carriers of SCA2 mutation (Inagaki et al., 2005).

Several PET and SPECT studies have demonstrated nigrostriatal dopaminergic abnormalities in SCA2 patients (Boesch et al., 2004; Furtado et al., 2002; Shan et al., 2001; Varrone et al., 2004; Wang et al., 2009; Wüllner et al., 2005). Dopamine transporter decrease is already present in asymptomatic carriers of the SCA2 gene (Kim et al., 2007), whereas D2 postsynaptic receptors showed increased or decreased activity (Boesch et al., 2004; Furtado et al., 2002; Kim et al., 2007).

One study reported correlation between volume loss and abnormal CAG repeat length (Reetz et al., 2018) that was not observed in a prior study (Goel et al., 2011).

3.2.3.3 SCA3

SCA3 (Machado-Joseph disease) is the most frequent autosomal dominant inherited ataxia in the world.

The atrophy pattern in SCA3 disease seems heterogeneous. In German families atrophy of the cerebellum including the dentate, of the brainstem and cerebellar peduncels is associated with a significant decrease of the putamen, caudate and thalamus volume (Eichler et al., 2011; Kang et al., 2014;

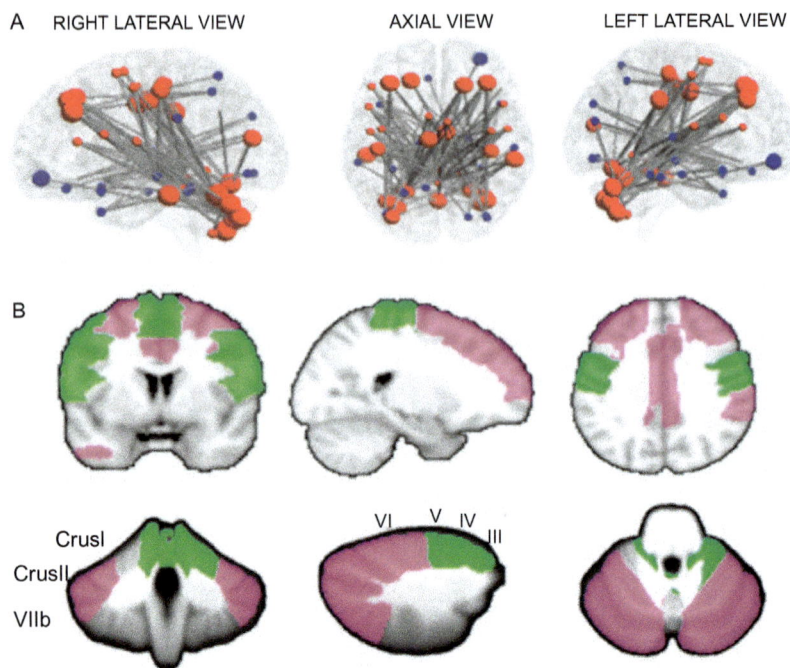

Fig. 5 (A and B) Network of significantly decreased functional connectivity in SCA2 patients as assessed by Network Based Statistics analysis (A). The regions of the cerebello-cortical modules are shown in red and of the cortico-cortical modules in blue. Bigger nodes correspond to cerebellar and cortical regions relevant to cognition and emotion; smaller nodes correspond to cerebellar and cortical regions relevant to motor control. Anatomical representations of cognitive (violet) and motor (green) nodes in the cerebellum and cerebral cortex showing underconnectivity between each other (B). *Reproduced with permission from Olivito, G., Cercignani, M., Lupo, M., Iacobacci, C., Clausi, S., Romano, S., et al. (2017). Neural substrates of motor and cognitive dysfunctions in SCA2 patients: A network based statistics analysis.* NeuroImage. Clinical, *14, 719–725. https://doi.org/10.1016/j.nicl.2017.03.009.*

Klockgether et al., 1998; Lukas et al., 2006; Schulz et al., 2010; Stefanescu et al., 2015). In Japanese families atrophy involves the brainstem and cerebellum and the superior cerebellar peduncles (Murata et al., 1998; Tokumaru et al., 2003).

In a voxel based morphometry study from India, gray matter loss in the vermis and cerebellar hemisphere and frontal and temporal cortex was observed without white matter loss (Goel et al., 2011). In SCA3 patients from South America atrophy of the pons, inferior olives, cerebellum and frontal lobes was reported (Etchebehere et al., 2001; Guimarães et al., 2013).

Also the size of the upper spinal cord is significantly reduced in SCA3 patients (Lukas et al., 2008).

The cerebellum of SCA3 patients showed significantly lower fractal dimension was compared to controls (Huang et al., 2017).

Diffusion tensor imaging studies showed widespread microstructural changes of the white matter in the cerebellum, brainstem, and bilaterally in thalamus and the cerebral hemisphere (Guimarães et al., 2013; Kang et al., 2014).

Proton MR spectroscopy studies have demonstrated lower NAA/Cr and total NAA/mI, total NAA and Glu with increase of total Cr and mI in the pons, cerebellar hemispheres and vermis and in the cerebral white matter in SCA3 patients, including presymptomatic and early symptomatic gene carriers, compared to controls (Adanyeguh et al., 2015; D'Abreu et al., 2009; Huang et al., 2017; Joers et al., 2018; Lirng et al., 2012).

Activation in the cerebellar cortex and nuclei during simple hand movements was not significantly reduced in SCA3 patients compared to controls (Stefanescu et al., 2015). In a further study involving a bilateral audiopaced thumb movements paradigm, SCA3 patients showed a diminished movement synchronization consistent with functional reorganization of the motor network (Duarte et al., 2016).

Many SPECT and PET studies investigated SCA3 patients.

Brain SPECT studies have demonstrated hypoperfusion of the cerebellum and cerebral regions including basal ganglia and frontal-temporal cortices in patients with SCA3 (Braga-Neto et al., 2016; Etchebehere et al., 2001).

FDG-PET studies in SCA3 demonstrated a decrease of glucose metabolism in the brainstem, cerebellum, putamen, thalamus and and limbic and occipital cortex in symptomatic and pre-symptomatic carriers of the abnormal gene (Soong et al., 1997; Soong & Liu, 1998; Wang et al., 2007; Wüllner et al., 2005) (Fig. 4). Increased FDG metabolism in the parietal and temporal cortices was also present in asymptomatic gene carriers (Soong & Liu, 1998).

Symptomatic and asymptomatic SCA3 gene carriers showed decreased F-dopa uptake and dopamine trasporter binding in the striatum (Shinotoh et al., 1997; Wang et al., 2009; Wüllner et al., 2005; Yen et al., 2000, 2002). D2 receptor binding was normal in a [^{11}C]raclopride PET study (Shinotoh et al., 1997).

A SPECT study showed that benzodiazepine receptor binding was decreased throughout the cerebral cortex and cerebellum in SCA3 patients compared to controls suggesting a diffuse GABAergic dysfunction in this disease (Ishibashi et al., 1998).

A PET study using ^{11}C-N-methylpiperidin-4-yl propionate (^{11}C-PMP) revelaed a reduction of acetylcholinesterase activity in the thalamus in SCA3 patients (Hirano et al., 2008).

The number of abnormal SCA3 triplet did not correlate with volumetry, diffusion or proton MR spectroscopy data (Adanyeguh et al., 2015; Guimaraes et al., 2013; Huang et al., 2017; Kang et al., 2014; Stefanescu et al., 2015).

3.2.3.4 Multi System Atrophy

Clinical presentation enables distinction of two variants of Multi System Atrophy, namely, cerebellar-type with predominance of ataxia and parkinsonian type with predominance of extra-pyramidal signs, which show partially different MRI and nuclear medicine features (Mascalchi, Vella, & Ceravolo, 2012). In patients with advanced disease the neuroimaging features tend to overlap.

Voxel based morphometry studies showed a decrease of the gray matter in the brainstem and cerebellum and in frontal, temporal and insular cortices and of the white matter in the middle cerebellar peduncle, cerebellum, and brainstem in patients with Multiple System Atrophy cerebellar type (Brenneis et al., 2006; Specht et al., 2005).

The relaxation rate R2 showed a significant reduction in the brainstem and cerebellum, but it was increased in the putamen (Specht et al., 2005).

Patients with Multiple Systema Atrophy cerebellar-type showed decreased fractal dimension in the cerebellum in compared to controls (Wu et al., 2010).

Diffusion weighted imaging and diffusion tensor imaging studies have shown involvement of the cerebellar peduncles, cerebellum and putamen, but also of extensive cerebral white matter and gray matter structures in patients with Multiple System Atrophy cerebellar-type compared to controls (Pellecchia et al., 2011; Prakash et al., 2009; Taoka et al., 2007; Tha et al., 2010).

Proton MR spectroscopy studies have demonstrated a decrease of the NAA/Cr and Cho/Cr ratios in the cerebellum, pons, putamen, and frontal cortex (Boesch et al., 2007; Lirng et al., 2012; Mascalchi et al., 2002; Terakawa et al., 1999). The decrease of the NAA/Cr ratio in the putamen differentiated patients with Multiple System Atrophy from those with idiopathic late onset cerebellar ataxia (Terakawa et al., 1999).

In a resting state functional MRI study the regional homogeneity in several cerebral cortical regions was increased or decreased in patients with multiple system atrophy (11 cerebellar type and 9 parkinsonian type) compared to controls indicating a widespread motor circuit dysfunction (You et al., 2011).

SPECT studies have shown hypoperfusion of the cerebellum, middle cerebellar peduncles, pons and putamen in patients with Multiple System Atrophy cerebellar type (Cilia, Marotta, Benti, Pezzoli, & Antonini, 2005; Kimura et al., 2009a; Waragai et al., 2007).

FDG-PET study demonstrated widespread areas of decreased metabolism in the frontal and parietal cortex, cerebellum, pons and striata in Multiple System Atrophy cerebellar type suggesting that all these areas may be associated with the primary disease process (Kwon, Choi, Kim, Lee, & Chung, 2007; Lee et al., 2008; Lyoo et al., 2008). A global cerebellar involvement with some asymmetry was noted in patients with Multiple System Atrophy cerebellar type (Oh et al., 2017).

PET and SPECT with a variety of radiotracers show in patients with Multiple System Atrophy a presynaptic dysfunction that is typically combined with a decrease of striatal D2 dopamine receptors (see Mascalchi et al., 2012).

A PET study with ^{11}C-flumazenil showed decreased binding of the gamma-aminobutyric acid type A/benzodiazepine in the cerebellum and brainstem of patients with Multiple System Atrophy cerebellar type (Gilman, 2001).

A PET study with ^{11}C-PMP demonstrated a significant decrease of binding in the thalamus and in the posterior cerebellar cortex in patients with Multiple System Atrophy cerebellar type, consistent with cholinergic dysfunction and suggesting that cholinergic modulating drugs may have a role in the treatment (Hirano et al., 2008). A SPECT study using a tracer($[^{123}I]$-iodobenzovesamicol) that binds to the vesicular acetylcholine transporter expression showed a decreased uptake in pedunculopontine-laterodorsal nuclei and the thalamus of Multiple System Atrophy patients as compared to controls, confirming that dysfunction of some cholinergic pathways occurs in this disease (Mazere et al., 2013).

Table 3 summarizes the neuroimaging features which have increased understanting of physiopathology in the most common chronic ataxias.

Importantly, the recent ENIGMA Ataxia Consortium initiative with collection and cumulative analysis of a large (hundreds) number of cases

of inherited ataxias examined with MRI worldwide (http://enigma.
ini.usc.edu/ongoing/enigma-ataxia) is expected to overcome the
fragmentary nature of single center studies and help to establish the physiopathology features of the common inherited and sporadic chronic
ataxias.

3.3 Markers of Disease Progression

In view of future gene-based therapies for inherited ataxias, there is a critical
need to identify biomarkers of disease progression with effect sizes greater
than clinical scores and capable to detect abnormalities in individuals before
manifest disease (Baldarçara et al., 2015). Neuroimaging instruments are well
suited for such a purpose.

Cross-sectional studies

Inferences on the natural history and progression of chronic ataxias can
certainly be made based on the results of cross-sectional neuroimaging
studies in which a given semiquantitative or quantitative measure correlates in the patients group with parametric variables as disease duration
or scores of clinical cerebellar and extra-cerebellar deficits. Moreover in
case of autosomal dominant diseases, further insight can be provided by
examining asymptomatic gene carriers discovered with molecular genetic
investigations.

Overall substantial, although variable, correlations have been reported
between the semiquantitative and quantitative neuroimaging features
assessing structural, microstructural and functional changes of the brain
reported above and clinical features and disease duration in chronic ataxias.
These include Friedreich ataxia (see Akhlaghi et al., 2014; Bonilha da Silva
et al., 2014; Chevis et al., 2013; Clemm von Hohenberg et al., 2013; Dogan
et al., 2018; Gramegna et al., 2017; Harding et al., 2016; Mascalchi & Vella,
2012; Pagani et al., 2010; Rizzo et al., 2011; Selvadurai et al., 2016;
Stefanescu et al., 2015), SCA6 (Eichler et al., 2011; Falcon et al., 2016;
Joers et al., 2018; Kang et al., 2017; Richter et al., 2005; Schulz et al.,
2010; Stefanescu et al., 2015), SCA17 (Brockmann et al., 2012; Lasek
et al., 2006), Sporadic Adult Onset Ataxia/Idiopathic Late Onset Cerebellar
Ataxia (Richter et al., 2005), SCA1 (see Adanyeguh et al., 2015; Goel et al.,
2011; Jacobi et al., 2013; Joers et al., 2018; Martins et al., 2017; Mascalchi &
Vella, 2012; Oz et al., 2010; Schulz et al., 2010; Solodkin et al., 2011;
Ying et al., 2006), SCA2 (see Adanyeguh et al., 2015; D'Agata et al., 2011;
Goel et al., 2011; Hernandez-Castillo et al., 2015; Joers et al., 2018;

Jung, Choi, Du, Cuzzocreo, Ying, et al., 2012b; Mascalchi & Vella, 2012; Olivito, Lupo, et al., 2017; Reetz et al., 2018; Salvatore et al., 2014), SCA3 (Adanyeguh et al., 2015; Eichler et al., 2011; Guimarães et al., 2013; Jacobi et al., 2013; Kang et al., 2014; Schulz et al., 2010) and Multi System Atrophy (Lyoo et al., 2008; Mazere et al., 2013; Prakash et al., 2009; Tha et al., 2010).

Longitudinal studies

Definitive information on the capability of neuroimaging techniques to track progression of disease requires longitudinal (serial) studies which however are still relatively few and with small sample sizes of patients and controls.

3.3.1 Spinal Atrophy (Friedreich Ataxia)

Two longitudinal small size studies using T1 weighted images and diffusion tensor imaging showed an accelerated progression of microstructural changes in terms of decreased fractional anisotropy and axial diffusivity in the cerebral white matter (pyramidal tracts, distal tracts of the superior cerebellar peduncles) and corpus callosum of patients with Friedreich ataxia compared to controls subjects (Mascalchi et al., 2016; Rezende et al., 2016). No correlation of diffusivity changes with worsening of the clinical deficit was observed. As well, no significant changes in the microstructure of the brainstem and cerebellum were detected. Additionally, the two studies using voxel based morphometry and tensor based morphometry, respectively, failed to reveal significant changes in the regional volumes including those of the brainstem and cerebellum (Mascalchi et al., 2016; Rezende et al., 2016).

Using manual measurements, lack of significant differences in the decline of the cross-section of the lower medulla in patients with Friedreich ataxia and controls over a 4 year interval was reported (Mascalchi et al., 2017), supporting the existence of a maldevelopmental component underlying the lower size of some central nervous system structures, namely, spinal cord and medulla, in Friedreich ataxia (Koeppen, Becker, Qian, & Feustel, 2017).

A study of texture analysis showed no progression of medulla abnormalities in patients with Freidreich ataxia after 1 year of follow-up (Santos et al., 2015).

On the other hand after 1 year follow-up a significant reduction of T2 in the dentate nuclei of patients with Friedreich ataxia compared to controls was observed that correlated with GAA expansions and clinical deterioration (Bonilha da Silva et al., 2014).

3.3.2 Cortical Cerebellar Atrophy (SCA6, SCA17)

3.3.2.1 SCA6

In one longitudinal volumetry and voxel based morphometry study patients with SCA6 showed accelerated loss of brainstem, cerebellar and basal ganglia volume, more pronounced in the caudate, compared to controls, with a greater effect size of morphometry biomarkers as compared to clinical scales (Reetz et al., 2013).

3.3.2.2 SCA17

A longitudinal voxel based morphometry showed greater atrophy in the cerebellum and in the limbic system and parietal precuneus in SCA17 patients compared to controls in follow-up MR obtained 18 months after baseline (Reetz et al., 2010).

Progression of motor deficit correlated with reduction of gray matter in the cerebellum and cerebral cortical motor areas, worsening of psychiatric symptoms correlated with widespread reduction of gray matter in the cerebral and cerebellar cortex, and a global functioning score correlated with reduction of the gray matter in the hippo- and parahippocampus bilaterally. Moreover, in a further longitudinal voxel based morphometry study CAG repeats were inversely correlated with the rate of cerebellar atrophy in SCA17 (Reetz et al., 2011).

3.3.3 Olivopontocerebellar Atrophy (SCA1, SCA2, SCA3, Multi System Atrophy)

3.3.3.1 SCA1

Longitudinal volumetry and voxel based morphometry studies revealed accelerated loss of bulk of the brainstem, especially the pons, cerebellum and basal ganglia in SCA1 compared to controls with a greater effect size of morphometry biomarkers as compared to clinical scales (Adanyeguh et al., 2018; Reetz et al., 2013). Loss of the bilateral cerebellum and the pons showed a mild correlation with CAG repeat length (Reetz et al., 2013).

3.3.3.2 SCA2

Tensor based morphometry and volumetry revealed accelerated volume loss of the brainstem and cerebellum in SCA2 patients compared to controls (Adanyeguh et al., 2018; Mascalchi et al., 2014) (Fig. 6) with a greater effect size compared to clinical scores (Adanyeguh et al., 2018).

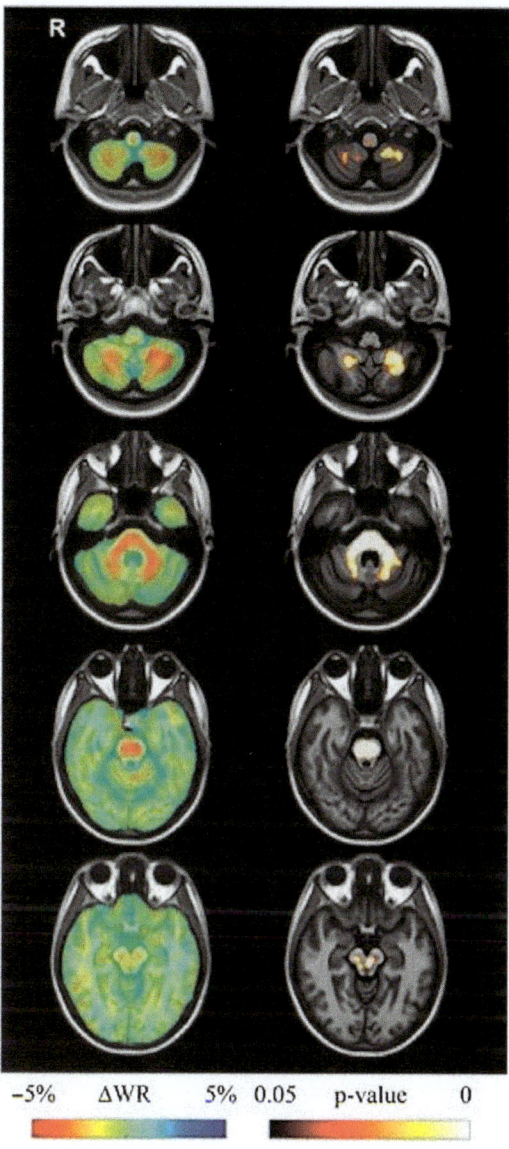

−5% ΔWR 5% 0.05 p-value 0

Fig. 6 Results of longitudinal Tensor Based Morphometry analysis comparing SCA2 patients and controls. Left column shows selected axial views of the difference in average longitudinal warp rate ΔWR maps between SCA2 patients and controls, where red indicates local atrophy and blue indicates local enlargement. Right column shows the corresponding voxel-wise corrected P-value maps testing the null hypothesis of zero differences in warp rate between SCA2 patients and healthy controls. Highlighted clusters indicate significantly (P < 0.05) more pronounced mean atrophy in SCA2 patients when compared to healthy controls. SCA2 patients exhibit significant higher atrophy

Histogram was more sensitive than voxel-wise analysis of diffusion tensor imaging data in revealing accelerated progression of microstructural changes in the brainstem and cerebellum (Mascalchi et al., 2018, 2015).

A longitudinal study of fractal dimension did not reveal accelerated decrease of structural organization of the cerebellum and cerebral cortex in SCA2 compared to controls (Marzi et al., 2018).

3.3.3.3 SCA3
Longitudinal volumetry and voxel based morphometry studies have revealed accelerated loss of brainstem, cerebellar and especially the basal ganglia volume in SCA3 patients compared to controls with a greater effect size of morphometry biomarkers as compared to clinical scales (Adanyeguh et al., 2018; Reetz et al., 2013).

3.3.3.4 Multi System Atrophy
A longitudinal study performed with a 2 year interval reported a significantly accelerated loss of volume in putamen and cerebellum of 14 patients with Multiple System Atrophy cerebellar type as compared to controls (Hauser et al., 2006). A 1 year interval voxel based morphometry demonstrated accelerated tissue loss in the middle cerebellar peduncles and corpus callosum in 14 patients with multiple system atrophy (10 cerebellar type and 4 parkinsonian type) as compared to controls (Minnerop et al., 2010).

A region of interest approach demonstrated the capability of diffusion weighted imaging to track progression of microstrutural damage in putamen, pons, cerebellar white matter, thalamus and frontal white matter over a 11 months interval in patients with Multi System Atrophy (Pellecchia et al., 2011). There was no correlation with progression of clinical deficit.

Table 3 summarizes the neuroimaging features which are capable to track neurodegeneration in the most common chronic ataxias.

---

rates with respect to controls in the midbrain (substantia nigra and medial lemniscus, bilaterally, right lateral lemniscus and central region corresponding to decussation of the superior cerebellar peduncles), the entire basis pontis, the middle cerebellar peduncles and posterior medulla corresponding to the gracilis and cuneatus tracts and nuclei. The cerebellum shows loss of white matter in the hemispheric and peridentate region and of gray matter in the cerebellar cortex of the inferior portions of the cerebellar hemisphers. *Reproduced with permission from Mascalchi, M., Diciotti, S., Giannelli, M., Ginestroni, A., Soricelli, A., Nicolai, E., et al. (2014). Progression of brain atrophy in spinocerebellar ataxia type 2: A longitudinal tensor-based morphometry study. PloS One, 9(2), e89410. https://doi.org/10.1371/journal.pone.0089410.*

To comparatively evaluate the capability of MRI features to track progression of neurodegeneration in chonic ataxias is one of the aims of the ENIGMA Ataxia Consortium initiative (http://enigma.ini.usc.edu/ongoing/enigma-ataxia).

3.4 Surrogate Markers of Treatments Effects

The quantitative nature of the data that can be extracted by MRI, SPECT and PET make them potentially useful to evaluate efficacy of new treatments. Due to the relative insensitivity of clinical scales (Schulte & Schols, 2002) and the possibility of a "ceiling effect" on the clinical ground hindering reversal of deficits in patients with advanced disease, inclusion of neuroimaging data as surrogate markers would theoretically allow to conduct an efficacy study with smaller number of patients (Baldarçara et al., 2015).

So far the use of neuroimaging techniques as surrogate markers in patients with chronic ataxias has been limited to few exploratory observational single center studies with few patients.

3.4.1 Spinal Atrophy (Freidreich Ataxia)

An Austrian group performed a voxel based morphometry study and a diffusion tensor imaging study of the efficacy of recombinant human erythropoietin in a sample of 9 patients with Friedreich ataxia who underwent MRI twice 8 months apart.

In the voxel based morphometry study (Santner et al., 2014) they observed an increase of gray matter volume bilaterally in the pulvinar and the posterior parietal cortex when comparing follow-up scans after recombinant human erythropoietin treatment with baseline scans. The increased volume in the pulvinar correlated with clinical improvement. In the diffusion tensor imaging study (Egger et al., 2014) they detected widespread longitudinal increase of fractional anisotropy and axial diffusivity thought to be consistent with reversal of the white matter microstructural damage in cerebral hemispheres bilaterally, while no changes were observed within the cerebellum, medulla oblongata, and pons.

3.4.2 Cortical Cerebellar Atrophy (Ataxia Telangectasia, SCA6, Gluten Ataxia, Sporadic Adult Onset Ataxia)

A functional MRI study evaluated with a simple motor task two children with Ataxia Telangectasia before and after a 10-days cycle of oral betamethasone (Quarantelli et al., 2013). The number of activated voxels within the motor cortex under the on-therapy condition increased as compared with the cortical activity under baseline condition (Fig. 7). This result suggests that steroid treatment could improve motor performance in patients with Ataxia Telangectasia by facilitating cortical compensatory mechanisms.

Kimura et al. (2009b) investigated with perfusion SPECT the effect of thyrotropin releasing hormone in 5 patients with late-onset cortical cerebellar atrophy, 2 patients with SCA6 and 3 patients with autosomal dominant cerebellar ataxia not otherwise characterized. They observed that, after 14 days of intravenous therapy, thyrotropin releasing hormone significantly improved the clinical cerebellar deficit and increased the perfusion in the callosomarginal region and cerebellum.

Perfusion SPECT was also applied to evaluate the efficacy of intravenous immunoglobulin therapy in patients with autoantibody-positive cerebellar ataxia, including patients with antigliadin antibodies (Nanri et al., 2009) with an increased cerebellar perfusion after treatment.

A study evaluated with proton MR spectroscopy of the cerebellum patients with gluten ataxia before and after gluten free diet (Hadjivassiliou et al., 2017). The NAA/Cr area ratio in the cerebellar vermis increased in 62 (98%) out of 63 patients on strict gluten free diet, in 9 (26%) of 35 patients on gluten free diet but still positive antibodies, and in only 1 (5%) of 19 control gluten ataxia patients who were not on gluten free diet. These results strengthen previous findings of clinical improvement in patients with gluten ataxia after introduction of gluten free diet.

3.4.3 Olivopontocerebellar Atrophy (Multi System Atrophy)

Kimura et al. (2011) investigated with perfusion SPECT the effect of thyrotropin releasing hormone on regional cerebral blood flow in 7 patients with Multi System Atrophy cerebellar type and observed that, after 14 days of intravenous therapy, the cerebellar regional blood flow was reduced and the cerebellar clinical deficit was unchanged.

Table 3 summarizes the neuroimaging features which have been used as surrogate markers to assess treatments effects in the most common chronic ataxias.

Overall the observational and single center type of the available studies represents major methodological limitations for establishing the validity of neuroimaging techniques as surrogate biomarkers. In view of randomized clinical trials propelled by gene-based therapies in inherited ataxias, a multicentric approach with protocol sharing, power analyses and data pooling is strongly recommended.

Patient 1 (male, 8 yo)

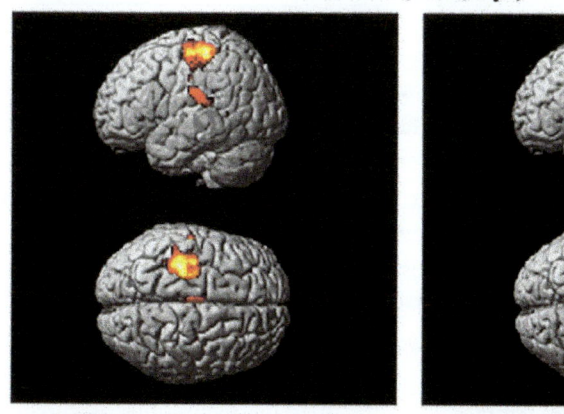
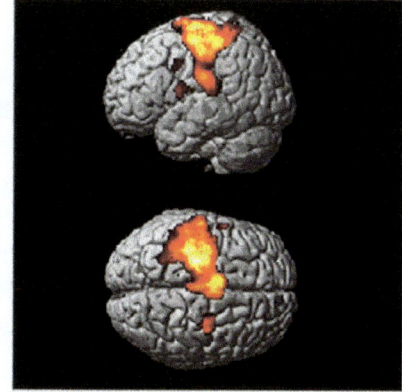

off-therapy (Total SARA score 19)   on-therapy (Total SARA score 8)

Patient 2 (male, 17yo)

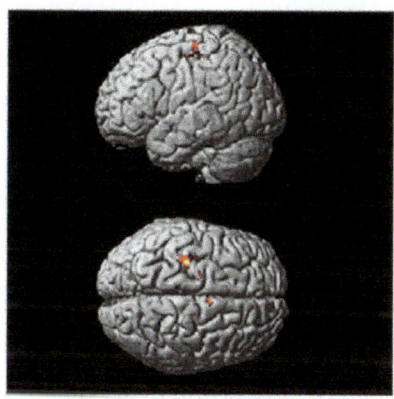
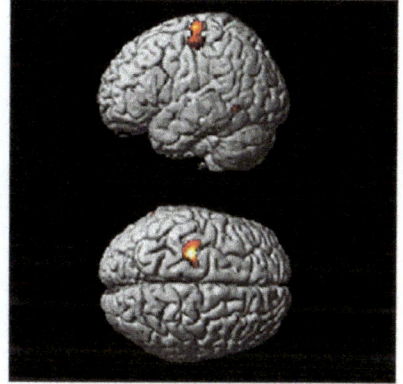

off-therapy (Total SARA score 29)   on-therapy (Total SARA score 24)

Fig. 7 Results of functional MRI during a simple motor task, pronosupination of the dominant (right) hand, before (off-therapy) and after a 10-days cycle (on-therapy) of oral betamethasone at 0.03 mg/kg/day in two patients with ataxia telangectasia. In both cases, the cluster of activation in the left motor cortex shows a substantial enlargement and increase in the strength of significance in the on-therapy study, as compared to the basal pattern. Within brackets the score of the Scale for the Assessment and Rating of Ataxia (SARA) score is indicated with a decrease in the on-therapy panels corresponding to clinical improvement. *Reproduced with permission from Quarantelli, M., Giardino, G., Prinster, A., Aloj, G., Carotenuto, B., Cirillo, E., et al. (2013). Steroid treatment in Ataxia-Telangiectasia induces alterations of functional magnetic resonance imaging during prono-supination task.* European Journal of Paediatric Neurology: EJPN: Official Journal of the European Paediatric Neurology Society, *17(2), 135–140. https://doi.org/10.1016/j. ejpn.2012.06.002.*

4. CONCLUSIONS

MRI is the fundamental diagnostic marker of chronic ataxias, especially in case of recessive or X-linked inheritance and of sporadic occurrence. In fact, visual assessment in the single patient enables appreciation of patterns of atrophy and areas exhibiting increased or decreased T2 signal which are characteristic of these diseases. Semiquantitative or quantitative MRI, SPECT and PET data describing structural, microstructural and functional changes of the cerebellum, brainstem, cerebrum and spinal cord are useful physiopathological markers. They have been also employed as progression markers to track neurodegeneration with a greater sensitivity than clinical scales. So far a few small-size single center studies employed neuroimaging techniques as surrogate markers of treatment effects in chronic ataxias. Recent inititatives aimed to promote data sharing and constitution of large data sets of MRI examinations are expected to overcome the fragmentary nature of single center studies which have predominatly been performed in these uncommon diseases.

REFERENCES

Adanyeguh, I. M., Henry, P.-G., Nguyen, T. M., Rinaldi, D., Jauffret, C., Valabregue, R., et al. (2015). In vivo neurometabolic profiling in patients with spinocerebellar ataxia types 1, 2, 3, and 7. *Movement Disorders: Official Journal of the Movement Disorder Society,* *30*(5), 662–670. https://doi.org/10.1002/mds.26181.

Adanyeguh, I. M., Perlbarg, V., Henry, P.-G., Rinaldi, D., Petit, E., Valabregue, R., et al. (2018). Autosomal dominant cerebellar ataxias: Imaging biomarkers with high effect sizes. *NeuroImage. Clinical, 19,* 858–867. https://doi.org/10.1016/j.nicl.2018.06.011.

Akhlaghi, H., Corben, L., Georgiou-Karistianis, N., Bradshaw, J., Delatycki, M. B., Storey, E., et al. (2012). A functional MRI study of motor dysfunction in Friedreich's ataxia. *Brain Research, 1471,* 138–154. https://doi.org/10.1016/j.brainres.2012.06.035. Oct 16 [Epub ahead of print]. https://doi.org/10.1111/ene.13843.

Akhlaghi, H., Corben, L., Georgiou-Karistianis, N., Bradshaw, J., Storey, E., Delatycki, M. B., et al. (2011). Superior cerebellar peduncle atrophy in Friedreich's ataxia correlates with disease symptoms. *Cerebellum (London, England), 10*(1), 81–87. https://doi.org/10.1007/s12311-010-0232-3.

Akhlaghi, H., Yu, J., Corben, L., Georgiou-Karistianis, N., Bradshaw, J. L., Storey, E., et al. (2014). Cognitive deficits in Friedreich ataxia correlate with micro-structural changes in dentatorubral tract. *Cerebellum (London, England), 13*(2), 187–198. https://doi.org/10.1007/s12311-013-0525-4.

Alves, C. A. P. F., Fragoso, D. C., Gonçalves, F. G., Marussi, V. H., & Amaral, L. L. F. d. (2018). Cerebellar ataxia in children: A clinical and MRI approach to the differential diagnosis. *Topics in Magnetic Resonance Imaging: TMRI, 27*(4), 275–302. https://doi.org/10.1097/RMR.0000000000000175.

Baldarçara, L., Currie, S., Hadjivassiliou, M., Hoggard, N., Jack, A., Jackowski, A. P., et al. (2015). Consensus paper: Radiological biomarkers of cerebellar diseases. *Cerebellum (London, England), 14*(2), 175–196. https://doi.org/10.1007/s12311-014-0610-3.

Boesch, S. M., Donnemiller, E., Müller, J., Seppi, K., Weirich-Schwaiger, H., Poewe, W., et al. (2004). Abnormalities of dopaminergic neurotransmission in SCA2: A combined 123I-betaCIT and 123I-IBZM SPECT study. *Movement Disorders: Official Journal of the Movement Disorder Society, 19*(11), 1320–1325. https://doi.org/10.1002/mds.20159.

Boesch, S. M., Schocke, M., Bürk, K., Hollosi, P., Fornai, F., Aichner, F. T., et al. (2001). Proton magnetic resonance spectroscopic imaging reveals differences in spinocerebellar ataxia types 2 and 6. *Journal of Magnetic Resonance Imaging: JMRI, 13*(4), 553–559.

Boesch, S. M., Wolf, C., Seppi, K., Felber, S., Wenning, G. K., & Schocke, M. (2007). Differentiation of SCA2 from MSA-C using proton magnetic resonance spectroscopic imaging. *Journal of Magnetic Resonance Imaging: JMRI, 25*(3), 564–569. https://doi.org/10.1002/jmri.20846.

Bonilha da Silva, C., Bergo, F. P. G., D'Abreu, A., Cendes, F., Lopes-Cendes, I., & França, M. C. (2014). Dentate nuclei T2 relaxometry is a reliable neuroimaging marker in Friedreich's ataxia. *European Journal of Neurology, 21*(8), 1131–1136. https://doi.org/10.1111/ene.12448.

Braga-Neto, P., Pedroso, J. L., Gadelha, A., Laureano, M. R., de Souza Noto, C., Garrido, G. J., et al. (2016). Psychosis in Machado-Joseph disease: Clinical correlates, pathophysiological discussion, and functional brain imaging. Expanding the cerebellar cognitive affective syndrome. *Cerebellum (London, England), 15*(4), 483–490. https://doi.org/10.1007/s12311-015-0716-2.

Brenneis, C., Boesch, S. M., Egger, K. E., Seppi, K., Scherfler, C., Schocke, M., et al. (2006). Cortical atrophy in the cerebellar variant of multiple system atrophy: A voxel-based morphometry study. *Movement Disorders: Official Journal of the Movement Disorder Society, 21*(2), 159–165. https://doi.org/10.1002/mds.20656.

Brenneis, C., Bösch, S. M., Schocke, M., Wenning, G. K., & Poewe, W. (2003). Atrophy pattern in SCA2 determined by voxel-based morphometry. *Neuroreport, 14*(14), 1799–1802. https://doi.org/10.1097/01.wnr.0000094105.16607.18.

Brockmann, K., Reimold, M., Globas, C., Hauser, T. K., Walter, U., Machulla, H.-J., et al. (2012). PET and MRI reveal early evidence of neurodegeneration in spinocerebellar ataxia type 17. *Journal of Nuclear Medicine: Official Publication, Society of Nuclear Medicine, 53*(7), 1074–1080. https://doi.org/10.2967/jnumed.111.101543.

Bürk, K., Bösch, S., Müller, C. A., Melms, A., Zühlke, C., Stern, M., et al. (2001). Sporadic cerebellar ataxia associated with gluten sensitivity. *Brain: A Journal of Neurology, 124*(Pt. 5), 1013–1019.

Bürk, K., Globas, C., Wahl, T., Bühring, U., Dietz, K., Zuhlke, C., et al. (2004). MRI-based volumetric differentiation of sporadic cerebellar ataxia. *Brain: A Journal of Neurology, 127*(Pt. 1), 175–181. https://doi.org/10.1093/brain/awh013.

Chavoix, C., Samson, Y., Pappata, S., Prenant, C., Mazière, M., Seck, A., et al. (1990). Positron emission tomography study of brain benzodiazepine receptors in Friedreich's ataxia. *The Canadian Journal of Neurological Sciences. Le Journal Canadien des Sciences Neurologiques, 17*(4), 404–409.

Chevis, C. F., da Silva, C. B., D'Abreu, A., Lopes-Cendes, I., Cendes, F., Bergo, F. P. G., et al. (2013). Spinal cord atrophy correlates with disability in Friedreich's ataxia. *Cerebellum (London, England), 12*(1), 43–47. https://doi.org/10.1007/s12311-012-0390-6.

Cilia, R., Marotta, G., Benti, R., Pezzoli, G., & Antonini, A. (2005). Brain SPECT imaging in multiple system atrophy. *Journal of Neural Transmission (Vienna, Austria: 1996), 112*(12), 1635–1645. https://doi.org/10.1007/s00702-005-0382-5.

Clemm von Hohenberg, C., Schocke, M. F., Wigand, M. C., Nachbauer, W., Guttmann, C. R. G., Kubicki, M., et al. (2013). Radial diffusivity in the cerebellar peduncles correlates with clinical severity in Friedreich ataxia. *Neurological Sciences: Official Journal of the Italian Neurological Society and of the Italian Society of Clinical Neurophysiology, 34*(8), 1459–1462. https://doi.org/10.1007/s10072-013-1402-0.

Cocozza, S., Constabile, T., Tedeschi, E., Abate, F., Russo, C., Liguori, A., et al. (2018). Cognitive and functional connectivity alterations in Friedreich's ataxia. *Annals of Clinical and Translational Neurology, 5,* 677–686. https://doi.org/10.1002/acn3.555.

Cocozza, S., Saccà, F., Cervo, A., Marsili, A., Russo, C. V., Giorgio, S. M. D. A., et al. (2015). Modifications of resting state networks in spinocerebellar ataxia type 2. *Movement Disorders: Official Journal of the Movement Disorder Society, 30*(10), 1382–1390. https://doi.org/10.1002/mds.26284.

D'Abreu, A., França, M., Appenzeller, S., Lopes-Cendes, I., & Cendes, F. (2009). Axonal dysfunction in the deep white matter in Machado-Joseph disease. *Journal of Neuroimaging: Official Journal of the American Society of Neuroimaging, 19*(1), 9–12. https://doi.org/10.1111/j.1552-6569.2008.00260.x.

D'Agata, F., Caroppo, P., Boghi, A., Coriasco, M., Caglio, M., Baudino, B., et al. (2011). Linking coordinative and executive dysfunctions to atrophy in spinocerebellar ataxia 2 patients. *Brain Structure & Function, 216*(3), 275–288. https://doi.org/10.1007/s00429-011-0310-4.

Déjérine, J., & Thomas, A. (1900). L'atrophie olivopontocérébelleuse. In *Nouvelle Iconographie Salpêtrière: Vol. 13.* (pp. 330–370).

Della Nave, R., Foresti, S., Tessa, C., Moretti, M., Ginestroni, A., Gavazzi, C., et al. (2004). ADC mapping of neurodegeneration in the brainstem and cerebellum of patients with progressive ataxias. *NeuroImage, 22*(2), 698–705. https://doi.org/10.1016/j.neuroimage.2004.01.035.

Della Nave, R., Ginestroni, A., Diciotti, S., Salvatore, E., Soricelli, A., & Mascalchi, M. (2011). Axial diffusivity is increased in the degenerating superior cerebellar peduncles of Friedreich's ataxia. *Neuroradiology, 53*(5), 367–372. https://doi.org/10.1007/s00234-010-0807-1.

Della Nave, R., Ginestroni, A., Giannelli, M., Tessa, C., Salvatore, E., Salvi, F., et al. (2008). Brain structural damage in Friedreich's ataxia. *Journal of Neurology, Neurosurgery, and Psychiatry, 79*(1), 82–85. https://doi.org/10.1136/jnnp.2007.124297.

Della Nave, R., Ginestroni, A., Tessa, C., Cosottini, M., Giannelli, M., Salvatore, E., et al. (2008). Brain structural damage in spinocerebellar ataxia type 2. A voxel-based morphometry study. *Movement Disorders: Official Journal of the Movement Disorder Society, 23*(6), 899–903. https://doi.org/10.1002/mds.21982.

Della Nave, R., Ginestroni, A., Tessa, C., Salvatore, E., Bartolomei, I., Salvi, F., et al. (2008). Brain white matter tracts degeneration in Friedreich ataxia. An in vivo MRI study using tract-based spatial statistics and voxel-based morphometry. *NeuroImage, 40*(1), 19–25. https://doi.org/10.1016/j.neuroimage.2007.11.050.

Della Nave, R., Ginestroni, A., Tessa, C., Salvatore, E., De Grandis, D., Plasmati, R., et al. (2008). Brain white matter damage in SCA1 and SCA2. An in vivo study using voxel-based morphometry, histogram analysis of mean diffusivity and tract-based spatial statistics. *NeuroImage, 43*(1), 10–19. https://doi.org/10.1016/j.neuroimage.2008.06.036.

Dogan, I., Romanzetti, S., Didszun, C., Mirzazade, S., Timmann, D., Saft, C., et al. (2018). Structural characteristics of the central nervous system in Friedreich ataxia: An in vivo spinal cord and brain MRI study. *Journal of Neurology, Neurosurgery, and Psychiatry.* https://doi.org/10.1136/jnnp-2018-318422, [Epub ahead of print].

Dogan, I., Tinnemann, E., Romanzetti, S., Mirzazade, S., Costa, A. S., Werner, C. J., et al. (2016). Cognition in Friedreich's ataxia: A behavioral and multimodal imaging study. *Annals of Clinical and Translational Neurology, 3*(8), 572–587. https://doi.org/10.1002/acn3.315.

Duarte, J. V., Faustino, R., Lobo, M., Cunha, G., Nunes, C., Ferreira, C., et al. (2016). Parametric fMRI of paced motor responses uncovers novel whole-brain imaging biomarkers in spinocerebellar ataxia type 3. *Human Brain Mapping, 37*(10), 3656–3668. https://doi.org/10.1002/hbm.23266.

Egger, K., Clemm von Hohenberg, C., Schocke, M. F., Guttmann, C. R. G., Wassermann, D., Wigand, M. C., et al. (2014). White matter changes in patients with friedreich ataxia after treatment with erythropoietin. *Journal of Neuroimaging: Official Journal of the American Society of Neuroimaging, 24*(5), 504–508. https://doi.org/10.1111/jon. 12050.

Eichler, L., Bellenberg, B., Hahn, H. K., Köster, O., Schöls, L., & Lukas, C. (2011). Quantitative assessment of brain stem and cerebellar atrophy in spinocerebellar ataxia types 3 and 6: Impact on clinical status. *AJNR American Journal of Neuroradiology, 32*(5), 890–897. https://doi.org/10.3174/ajnr.A2387.

Etchebehere, E. C., Cendes, F., Lopes-Cendes, I., Pereira, J. A., Lima, M. C., Sansana, C. R., et al. (2001). Brain single-photon emission computed tomography and magnetic resonance imaging in Machado-Joseph disease. *Archives of Neurology, 58*(8), 1257–1263.

Falcon, M. I., Gomez, C. M., Chen, E. E., Shereen, A., & Solodkin, A. (2016). Early cerebellar network shifting in spinocerebellar ataxia type 6. *Cerebral Cortex (New York, N.Y.: 1991), 26*(7), 3205–3218. https://doi.org/10.1093/cercor/bhv154.

Firat, A. K., Karakaş, H. M., Firat, Y., & Yakinci, C. (2005). Quantitative evaluation of brain involvement in ataxia telangiectasia by diffusion weighted MR imaging. *European Journal of Radiology, 56*(2), 192–196. https://doi.org/10.1016/j.ejrad.2005.04.009.

Fortuna, F., Barboni, P., Liguori, R., Valentino, M. L., Savini, G., Gellera, C., et al. (2009). Visual system involvement in patients with Friedreich's ataxia. *Brain: A Journal of Neurology, 132*(Pt. 1), 116–123. https://doi.org/10.1093/brain/awn269.

França, M. C., D'Abreu, A., Yasuda, C. L., Bonadia, L. C., Santos da Silva, M., Nucci, A., et al. (2009). A combined voxel-based morphometry and 1H-MRS study in patients with Friedreich's ataxia. *Journal of Neurology, 256*(7), 1114–1120. https://doi.org/10.1007/s00415-009-5079-5.

Friedreich, N. (1877). Ueber Ataxie mit besonderer Berucksichtigung der hereditaren Formen. Nachtrag *Virchows Archives fu Pathologische Anatomie und Physiologie und fur Klinische Medizin: Vol. 70.* (pp. 140–152).

Furtado, S., Farrer, M., Tsuboi, Y., Klimek, M. L., de la Fuente-Fernández, R., Hussey, J., et al. (2002). SCA-2 presenting as parkinsonism in an Alberta family: Clinical, genetic, and PET findings. *Neurology, 59*(10), 1625–1627.

Gilman, S. (2001). Biochemical changes in multiple system atrophy detected with positron emission tomography. *Parkinsonism & Related Disorders, 7*, 253–256.

Gilman, S., Junck, L., Markel, D. S., Koeppe, R. A., & Kluin, K. J. (1990). Cerebral glucose hypermetabolism in Friedreich's ataxia detected with positron emission tomography. *Annals of Neurology, 28*(6), 750–757. https://doi.org/10.1002/ana.410280605.

Gilman, S., Sima, A. A., Junck, L., Kluin, K. J., Koeppe, R. A., Lohman, M. E., et al. (1996). Spinocerebellar ataxia type 1 with multiple system degeneration and glial cytoplasmic inclusions. *Annals of Neurology, 39*(2), 241–255. https://doi.org/10.1002/ana.410390214.

Ginestroni, A., Della Nave, R., Tessa, C., Giannelli, M., De Grandis, D., Plasmati, R., et al. (2008). Brain structural damage in spinocerebellar ataxia type 1: A VBM study. *Journal of Neurology, 255*(8), 1153–1158. https://doi.org/10.1007/s00415-008-0860-4.

Ginestroni, A., Diciotti, S., Cecchi, P., Pesaresi, I., Tessa, C., Giannelli, M., et al. (2012). Neurodegeneration in friedreich's ataxia is associated with a mixed activation pattern of the brain. A fMRI study. *Human Brain Mapping, 33*(8), 1780–1791. https://doi.org/10.1002/hbm.21319.

Goel, G., Pal, P. K., Ravishankar, S., Venkatasubramanian, G., Jayakumar, P. N., Krishna, N., et al. (2011). Gray matter volume deficits in spinocerebellar ataxia: An optimized voxel based morphometric study. *Parkinsonism & Related Disorders, 17*(7), 521–527. https://doi.org/10.1016/j.parkreldis.2011.04.008.

Gramegna, L. L., Tonon, C., Manners, D. N., Pini, A., Rinaldi, R., Zanigni, S., et al. (2017). Combined cerebellar proton MR spectroscopy and DWI study of patients with

Friedreich's ataxia. *Cerebellum (London, England)*, *16*(1), 82–88. https://doi.org/10.1007/s12311-016-0767-z.

Guerrini, L., Lolli, F., Ginestroni, A., Belli, G., Della Nave, R., Tessa, C., et al. (2004). Brainstem neurodegeneration correlates with clinical dysfunction in SCA1 but not in SCA2. A quantitative volumetric, diffusion and proton spectroscopy MR study. *Brain: A Journal of Neurology*, *127*(Pt. 8), 1785–1795. https://doi.org/10.1093/brain/awh201.

Guimarães, R. P., D'Abreu, A., Yasuda, C. L., França, M. C., Silva, B. H. B., Cappabianco, F. A. M., et al. (2013). A multimodal evaluation of microstructural white matter damage in spinocerebellar ataxia type 3. *Movement Disorders: Official Journal of the Movement Disorder Society*, *28*(8), 1125–1132. https://doi.org/10.1002/mds.25451.

Hadjivassiliou, M., Grünewald, R. A., Sanders, D. S., Shanmugarajah, P., & Hoggard, N. (2017). Effect of gluten-free diet on cerebellar MR spectroscopy in gluten ataxia. *Neurology*, *89*(7), 705–709. https://doi.org/10.1212/WNL.0000000000004237.

Hadjivassiliou, M., Wallis, L. I., Hoggard, N., Grünewald, R. A., Griffiths, P. D., & Wilkinson, I. D. (2012). MR spectroscopy and atrophy in Gluten, Friedreich's and SCA6 ataxias. *Acta Neurologica Scandinavica*, *126*(2), 138–143. https://doi.org/10.1111/j.1600-0404.2011.01620.x.

Harding, I. H., Corben, L. A., Storey, E., Egan, G. F., Stagnitti, M. R., Poudel, G. R., et al. (2016). Fronto-cerebellar dysfunction and dysconnectivity underlying cognition in friedreich ataxia: The IMAGE-FRDA study. *Human Brain Mapping*, *37*(1), 338–350. https://doi.org/10.1002/hbm.23034.

Hauser, T.-K., Luft, A., Skalej, M., Nägele, T., Kircher, T. T. J., Leube, D. T., et al. (2006). Visualization and quantification of disease progression in multiple system atrophy. *Movement Disorders: Official Journal of the Movement Disorder Society*, *21*(10), 1674–1681. https://doi.org/10.1002/mds.21032.

Hernandez-Castillo, C. R., Galvez, V., Mercadillo, R. E., Díaz, R., Yescas, P., Martinez, L., et al. (2015). Functional connectivity changes related to cognitive and motor performance in spinocerebellar ataxia type 2. *Movement Disorders: Official Journal of the Movement Disorder Society*, *30*(10), 1391–1399. https://doi.org/10.1002/mds.26320.

Hirano, S., Shinotoh, H., Arai, K., Aotsuka, A., Yasuno, F., Tanaka, N., et al. (2008). PET study of brain acetylcholinesterase in cerebellar degenerative disorders. *Movement Disorders: Official Journal of the Movement Disorder Society*, *23*(8), 1154–1160. https://doi.org/10.1002/mds.22056.

Honjo, K., Ohshita, T., Kawakami, H., Naka, H., Imon, Y., Maruyama, H., et al. (2004). Quantitative assessment of cerebral blood flow in genetically confirmed spinocerebellar ataxia type 6. *Archives of Neurology*, *61*(6), 933–937. https://doi.org/10.1001/archneur.61.6.933.

Huang, S.-R., Wu, Y.-T., Jao, C.-W., Soong, B.-W., Lirng, J.-F., Wu, H.-M., et al. (2017). CAG repeat length does not associate with the rate of cerebellar degeneration in spinocerebellar ataxia type 3. *NeuroImage. Clinical*, *13*, 97–105. https://doi.org/10.1016/j.nicl.2016.11.007.

Inagaki, A., Iida, A., Matsubara, M., & Inagaki, H. (2005). Positron emission tomography and magnetic resonance imaging in spinocerebellar ataxia type 2: A study of symptomatic and asymptomatic individuals. *European Journal of Neurology*, *12*(9), 725–728. https://doi.org/10.1111/j.1468-1331.2005.01011.x.

Ishibashi, M., Sakai, T., Matsuishi, T., Yonekura, Y., Yamashita, Y., Abe, T., et al. (1998). Decreased benzodiazepine receptor binding in Machado-Joseph disease. *Journal of Nuclear Medicine: Official Publication, Society of Nuclear Medicine*, *39*(9), 1518–1520.

Jacobi, H., Reetz, K., du Montcel, S. T., Bauer, P., Mariotti, C., Nanetti, L., et al. (2013). Biological and clinical characteristics of individuals at risk for spinocerebellar ataxia types

1, 2, 3, and 6 in the longitudinal RISCA study: Analysis of baseline data. *The Lancet. Neurology, 12*(7), 650–658. https://doi.org/10.1016/S1474-4422(13)70104-2.

Jayakumar, P. N., Desai, S., Pal, P. K., Balivada, S., Ellika, S., & Kalladka, D. (2008). Functional correlates of incoordination in patients with spinocerebellar ataxia 1: A preliminary fMRI study. *Journal of Clinical Neuroscience: Official Journal of the Neurosurgical Society of Australasia, 15*(3), 269–277. https://doi.org/10.1016/j.jocn.2007.06.021.

Joers, J. M., Deelchand, D. K., Lyu, T., Emir, U. E., Hutter, D., Gomez, C. M., et al. (2018). Neurochemical abnormalities in premanifest and early spinocerebellar ataxias. *Annals of Neurology, 83*(4), 816–829. https://doi.org/10.1002/ana.25212.

Jung, B. C., Choi, S. I., Du, A. X., Cuzzocreo, J. L., Geng, Z. Z., Ying, H. S., et al. (2012a). Principal component analysis of cerebellar shape on MRI separates SCA types 2 and 6 into two archetypal modes of degeneration. *Cerebellum (London, England), 11*(4), 887–895. https://doi.org/10.1007/s12311-011-0334-6.

Jung, B. C., Choi, S. I., Du, A. X., Cuzzocreo, J. L., Ying, H. S., Landman, B. A., et al. (2012b). MRI shows a region-specific pattern of atrophy in spinocerebellar ataxia type 2. *Cerebellum (London, England), 11*(1), 272–279. https://doi.org/10.1007/s12311-011-0308-8.

Kang, N., Christou, E. A., Burciu, R. G., Chung, J. W., DeSimone, J. C., Ofori, E., et al. (2017). Sensory and motor cortex function contributes to symptom severity in spinocerebellar ataxia type 6. *Brain Structure & Function, 222*(2), 1039–1052. https://doi.org/10.1007/s00429-016-1263-4.

Kang, J.-S., Klein, J. C., Baudrexel, S., Deichmann, R., Nolte, D., & Hilker, R. (2014). White matter damage is related to ataxia severity in SCA3. *Journal of Neurology, 261*(2), 291–299. https://doi.org/10.1007/s00415-013-7186-6.

Kawai, Y., Suenaga, M., Watanabe, H., Ito, M., Kato, K., Kato, T., et al. (2008). Prefrontal hypoperfusion and cognitive dysfunction correlates in spinocerebellar ataxia type 6. *Journal of the Neurological Sciences, 271*(1–2), 68–74. https://doi.org/10.1016/j.jns.2008.03.018.

Kim, J.-M., Hong, S., Kim, G. P., Choi, Y. J., Kim, Y. K., Park, S. S., et al. (2007). Importance of low-range CAG expansion and CAA interruption in SCA2 parkinsonism. *Archives of Neurology, 64*(10), 1510–1518. https://doi.org/10.1001/archneur.64.10.1510.

Kim, J.-Y., Kim, S. Y., Kim, J.-M., Kim, Y. K., Yoon, K.-Y., Kim, J. Y., et al. (2009). Spinocerebellar ataxia type 17 mutation as a causative and susceptibility gene in parkinsonism. *Neurology, 72*(16), 1385–1389. https://doi.org/10.1212/WNL.0b013e3181a18876.

Kim, J.-M., Lee, J.-Y., Kim, H. J., Kim, J. S., Kim, Y. K., Park, S. S., et al. (2010). The wide clinical spectrum and nigrostriatal dopaminergic damage in spinocerebellar ataxia type 6. *Journal of Neurology, Neurosurgery, and Psychiatry, 81*(5), 529–532. https://doi.org/10.1136/jnnp.2008.166728.

Kimura, N., Kumamoto, T., Masuda, T., Nomura, Y., Hanaoka, T., Hazama, Y., et al. (2009a). Evaluation of regional cerebral blood flow in cerebellar variant of multiple system atrophy using FineSRT. *Clinical Neurology and Neurosurgery, 111*(10), 829–834. https://doi.org/10.1016/j.clineuro.2009.08.014.

Kimura, N., Kumamoto, T., Masuda, T., Nomura, Y., Hanaoka, T., Hazama, Y., et al. (2011). Evaluation of the effects of thyrotropin releasing hormone (TRH) therapy on regional cerebral blood flow in the cerebellar variant of multiple system atrophy using 3DSRT. *Journal of Neuroimaging: Official Journal of the American Society of Neuroimaging, 21*(2), 132–137. https://doi.org/10.1111/j.1552-6569.2009.00411.x.

Kimura, N., Kumamoto, T., Masuda, T., Nomura, Y., Hanaoka, T., Hazama, Y., et al. (2009b). Evaluation of the effect of thyrotropin releasing hormone (TRH) on regional cerebral blood flow in spinocerebellar degeneration using 3DSRT. *Journal of the Neurological Sciences, 281*(1–2), 93–98. https://doi.org/10.1016/j.jns.2009.01.023.

Klockgether, T., Skalej, M., Wedekind, D., Luft, A. R., Welte, D., Schulz, J. B., et al. (1998). Autosomal dominant cerebellar ataxia type I. MRI-based volumetry of posterior fossa structures and basal ganglia in spinocerebellar ataxia types 1, 2 and 3. *Brain: A Journal of Neurology, 121*(Pt. 9), 1687–1693.

Koepp, M., Schelosky, L., Cordes, I., Cordes, M., & Poewe, W. (1994). Dystonia in ataxia telangiectasia: Report of a case with putaminal lesions and decreased striatal [123I] iodobenzamide binding. *Movement Disorders: Official Journal of the Movement Disorder Society, 9*(4), 455–459. https://doi.org/10.1002/mds.870090414.

Koeppen, A. H., Becker, A. B., Qian, J., & Feustel, P. J. (2017). Friedreich ataxia: Hypoplasia of spinal cord and dorsal root ganglia. *Journal of Neuropathology and Experimental Neurology, 76*(2), 101–108. https://doi.org/10.1093/jnen/nlw111.

Kono, S., Terada, T., Ouchi, Y., & Miyajima, H. (2014). An altered GABA-A receptor function in spinocerebellar ataxia type 6 and familial hemiplegic migraine type 1 associated with the CACNA1A gene mutation. *BBA Clinical, 2*, 56–61. https://doi.org/10.1016/j.bbacli.2014.09.005.

Kwon, K.-Y., Choi, C. G., Kim, J. S., Lee, M. C., & Chung, S. J. (2007). Comparison of brain MRI and 18F-FDG PET in the differential diagnosis of multiple system atrophy from Parkinson's disease. *Movement Disorders: Official Journal of the Movement Disorder Society, 22*(16), 2352–2358. https://doi.org/10.1002/mds.21714.

Lasek, K., Lencer, R., Gaser, C., Hagenah, J., Walter, U., Wolters, A., et al. (2006). Morphological basis for the spectrum of clinical deficits in spinocerebellar ataxia 17 (SCA17). *Brain: A Journal of Neurology, 129*(Pt. 9), 2341–2352. https://doi.org/10.1093/brain/awl148.

Lee, P. H., An, Y.-S., Yong, S. W., & Yoon, S. N. (2008). Cortical metabolic changes in the cerebellar variant of multiple system atrophy: A voxel-based FDG-PET study in 41 patients. *NeuroImage, 40*(2), 796–801. https://doi.org/10.1016/j.neuroimage.2007.11.055.

Lin, D. D. M., Crawford, T. O., Lederman, H. M., & Barker, P. B. (2006). Proton MR spectroscopic imaging in ataxia-telangiectasia. *Neuropediatrics, 37*(4), 241–246. https://doi.org/10.1055/s-2006-924722.

Lirng, J.-F., Wang, P.-S., Chen, H.-C., Soong, B.-W., Guo, W. Y., Wu, H.-M., et al. (2012). Differences between spinocerebellar ataxias and multiple system atrophy-cerebellar type on proton magnetic resonance spectroscopy. *PLoS One, 7*(10), e47925. https://doi.org/10.1371/journal.pone.0047925.

Lukas, C., Hahn, H. K., Bellenberg, B., Hellwig, K., Globas, C., Schimrigk, S. K., et al. (2008). Spinal cord atrophy in spinocerebellar ataxia type 3 and 6: Impact on clinical disability. *Journal of Neurology, 255*(8), 1244–1249. https://doi.org/10.1007/s00415-008-0907-6.

Lukas, C., Schöls, L., Bellenberg, B., Rüb, U., Przuntek, H., Schmid, G., et al. (2006). Dissociation of grey and white matter reduction in spinocerebellar ataxia type 3 and 6: A voxel-based morphometry study. *Neuroscience Letters, 408*(3), 230–235. https://doi.org/10.1016/j.neulet.2006.09.007.

Lyoo, C. H., Jeong, Y., Ryu, Y. H., Lee, S. Y., Song, T. J., Lee, J. H., et al. (2008). Effects of disease duration on the clinical features and brain glucose metabolism in patients with mixed type multiple system atrophy. *Brain: A Journal of Neurology, 131*(Pt. 2), 438–446. https://doi.org/10.1093/brain/awm328.

Mandelli, M. L., De Simone, T., Minati, L., Bruzzone, M. G., Mariotti, C., Fancellu, R., et al. (2007). Diffusion tensor imaging of spinocerebellar ataxias types 1 and 2. *AJNR. American Journal of Neuroradiology, 28*(10), 1996–2000. https://doi.org/10.3174/ajnr.A0716.

Marie, P., Foix, C., & Alajouanine, T. (1922). De l'atrophie cerebelleuse tardive a predominance corticale. *Revue Neurologique, 38*(849–885), 1082–1111.

Martins, C. R., Martinez, A. R. M., de Rezende, T. J. R., Branco, L. M. T., Pedroso, J. L., Barsottini, O. G. P., et al. (2017). Spinal cord damage in spinocerebellar ataxia type 1. *Cerebellum (London, England)*, *16*(4), 792–796. https://doi.org/10.1007/s12311-017-0854-9.

Marzi, C., Ciulli, S., Giannelli, M., Ginestroni, A., Tessa, C., Mascalchi, M., et al. (2018). Structural complexity of the cerebellum and cerebral cortex is reduced in spinocerebellar ataxia type 2. *Journal of Neuroimaging: Official Journal of the American Society of Neuroimaging*. https://doi.org/10.1111/jon.12534, [Epub ahead of print].

Mascalchi, M., Bianchi, A., Ciulli, S., Ginestroni, A., Aiello, M., Dotti, M. T., et al. (2017). Lower medulla hypoplasia in Friedreich ataxia: MR imaging confirmation 140 years later. *Journal of Neurology*, *264*(7), 1526–1528. https://doi.org/10.1007/s00415-017-8542-8.

Mascalchi, M., Cosottini, M., Lolli, F., Salvi, F., Tessa, C., Macucci, M., et al. (2002). Proton MR spectroscopy of the cerebellum and pons in patients with degenerative ataxia. *Radiology*, *223*(2), 371–378. https://doi.org/10.1148/radiol.2232010722.

Mascalchi, M., Diciotti, S., Giannelli, M., Ginestroni, A., Soricelli, A., Nicolai, E., et al. (2014). Progression of brain atrophy in spinocerebellar ataxia type 2: A longitudinal tensor-based morphometry study. *PLoS One*, *9*(2), e89410. https://doi.org/10.1371/journal.pone.0089410.

Mascalchi, M., Marzi, C., Giannelli, M., Ciulli, S., Bianchi, A., Ginestroni, A., et al. (2018). Histogram analysis of DTI-derived indices reveals pontocerebellar degeneration and its progression in SCA2. *PLoS One*, *13*(7), e0200258. https://doi.org/10.1371/journal.pone.0200258.

Mascalchi, M., Salvi, F., Piacentini, S., & Bartolozzi, C. (1994). Friedreich's ataxia: MR findings involving the cervical portion of the spinal cord. *AJR. American Journal of Roentgenology*, *163*(1), 187–191. https://doi.org/10.2214/ajr.163.1.8010211.

Mascalchi, M., Toschi, N., Giannelli, M., Ginestroni, A., Della Nave, R., Nicolai, E., et al. (2015). Progression of microstructural damage in spinocerebellar ataxia type 2: A longitudinal DTI study. *AJNR. American Journal of Neuroradiology*, *36*(6), 1096–1101. https://doi.org/10.3174/ajnr.A4343.

Mascalchi, M., Toschi, N., Giannelli, M., Ginestroni, A., Della Nave, R., Tessa, C., et al. (2016). Regional cerebral disease progression in Friedreich's ataxia: A longitudinal diffusion tensor imaging study. *Journal of Neuroimaging: Official Journal of the American Society of Neuroimaging*, *26*(2), 197–200. https://doi.org/10.1111/jon.12270.

Mascalchi, M., Tosetti, M., Plasmati, R., Bianchi, M. C., Tessa, C., Salvi, F., et al. (1998). Proton magnetic resonance spectroscopy in an Italian family with spinocerebellar ataxia type 1. *Annals of Neurology*, *43*(2), 244–252. https://doi.org/10.1002/ana.410430215.

Mascalchi, M., & Vella, A. (2012). Magnetic resonance and nuclear medicine imaging in ataxias. *Handbook of Clinical Neurology*, *103*, 85–110. https://doi.org/10.1016/B978-0-444-51892-7.00004-8.

Mascalchi, M., Vella, A., & Ceravolo, R. (2012). Movement disorders: Role of imaging in diagnosis. *Journal of Magnetic Resonance Imaging: JMRI*, *35*(2), 239–256. https://doi.org/10.1002/jmri.22825.

Mazere, J., Meissner, W. G., Sibon, I., Lamare, F., Tison, F., Allard, M., et al. (2013). [(123) I]-IBVM SPECT imaging of cholinergic systems in multiple system atrophy: A specific alteration of the ponto-thalamic cholinergic pathways (Ch5-Ch6). *NeuroImage. Clinical*, *3*, 212–217. https://doi.org/10.1016/j.nicl.2013.07.012.

Minnerop, M., Joe, A., Lutz, M., Bauer, P., Urbach, H., Helmstaedter, C., et al. (2005). Putamen dopamine transporter and glucose metabolism are reduced in SCA17. *Annals of Neurology*, *58*(3), 490–491. https://doi.org/10.1002/ana.20609.

Minnerop, M., Lüders, E., Specht, K., Ruhlmann, J., Schimke, N., Thompson, P. M., et al. (2010). Callosal tissue loss in multiple system atrophy—A one-year follow-up study. *Movement Disorders: Official Journal of the Movement Disorder Society, 25*(15), 2613–2620. https://doi.org/10.1002/mds.23318.

Murata, Y., Yamaguchi, S., Kawakami, H., Imon, Y., Maruyama, H., Sakai, T., et al. (1998). Characteristic magnetic resonance imaging findings in Machado-Joseph disease. *Archives of Neurology, 55*(1), 33–37.

Nanri, K., Okita, M., Takeguchi, M., Taguchi, T., Ishiko, T., Saito, H., et al. (2009). Intravenous immunoglobulin therapy for autoantibody-positive cerebellar ataxia. *Internal Medicine (Tokyo, Japan), 48*(10), 783–790.

Oh, M., Kim, J. S., Oh, J. S., Lee, C. S., & Chung, S. J. (2017). Different subregional metabolism patterns in patients with cerebellar ataxia by 18F-fluorodeoxyglucose positron emission tomography. *PLoS One, 12*(3), e0173275. https://doi.org/10.1371/journal.pone.0173275.

Olivito, G., Cercignani, M., Lupo, M., Iacobacci, C., Clausi, S., Romano, S., et al. (2017). Neural substrates of motor and cognitive dysfunctions in SCA2 patients: A network based statistics analysis. *NeuroImage. Clinical, 14*, 719–725. https://doi.org/10.1016/j.nicl.2017.03.009.

Olivito, G., Lupo, M., Iacobacci, C., Clausi, S., Romano, S., Masciullo, M., et al. (2017). Microstructural MRI basis of the cognitive functions in patients with spinocerebellar ataxia type 2. *Neuroscience, 366*, 44–53. https://doi.org/10.1016/j.neuroscience.2017.10.007.

Oz, G., Hutter, D., Tkác, I., Clark, H. B., Gross, M. D., Jiang, H., et al. (2010). Neurochemical alterations in spinocerebellar ataxia type 1 and their correlations with clinical status. *Movement Disorders: Official Journal of the Movement Disorder Society, 25*(9), 1253–1261. https://doi.org/10.1002/mds.23067.

Pagani, E., Ginestroni, A., Della Nave, R., Agosta, F., Salvi, F., De Michele, G., et al. (2010). Assessment of brain white matter fiber bundle atrophy in patients with Friedreich ataxia. *Radiology, 255*(3), 882–889. https://doi.org/10.1148/radiol.10091742.

Pellecchia, M. T., Barone, P., Vicidomini, C., Mollica, C., Salvatore, E., Ianniciello, M., et al. (2011). Progression of striatal and extrastriatal degeneration in multiple system atrophy: A longitudinal diffusion-weighted MR study. *Movement Disorders: Official Journal of the Movement Disorder Society, 26*(7), 1303–1309. https://doi.org/10.1002/mds.23601.

Prakash, N., Hageman, N., Hua, X., Toga, A. W., Perlman, S. L., & Salamon, N. (2009). Patterns of fractional anisotropy changes in white matter of cerebellar peduncles distinguish spinocerebellar ataxia-1 from multiple system atrophy and other ataxia syndromes. *NeuroImage, 47*(Suppl. 2), T72–T81. https://doi.org/10.1016/j.neuroimage.2009.05.013.

Quarantelli, M., Giardino, G., Prinster, A., Aloj, G., Carotenuto, B., Cirillo, E., et al. (2013). Steroid treatment in ataxia-telangiectasia induces alterations of functional magnetic resonance imaging during prono-supination task. *European Journal of Paediatric Neurology: EJPN: Official Journal of the European Paediatric Neurology Society, 17*(2), 135–140. https://doi.org/10.1016/j.ejpn.2012.06.002.

Reetz, K., Costa, A. S., Mirzazade, S., Lehmann, A., Juzek, A., Rakowicz, M., et al. (2013). Genotype-specific patterns of atrophy progression are more sensitive than clinical decline in SCA1, SCA3 and SCA6. *Brain: A Journal of Neurology, 136*(Pt. 3), 905–917. https://doi.org/10.1093/brain/aws369.

Reetz, K., Kleiman, A., Klein, C., Lencer, R., Zuehlke, C., Brockmann, K., et al. (2011). CAG repeats determine brain atrophy in spinocerebellar ataxia 17: A VBM study. *PLoS One, 6*(1), e15125. https://doi.org/10.1371/journal.pone.0015125.

Reetz, K., Lencer, R., Hagenah, J. M., Gaser, C., Tadic, V., Walter, U., et al. (2010). Structural changes associated with progression of motor deficits in spinocerebellar ataxia 17. *Cerebellum (London, England)*, 9(2), 210–217. https://doi.org/10.1007/s12311-009-0150-4.

Reetz, K., Rodríguez-Labrada, R., Dogan, I., Mirzazade, S., Romanzetti, S., Schulz, J. B., et al. (2018). Brain atrophy measures in preclinical and manifest spinocerebellar ataxia type 2. *Annals of Clinical and Translational Neurology*, 5(2), 128–137. https://doi.org/10.1002/acn3.504.

Rezende, T. J. R., Martinez, A. R. M., Faber, I., Girotto, K., Martins, M. P., de Lima, F. D., et al. (2018). Developmental and neurodegenerative damage in Friedreich ataxia. *European Journal of Neurology*. https://doi.org/10.1111/ene.13843. October 16. [Epub ahead of print].

Rezende, T. J. R., Martinez, A. R. M., Faber, I., Girotto, K., Pedroso, J. L., Barsottini, O. G., et al. (2017). Structural signature of classical versus late-onset friedreich's ataxia by multimodality brain MRI. *Human Brain Mapping*, 38(8), 4157–4168. https://doi.org/10.1002/hbm.23655.

Rezende, T. J. R., Silva, C. B., Yassuda, C. L., Campos, B. M., D'Abreu, A., Cendes, F., et al. (2016). Longitudinal magnetic resonance imaging study shows progressive pyramidal and callosal damage in Friedreich's ataxia. *Movement Disorders: Official Journal of the Movement Disorder Society*, 31(1), 70–78. https://doi.org/10.1002/mds.26436.

Richter, S., Dimitrova, A., Maschke, M., Gizewski, E., Beck, A., Aurich, V., et al. (2005). Degree of cerebellar ataxia correlates with three-dimensional mri-based cerebellar volume in pure cerebellar degeneration. *European Neurology*, 54(1), 23–27. https://doi.org/10.1159/000087241.

Rizzo, G., Tonon, C., Valentino, M. L., Manners, D., Fortuna, F., Gellera, C., et al. (2011). Brain diffusion-weighted imaging in Friedreich's ataxia. *Movement Disorders: Official Journal of the Movement Disorder Society*, 26(4), 705–712. https://doi.org/10.1002/mds.23518.

Salvatore, E., Tedeschi, E., Mollica, C., Vicidomini, C., Varrone, A., Coda, A. R. D., et al. (2014). Supratentorial and infratentorial damage in spinocerebellar ataxia 2: A diffusion-weighted MRI study. *Movement Disorders: Official Journal of the Movement Disorder Society*, 29(6), 780–786. https://doi.org/10.1002/mds.25757.

Santner, W., Schocke, M., Boesch, S., Nachbauer, W., & Egger, K. (2014). A longitudinal VBM study monitoring treatment with erythropoietin in patients with Friedreich ataxia. *Acta Radiologica Short Reports*, 3(4) 2047981614531573. https://doi.org/10.1177/2047981614531573.

Santos, T. A., Maistro, C. E. B., Silva, C. B., Oliveira, M. S., França, M. C., & Castellano, G. (2015). MRI texture analysis reveals bulbar abnormalities in Friedreich ataxia. *AJNR. American Journal of Neuroradiology*, 36(12), 2214–2218. https://doi.org/10.3174/ajnr.A4455.

Schöls, L., Amoiridis, G., Przuntek, H., Frank, G., Epplen, J. T., & Epplen, C. (1997). Friedreich's ataxia. Revision of the phenotype according to molecular genetics. *Brain: A Journal of Neurology*, 120(Pt. 12), 2131–2140.

Schulte, T., & Schols, L. (2002). The use of quantitative methods in clinical trials for spinocerebellar ataxia [letter]. *Archives of Neurology*, 59, 1044–1045.

Schulz, J. B., Borkert, J., Wolf, S., Schmitz-Hübsch, T., Rakowicz, M., Mariotti, C., et al. (2010). Visualization, quantification and correlation of brain atrophy with clinical symptoms in spinocerebellar ataxia types 1, 3 and 6. *NeuroImage*, 49(1), 158–168. https://doi.org/10.1016/j.neuroimage.2009.07.027.

Selvadurai, L. P., Harding, I. H., Corben, L. A., & Georgiou Karistianis, N. (2018). Cerebral abnormalities in Friedreich ataxia: A review. *Neuroscience and Biobehavioral Reviews*, 84, 394–406. https://doi.org/10.1016/j.neubiorev.2017.08.006.

Selvadurai, L. P., Harding, I. H., Corben, L. A., Stagnitti, M. R., Storey, E., Egan, G. F., et al. (2016). Cerebral and cerebellar grey matter atrophy in Friedreich ataxia: The IMAGE-FRDA study. *Journal of Neurology, 263*(11), 2215–2223. https://doi.org/10. 1007/s00415-016-8252-7.

Shan, D. E., Soong, B. W., Sun, C. M., Lee, S. J., Liao, K. K., & Liu, R. S. (2001). Spinocerebellar ataxia type 2 presenting as familial levodopa-responsive parkinsonism. *Annals of Neurology, 50*(6), 812–815.

Shinotoh, H., Thiessen, B., Snow, B. J., Hashimoto, S., MacLeod, P., Silveira, I., et al. (1997). Fluorodopa and raclopride PET analysis of patients with Machado-Joseph disease. *Neurology, 49*(4), 1133–1136.

Solbach, K., Kraff, O., Minnerop, M., Beck, A., Schöls, L., Gizewski, E. R., et al. (2014). Cerebellar pathology in Friedreich's ataxia: Atrophied dentate nuclei with normal iron content. *NeuroImage. Clinical, 6*, 93–99. https://doi.org/10.1016/j.nicl. 2014.08.018.

Solodkin, A., Peri, E., Chen, E. E., Ben-Jacob, E., & Gomez, C. M. (2011). Loss of intrinsic organization of cerebellar networks in spinocerebellar ataxia type 1: Correlates with disease severity and duration. *Cerebellum (London, England), 10*(2), 218–232. https://doi. org/10.1007/s12311-010-0214-5.

Soong, B., Cheng, C., Liu, R., & Shan, D. (1997). Machado-Joseph disease: Clinical, molecular, and metabolic characterization in Chinese kindreds. *Annals of Neurology, 41*(4), 446–452. https://doi.org/10.1002/ana.410410407.

Soong, B. W., & Liu, R. S. (1998). Positron emission tomography in asymptomatic gene carriers of Machado-Joseph disease. *Journal of Neurology, Neurosurgery, and Psychiatry, 64*(4), 499–504.

Soong, B., Liu, R., Wu, L., Lu, Y., & Lee, H. (2001). Metabolic characterization of spinocerebellar ataxia type 6. *Archives of Neurology, 58*(2), 300–304.

Specht, K., Minnerop, M., Müller-Hübenthal, J., & Klockgether, T. (2005). Voxel-based analysis of multiple-system atrophy of cerebellar type: Complementary results by combining voxel-based morphometry and voxel-based relaxometry. *NeuroImage, 25*(1), 287–293. https://doi.org/10.1016/j.neuroimage.2004.11.022.

Stefanescu, M. R., Dohnalek, M., Maderwald, S., Thürling, M., Minnerop, M., Beck, A., et al. (2015). Structural and functional MRI abnormalities of cerebellar cortex and nuclei in SCA3, SCA6 and Friedreich's ataxia. *Brain: A Journal of Neurology, 138*(Pt. 5), 1182–1197. https://doi.org/10.1093/brain/awv064.

Tachibana, H., Kawabata, K., Tomino, Y., & Sugita, M. (1999). Prolonged P3 latency and decreased brain perfusion in cerebellar degeneration. *Acta Neurologica Scandinavica, 100*(5), 310–316.

Taoka, T., Kin, T., Nakagawa, H., Hirano, M., Sakamoto, M., Wada, T., et al. (2007). Diffusivity and diffusion anisotropy of cerebellar peduncles in cases of spinocerebellar degenerative disease. *NeuroImage, 37*(2), 387–393. https://doi.org/10.1016/j.neuroimage. 2007.05.028.

Terakawa, H., Abe, K., Watanabe, Y., Nakamura, M., Fujita, N., Hirabuki, N., et al. (1999). Proton magnetic resonance spectroscopy (1H MRS) in patients with sporadic cerebellar degeneration. *Journal of Neuroimaging: Official Journal of the American Society of Neuroimaging, 9*(2), 72–77.

Tha, K. K., Terae, S., Yabe, I., Miyamoto, T., Soma, H., Zaitsu, Y., et al. (2010). Microstructural white matter abnormalities of multiple system atrophy: In vivo topographic illustration by using diffusion-tensor MR imaging. *Radiology, 255*(2), 563–569. https://doi.org/10.1148/radiol.10090988.

Tokumaru, A. M., Kamakura, K., Maki, T., Murayama, S., Sakata, I., Kaji, T., et al. (2003). Magnetic resonance imaging findings of Machado-Joseph disease: Histopathologic correlation. *Journal of Computer Assisted Tomography, 27*(2), 241–248.

Varrone, A., Salvatore, E., De Michele, G., Barone, P., Sansone, V., Pellecchia, M. T., et al. (2004). Reduced striatal [123 I]FP-CIT binding in SCA2 patients without parkinsonism. *Annals of Neurology, 55*(3), 426–430. https://doi.org/10.1002/ana.20054.

Vella, A., & Mascalchi, M. (2018). Nuclear medicine of the cerebellum. *Handbook of Clinical Neurology, 154,* 251–266. https://doi.org/10.1016/B978-0-444-63956-1.00015-1.

Viau, M., & Boulanger, Y. (2004). Characterization of ataxias with magnetic resonance imaging and spectroscopy. *Parkinsonism & Related Disorders, 10*(6), 335–351. https://doi.org/10.1016/j.parkreldis.2004.02.006.

Vieira Karuta, S. C., Raskin, S., de Carvalho Neto, A., Gasparetto, E. L., Doring, T., & Teive, H. A. G. (2015). Diffusion tensor imaging and tract-based spatial statistics analysis in Friedreich's ataxia patients. *Parkinsonism & Related Disorders, 21*(5), 504–508. https://doi.org/10.1016/j.parkreldis.2015.02.021.

Volkow, N. D., Tomasi, D., Wang, G.-J., Studentsova, Y., Margus, B., & Crawford, T. O. (2014). Brain glucose metabolism in adults with ataxia-telangiectasia and their asymptomatic relatives. *Brain: A Journal of Neurology, 137*(Pt. 6), 1753–1761. https://doi.org/10.1093/brain/awu092.

Wallis, L. I., Griffiths, P. D., Ritchie, S. J., Romanowski, C. a. J., Darwent, G., & Wilkinson, I. D. (2007). Proton spectroscopy and imaging at 3T in ataxia-telangiectasia. *AJNR. American Journal of Neuroradiology, 28*(1), 79–83.

Wang, P.-S., Liu, R.-S., Yang, B.-H., & Soong, B.-W. (2007). Regional patterns of cerebral glucose metabolism in spinocerebellar ataxia type 2, 3 and 6: A voxel-based FDG-positron emission tomography analysis. *Journal of Neurology, 254*(7), 838–845. https://doi.org/10.1007/s00415-006-0383-9.

Wang, J.-L., Xiao, B., Cui, X.-X., Guo, J.-F., Lei, L.-F., Song, X.-X., et al. (2009). Analysis of SCA2 and SCA3/MJD repeats in Parkinson's disease in mainland China: Genetic, clinical, and positron emission tomography findings. *Movement Disorders: Official Journal of the Movement Disorder Society, 24*(13), 2007–2011. https://doi.org/10.1002/mds.22727.

Waragai, M., Yamada, T., & Matsuda, H. (2007). Evaluation of brain perfusion SPECT using an easy Z-score imaging system (eZIS) as an adjunct to early-diagnosis of neurodegenerative diseases. *Journal of the Neurological Sciences, 260*(1–2), 57–64. https://doi.org/10.1016/j.jns.2007.03.027.

Wilkinson, I. D., Hadjivassiliou, M., Dickson, J. M., Wallis, L., Grünewald, R. A., Coley, S. C., et al. (2005). Cerebellar abnormalities on proton MR spectroscopy in gluten ataxia. *Journal of Neurology, Neurosurgery, and Psychiatry, 76*(7), 1011–1013. https://doi.org/10.1136/jnnp.2004.049809.

Wu, Y.-T., Shyu, K.-K., Jao, C.-W., Wang, Z.-Y., Soong, B.-W., Wu, H.-M., et al. (2010). Fractal dimension analysis for quantifying cerebellar morphological change of multiple system atrophy of the cerebellar type (MSA-C). *NeuroImage, 49*(1), 539–551. https://doi.org/10.1016/j.neuroimage.2009.07.042.

Wu, T., Wang, C., Wang, J., Hallett, M., Zang, Y., & Chan, P. (2013). Preclinical and clinical neural network changes in SCA2 parkinsonism. *Parkinsonism & Related Disorders, 19,* 158–164. https://doi.org/10.1016/j.parkreldis.2012.08.011.

Wüllner, U., Klockgether, T., Petersen, D., Naegele, T., & Dichgans, J. (1993). Magnetic resonance imaging in hereditary and idiopathic ataxia. *Neurology, 43*(2), 318–325.

Wüllner, U., Reimold, M., Abele, M., Bürk, K., Minnerop, M., Dohmen, B.-M., et al. (2005). Dopamine transporter positron emission tomography in spinocerebellar ataxias type 1, 2, 3, and 6. *Archives of Neurology, 62*(8), 1280–1285. https://doi.org/10.1001/archneur.62.8.1280.

Yen, T. C., Lu, C. S., Tzen, K. Y., Wey, S. P., Chou, Y. H., Weng, Y. H., et al. (2000). Decreased dopamine transporter binding in Machado-Joseph disease. *Journal of Nuclear Medicine: Official Publication, Society of Nuclear Medicine, 41*(6), 994–998.

Yen, T.-C., Tzen, K.-Y., Chen, M.-C., Chou, Y.-H. W., Chen, R.-S., Chen, C.-J., et al. (2002). Dopamine transporter concentration is reduced in asymptomatic Machado-Joseph disease gene carriers. *Journal of Nuclear Medicine: Official Publication, Society of Nuclear Medicine, 43*(2), 153–159.

Ying, S. H., Choi, S. I., Perlman, S. L., Baloh, R. W., Zee, D. S., & Toga, A. W. (2006). Pontine and cerebellar atrophy correlate with clinical disability in SCA2. *Neurology, 66*(3), 424–426. https://doi.org/10.1212/01.wnl.0000196464.47508.00.

You, H., Wang, J., Wang, H., Zang, Y.-F., Zheng, F.-L., Meng, C.-L., et al. (2011). Altered regional homogeneity in motor cortices in patients with multiple system atrophy. *Neuroscience Letters, 502*(1), 18–23. https://doi.org/10.1016/j.neulet.2011.07.015.

Zalesky, A., Akhlaghi, H., Corben, L. A., Bradshaw, J. L., Delatycki, M. B., Storey, E., et al. (2014). Cerebello-cerebral connectivity deficits in Friedreich ataxia. *Brain Structure and Function, 219*(1), 969–981. https://doi.org/10.1007/s00429-013-0547-1.

CHAPTER FIVE

Neuroimaging Applications in Functional Movement Disorders

Luis Pedro Faria de Abreu*,†, Tiago Teodoro‡,§, Mark J. Edwards‡,1

*Departamento de Neurociências do Hospital de Santa Maria, Lisbon, Portugal
†Assistente Convidado da Faculdade de Medicina de Lisboa, Universidade de Lisboa, Lisbon, Portugal
‡Neuroscience Research Centre, Institute of Molecular and Clinical Sciences, St George's University of London, London, United Kingdom
§Instituto de Medicina Molecular, Faculdade de Medicina de Lisboa, Universidade de Lisboa, Lisbon, Portugal
1Corresponding author: e-mail address: medwards@sgul.ac.uk

Contents

Abstract

Functional (psychogenic) movement disorders are a subtype of functional neurological disorder, a common and disabling cause of neurological symptoms. Abnormal movement in people with functional movement disorders has specific characteristics (e.g., distractibility, variability, incongruence with deficits caused by neurological disease), allowing positive diagnosis and differentiation from other causes of movement disorder. Attempts to understand the pathophysiology of this disorder have previously focused mainly on the psychological level, emphasizing the importance of psychological trauma and adverse life events. However, the last two decades has seen a broadening of this approach to consider the neurobiological level, and brain imaging has formed a key part of this work. Here we review the available imaging evidence in functional movement disorders and explain how this evidence can help us understand more about the underlying pathophysiology of this common cause of abnormal movement control.

International Review of Neurobiology, Volume 143
ISSN 0074-7742
https://doi.org/10.1016/bs.irn.2018.10.001

163

1. INTRODUCTION

In the last decade, Functional Neurological Disorders (FND) have emerged as a subspecialty area within Neurology. This very recent development followed almost 50 years of relative stagnation in this area. This is surprising given that FND are a common problem, accounting for 16% of all referrals from primary care to Neurology outpatient clinics (Stone et al., 2010).

Functional Movement Disorders (FMD) have been at the center of the paradigm change in the understanding and management of FND which has moved away from a sole focus on devising a psychological formulation for the cause of the disorder. From a diagnostic perspective, emphasis has been put on establishing a positive diagnosis of FND, based on the demonstration of positive physical signs, particularly the modulation of functional symptoms and signs by attention (Carson et al., 2016). Functional motor symptoms are particularly amenable for direct assessment during examination, thus providing an excellent model for catalyzing this paradigm change.

Neuroimaging has two main applications in FND: clinical diagnosis (of co-morbidity) and understanding pathophysiology. Parsimony is highly praised in the neurological diagnostic process. We are taught to follow the principle of Occam's Razor, and to try to find the minimum number of diagnoses capable of explaining the whole clinical picture presented by patient. FND constitute a relative exception to this rule. This is because FND commonly coexist with organic neurological and non-neurological disorders (so-called "functional overlay"). Therefore, establishing a positive clinical diagnosis of FMD does not exclude the possibility of other simultaneous neurological disorders, including those associated with structural damage. In fact, having an organic neurological condition is a risk factor for developing functional neurological symptoms. In this context, conventional neuroimaging is often essential for the general diagnostic work-out, to rule out coexisting structural damage.

From a pathophysiological perspective, the field of FND is moving away from a sometimes reductionist approach framed by Freudian theories, which focused on the "cause" rather than on the "mechanism." The emerging management strategies for FND are informed by novel neurobiological models. Functional neurological symptoms have been related with abnormal attention, disease-related beliefs and expectations, abnormal agency, while also incorporating the role of stressors (understood in a broad sense—including acute medical illnesses, traumatic injuries and "psychological" stress) (Edwards, 2017). Functional neuroimaging techniques have been instrumental for testing new neurobiological theories.

2. THE ROLE OF NEUROIMAGING IN THE DIAGNOSTIC WORK-UP OF PATIENTS WITH SUSPECTED FUNCTIONAL MOVEMENT DISORDERS

As commented above, FND often coexist with "organic" neurological and non-neurological disorders. This is commonly called "functional overlay." Importantly, any form of "stress," including a physical injury (e.g., leg fracture), an acute medical (e.g., asthma attack) or neurological condition (e.g., migraine attack), psychological stress (e.g., past history of childhood abuse), psychiatric comorbidity (e.g., anxiety disorder) or a chronic medical condition (e.g., multiple sclerosis) can all trigger/increase the risk of developing an "overlay" of FNS.

The diagnosis of FMD is mainly clinical, based on detecting positive evidence of incongruency and inconsistency on history and examination. However, the presence of FMD does not exclude the possibility of other simultaneous neurological disorders, sometimes causing structural damage. This coexistence and interaction between functional and other disorders can sometimes generate a very complex clinical picture.

In this context, conventional neuroimaging is often essential for the general diagnostic work-out, to rule out coexisting, underlying structural damage.

Several case reports and case-series illustrate the usefulness of neuroimaging techniques in the management of patients with FND. For example, Wilson's disease (Elmalı, Gündüz, Poyraz, Kızıltan, & Ertan, 2017) and Hirayama's disease (Kwon, Kim, Yoon, & Park, 2018) (a rare focal form of motor neuron disease, causing upper limb weakness) have both been detected with the help of structural neuroimaging techniques, in patients with an *overlay* of functional tremor.

It has been reported that there is a high prevalence of functional neurological symptoms in patients with Parkinson's disease (Pareés et al., 2013). On the other hand, pure functional parkinsonism appears to be rare (Lang et al., 1995), but may be difficult to differentiate from Parkinson's disease based solely on clinical grounds.

Dopamine transporter-single-photon emission computer tomography (DAT-SPECT; DaTScan) imaging uses a radioactive ligand that binds to presynaptic dopamine transporters. This imaging method has high sensitivity to detect the nigrostriatal denervation characteristics of neurodegenerative parkinsonism (Ba & Martin, 2015; Kägi, Bhatia, & Tolosa, 2010).

DaTScans are normal in purely functional parkinsonism, and also in drug-induced parkinsonism, while degenerative forms of Parkinsonism, including idiopathic Parkinson's disease, show signs of presynaptic dopaminergic denervation (Tolosa, Coelho, & Gallardo, 2003). Hence, DaTScan can have a decisive influence in diagnosis and management (Seifert & Wiener, 2013).

Benaderette and colleagues performed clinical, electrophysiological, and [^{123} I]-FP-CIT SPECT studies in three patients with functional parkinsonism and six patients with a combination of functional parkinsonism and Parkinson's disease (Benaderette et al., 2006). This study supported the role of [^{123} I]-FP-CIT SPECT scan in improving diagnostic accuracy.

The usefulness of dopamine transporter imaging in patients with overlapping organic and functional features of parkinsonism was also pointed by Umeh, Szabo, Pontone and Mari (2013). In this study, imaging with DAT-SPECT was added to clinical assessment in three patients presenting with overlapping clinical features of functional and neurodegenerative Parkinsonism. Two patients had abnormal DAT-SPECT, leading to the diagnosis of Parkinson's disease and offering dopaminergic treatment. A third patient had normal imaging, supporting the diagnosis of functional parkinsonism.

Gaig and colleagues (Gaig et al., 2006) performed ^{123}I-Ioflipane SPECT in nine patients with suspected functional parkinsonism. The results were within normal limits in eight participants. In one patient there was bilateral decrease of striatal tracer uptake, supporting an alternative diagnosis (further investigations lead to the detection of a *Parkin* gene mutation).

These studies illustrate the usefulness of DaTScan for detecting underlying neurodegenerative Parkinsonism, particularly in patients with more atypical presentations, including an overlay of functional neurological symptoms.

3. BRAIN IMAGING IN THE DEVELOPMENT OF NOVEL NEUROBIOLOGICAL MODELS OF FUNCTIONAL MOVEMENT DISORDERS

3.1 Abnormal Movement Control and Its Relationship With Emotional Processing

Models of functional weakness have proposed abnormal motor intention and/or abnormal motor execution. It has been hypothesized that abnormal self-monitoring, limbic processing and/or top-down regulation from higher-order frontal regions could interfere with motor execution (Voon, Brezing, Gallea, & Hallett, 2011). Neuroimaging has provided insights into these theoretical models.

Voon and colleagues found that patients with functional weakness performing either self-paced or externally cued movement had reduced activity of left supplementary motor area (involved in movement initiation) and increased activity in right amygdala, left anterior insula, and bilateral posterior cingulate cortex, which participate in emotional processing. Patients with functional weakness also showed less functional connectivity between left supplementary motor area and bilateral dorsolateral prefrontal cortices in internally-generated as compared with externally-generated movements. Voon and colleagues proposed that, in an "arousing" context, abnormal learned motor programs may hijack the voluntary action selection system, which is hypoactive and disconnected from prefrontal top-down regulation, which could usually inhibit unwanted actions (Voon et al., 2011). The result is a movement that arises without a normal prediction of its sensory consequences (i.e., without an efference copy) and is therefore experienced by patients as arising spontaneously without will or self-control (Edwards, 2017; Voon et al., 2011).

There is a substantial body of research suggesting abnormal movement-related brain activity and a link with abnormal emotional processing in people with FMD. In 1997, Marshall and colleagues made a pioneering observation in a patient with functional lower limb paralysis (Marshall, Halligan, Fink, Wade, & Frackowiak, 1997). They recorded her brain activity [regional cerebral blood flow (rCBF)], while she was preparing to move and moving (or trying to move) her normal and her weak lower limbs. Preparing to move or moving her good leg, and also preparing to move her weak leg, activated motor and/or premotor areas previously described to be involved in motor preparation and execution. Interestingly, an attempt to move her weak limb failed to activate the primary motor cortex but was associated with an activation of the orbitofrontal and anterior cingulate cortices. Activation of these areas has been related with previous traumatic emotional experiences, and it was proposed that this activation prevented normal activation of the motor cortex.

Voon and colleagues investigated the relationship between functional motor symptoms and emotional processing of affective stimuli (Voon, Brezing, et al., 2010). As predicted, healthy subjects showed greater right amygdala activity on exposure to fearful versus neutral faces compared with happy versus neutral faces, while there were no differences in patients with functional weakness. In addition, patients with functional weakness showed a tendency for greater activity of right amygdala, as compared with healthy subjects. This pattern was interpreted as representing impaired amygdala habituation. They also identified a greater functional connectivity between the right amygdala and the right supplementary motor area in functional patients as compared with healthy controls, suggesting a possible mechanism for how psychological stressors could trigger or exacerbate functional symptoms.

Espay and colleagues investigated the fMRI neural correlates of emotional processing in patients with functional tremor (Espay et al., 2018a). They probed motor and emotional circuits of patients with functional tremor, essential tremor and healthy controls during performance of finger-tapping motor task, a basic-emotion task and an intense-emotion task. In the motor task, participants with functional tremor showed increased activation in the right cerebellum, as compared to participants with essential tremor. In the basic-emotion task, having functional tremor was associated with an increased activation in the paracingulate gyrus and left Heschl's gyrus (in comparison with healthy controls) and with decreased activation in the right precentral gyrus (in comparison with essential tremor). The differences were less clear for the intense-emotion task. The authors concluded that in response to emotional stimuli, participants with functional tremor had changes in the activation and functional connectivity in networks involved in emotion processing and theory of mind. How this related to the mechanism of the tremor itself was not resolved in this study.

A second study from the same group applied the same motor and emotional paradigms to patients with functional dystonia, primary organic dystonia and healthy controls (Espay et al., 2018b). In basic-emotion tasks, participants with functional dystonia showed decreased activation in the right middle temporal gyrus and bilateral precuneus, and increased activation in right inferior frontal gyrus, bilateral occipital cortex and fusiform gyrus, and bilateral cerebellum. In the intense-emotion task patients with functional dystonia showed a decreased activation in the left insular and motor cortices and increased activation in the left fusiform gyrus. There were no differences in the motor tasks. Espay and colleagues concluded that patients with functional dystonia patients showed altered patterns of activation in networks involved in motor preparation and execution, spatial cognition, and attentional control.

3.2 Abnormal Sense of Agency (SOA)

Functional motor symptoms show some characteristics usually regarded as typical of voluntary movements, such as distractibility and dual task phenomena. However, the patients report their abnormal movements as involuntary (i.e., lacking in self-agency). This contrast between motor abnormalities that look voluntary at first glance but are described as involuntary has been a source of mutual misunderstanding and distrust between patients, health professionals and people without FMD.

Advances in functional imaging have recently investigated the neurobiology of agency in FMD. Self-agency refers to the experience that one is the cause of one's own actions (Haggard, 2008). An impaired sense of agency may prevent patients with FMD experiencing agency over abnormal movements that otherwise share some common neurobiological pathways with normal voluntary movements (Kranick & Hallett, 2013).

Although intention is also linked to voluntary action, self of agency is in fact a "postintention" process of retrospective assessment of one's action consequences (Lau, Rogers, Haggard, & Passingham, 2004).

Indeed, a sense of agency has been proposed to emerge when the prediction of one's action consequences (feed-forward signal or efference copy) match with the actual sensory feedback resulting from the movement (action) (Blakemore, Goodbody, & Wolpert, 1998; Blakemore & Sirigu, 2003). The monitoring of the discrepancy between intended and resulting movement has been localized to the inferior temporo-parietal and prefrontal cortices and the cerebellum (Sirigu, Daprati, Pradat-Diehl, Franck, & Jeannerod, 1999; Voon, Gallea, et al., 2010).

Voon and colleagues compared the fMRI correlates of functional tremor and voluntarily mimicked tremor in the same patient group (Voon, Gallea, et al., 2010). They observed that functional tremor was associated with hypoactivity of the right temporo-parietal junction (rTPJ) and also a lower functional connectivity between rTPJ and sensorimotor regions (sensorimotor cortices and cerebellar vermis) and limbic regions. They proposed that these findings could reflect a lack of sensory prediction signals, which could lead to the perception that the abnormal movement was not self-generated.

Maurer and colleagues compared the functional rTPJ connectivity of patients with FMD and healthy controls in a resting-state fMRI imaging study (Maurer et al., 2016). Maurer also observed a decreased functional connectivity between rTPJ and sensorimotor areas bilaterally, including sensorimotor cortex (right), SMA (bilateral), insula (right) and cerebellum (bilateral).

This impaired functional connectivity was proposed to be related with impaired motor feed-forward and sensory feed-back signaling involving the rTPJ. Notably, SMA is proposed to be involved in sensorimotor integration.

Nahab and colleagues investigated the neurobiology of loss of agency by comparing the fMRI correlates of a group of patients with FMD and a previously assessed control group of healthy individuals, during performance of a validated virtual-reality movement paradigm that could modulate sense of agency over movement (Nahab et al., 2011; Nahab, Kundu, Maurer, Shen, & Hallett, 2017). Patients with FMD showed a dysfunction of dorsolateral prefrontal cortex and pre-supplementary motor area on the right, which did not respond differentially to the loss of movement control induced by the experimental paradigm (Nahab et al., 2017).

Baek and colleagues (Baek et al., 2017) compared the fMRI correlates of FND patients and healthy controls during performance of Libet's clock task. FND patients showed a significant reduction in the activity of the inferior parietal lobule within the rTPJ, when comparing intention versus movement trials. In addition, there was a reduction of resting-state connectivity between the right inferior parietal cortex and the frontal "control" regions (dorsolateral prefrontal cortex, anterior cingulate cortex), and increased functional connectivity with the premotor cortex and SMA.

Van Beilen and colleagues reported abnormal parietal activation in patients with functional weakness, along with reduced activity in the prefrontal cortex, supramarginal gyrus and precuneus (van Beilen, de Jong, Gieteling, Renken, & Leenders, 2011). They proposed that these regions may provide an interface between psychological mechanisms and impaired higher-order motor control. More specifically, decreased activity on the supramarginal gyrus was related with impaired interaction of bodily schema information and environmental cues, which would be predicted to impair movement initiation. Importantly, precuneus activity was reduced in functional patients but increased in controls asked to feign symptoms (van Beilen et al., 2011). The decreased activity of the precuneus in functional patients was related with a perception of unintentionality (lack of agency). Patients showed increased activation of the premotor cortex, while in feigners there was increased activation of pre-supplementary motor area (pre-SMA). Notably, increased activity in pre-SMA was previously related with a representation of motor intention (Lau et al., 2004). This is in accordance with the differences in intentionality between functionals and feigners. Finally, the increased activation of premotor cortex in people with functional weakness was proposed to reflect a greater preparatory effort (van Beilen et al., 2011).

Hassa and colleagues aimed to compare the differences in the neural correlates of motor inhibition between patients with functional motor symptoms and healthy subjects feigning weakness (Hassa, de Jel, Tuescher, Schmidt, & Schoenfeld, 2016). They observed that passive movement of the weak hand elicited activity in different regions of the inferior frontal gyrus. This was proposed to reflect a role of this area in motor inhibition in feigning controls and functional disorder patients. In addition, they observed an increase in the activity of the medial prefrontal cortex in functional weakness, in comparison with feigning (and non-feigning) healthy subjects. This was also proposed to be related to an altered sense of agency (Hassa et al., 2016).

Taken together, these findings suggest the impaired sense of agency in functional motor disorders may be related to dysfunction of a network including the rTPJ among other areas linked to sensorimotor integration and motor control.

3.3 Further Evidence of Abnormal Activity of Motor Circuits

Burgmer and colleagues analyzed brain activation using fMRI during observation and subsequent imitative execution of movements in patients with functional hand weakness and healthy controls (Burgmer et al., 2006). Patients with functional weakness showed a (side-specific) reduced activation of cortical hand areas during movement observation. In contrast, the authors found no activity suggesting movement inhibition. These observations were considered to support the existence of an upstream impairment of movement conceptualization.

Cojan and colleagues compared the fMRI correlate during performance of a Go/No-Go task in 1 patient with functional weakness and 30 healthy volunteers feigning paralysis (Cojan, Waber, Carruzzo, & Vuilleumier, 2009). Right motor cortex showed preparatory activation despite left-sided functional paralysis, suggesting preserved motor intention. The inability to move the affected left hand on go trials was associated with activations of precuneus and ventrolateral frontal gyrus. Interestingly, there was an enhancement of functional connectivity between the posterior cingulate cortex, precuneus and vmPFC. Inhibitory right frontal areas which became activate on no-go trials with the normal right hand, remained inactive for the left (weak) hand. In contrast, a group of healthy subjects feigning paralysis showed similar patterns of activity on no-go trials and left-go trials with feigned weakness. The authors argued that functional weakness does not directly result from the activity of motor inhibitory circuits, instead reflecting the selective activation of midline brain regions related with self-representation and emotional regulation.

Spence and colleagues used positron emission tomography (PET) to compare functional weakness of left upper limb in two patients with functional movement disorders and feigned weakness of the same limb in two healthy subjects (Spence, Crimlisk, Cope, Ron, & Grasby, 2000). During attempted movement of the affected hand, functional weakness was associated with hypoactivation of left dorsolateral prefrontal cortex, and feigned weakness with hypoactivation of the right anterior prefrontal cortex. These observations were proposed to reflect abnormal motor inhibition, driven by the prefrontal cortex (Spence et al., 2000).

Czarnecki and colleagues performed a SPECT-based study to compare the pattern of regional blood flow (rCBF) in patients with functional tremor, essential tremor and healthy controls, during performance of a tremor-inducing motor task and at rest (Czarnecki, Jones, Burnett, Mullan, & Matsumoto, 2011). During the motor task, patients with functional tremor showed reduced rCBF in anterior regions of the default mode network. At rest, there was increased rCBF in left inferior frontal gyrus and left insula. The authors related the functional tremor with abnormalities in the default mode network.

Finally, there is also evidence of subcortical involvement in functional neurological disorders. Vuilleumier and colleagues performed single positron emission computes tomography (SPECT) during hand vibratory stimulation on patients with functional unilateral sensorimotor impairment and weakness (Vuilleumier et al., 2001). They observed a consistent decreased in rCBF in the thalamus and basal ganglia (thalamus, caudate and putamen) contralateral to the functional deficit. Notably, they repeated their assessment a few months later, after symptom improvement, and there was also a recovery of rCBF. Vuilleumier proposed the contribution of low-level dysfunction to functional sensorimotor symptoms. Imaging normalization after symptom improvement suggests that the abnormalities on functional imaging are related with symptom neurobiology itself, and not simply disease-related traits.

Schrag and colleagues used positron emission tomography (PET) to compare functional dystonia with organic dystonia (Schrag et al., 2013). Participants were assessed at rest, during fixed posturing of the right leg and during paced ankle movements. Patients with organic dystonia showed increased activity in the primary motor cortex and thalamus and decreased activity in the cerebellum, in comparison with controls. In contrast, patients with functional dystonia showed signs of increased activity in the cerebellum and basal ganglia, and decreased activity in the primary motor cortex. On direct comparison, organic dystonia showed greater activity in the primary motor cortex, and functional dystonia in the cerebellum and basal ganglia. Both types of dystonia showed abnormal activation of right dorsolateral pre-frontal cortex during movement.

The authors concluded that abnormal activation of prefrontal cortex is not specific of functional dystonia, being also present in organic forms of dystonia.

3.4 Structural Imaging

After making a diagnosis of a functional neurological disorder, we usually explain to patients that this is a condition caused by malfunction of the brain without underlying structural damage. However, the reality appears to be less straightforward, with recent studies highlighting subtle structural changes in patients with FND.

MRI-based morphometric techniques have been applied to investigate subtle changes in cortical areas proposed to be involved in stress-mediated neuroplasticity (Perez, Matin, et al., 2017; Perez et al., 2018; Perez, Williams, et al., 2017). Adverse early-life events are considered risk factors for both post-traumatic stress disorder (PTSD) and functional neurological disorders. On the other hand, cingulo-insular structures are considered to be involved in the neurobiology of both conditions, and to be a target of stress-mediated neuroplasticity. In a subgroup of women with FND, Perez and colleagues observed an association between lower left anterior insula volumes and more severe functional symptoms and childhood abuse burden (Perez, Matin, et al., 2017). In contrast, in patients with PTSD, greater symptom severity was associated with lower volumes of dorsal anterior cingular cortex (ACC) and the magnitude of lifetime adverse events was also inversely related with left hippocampal volumes. These results suggest that the cingulo-insular complex appears to be involved in both FND and PTSD, but with a different profile of changes.

In another study from the same group a post-hoc stratified analysis found an association between higher levels of physical health impairment and reduced left anterior insular volumes in patients with FND, as compared with controls (Perez, Williams, et al., 2017). In addition, within-group analysis of patients with FND found an association between impaired mental health and increased trait anxiety and increased right amygdala volumes.

Finally, Perez and colleagues also found patients with functional motor symptoms and high levels of somatoform dissociation on Somatoform Dissociation Questionaire-20 to show reduced left caudal anterior cingulate cortex cortical thickness as compared with healthy controls. In addition, they reported a positive correlation between depersonalization/derealization scores and right lateral occipital thickness. Both findings retained statistical significance after controlling for trait anxiety and depression, PTSD, adverse life events and motor FND subtypes (Perez et al., 2018).

Aybek and colleagues (Aybek et al., 2014) used MRI voxel-based morphometry (VMB) and voxel-based cortical thickness (VBCT) analysis and found a significant increase in cortical thickness (VBCT) in the premotor cortices of patients with functional hemiparesis, in comparison with healthy controls. Nicholson and colleagues (Nicholson et al., 2014) observed a decrease in the volume of the left thalamus in a group of patients with functional motor symptoms, as compared with healthy participants. The authors discussed that this could be related with the primary disease process or could constitute a secondary effect of the disorder, for example, resulting from limb disuse.

In summary, there is evidence of subtle changes in cortical thickness and volumes among patient with functional movement disorders, with an possible involvement of cingular-insular structures and sensorimotor areas.

4. CONCLUSIONS

Functional neurological disorders in general, and functional movement disorders in particular have seen a growing research and clinical interest over the past 10–20 years. Imaging has a clinical application in the assessment of some people with FMD as a method to look for co-morbid organic disorders, for example, MS and Parkinson's disease.

More importantly, neuroimaging has found an application in helping to understand the underlying pathophysiology of FMD. Although functional imaging studies are very heterogenous regarding number of patients and methodological approaches, data from these studies are helping to build a picture of the neurobiology of FMD. These data support the difference between people with FMD and those feigning symptoms. The data also, while very heterogenous suggest failure to activate downstream motor areas in a normal fashion, both M1 and SMA, in contrast to abnormal activation of prefrontal and limbic regions and areas associated with agentic movement.

Future studies, perhaps in particular concentrating on difference between functional neuroimaging before and after successful treatment, will no doubt shed more light on the underlying mechanism of FMD.

REFERENCES

Aybek, S., Nicholson, T. R. J., Draganski, B., Daly, E., Murphy, D. G., David, A. S., et al. (2014). Grey matter changes in motor conversion disorder. *Journal of Neurology, Neurosurgery, and Psychiatry, 85*(2), 236–238.

Ba, F., & Martin, W. R. W. (2015). Dopamine transporter imaging as a diagnostic tool for parkinsonism and related disorders in clinical practice. *Parkinsonism & Related Disorders, 21*(2), 87–94.

Baek, K., Doñamayor, N., Morris, L. S., Strelchuk, D., Mitchell, S., Mikheenko, Y., et al. (2017). Impaired awareness of motor intention in functional neurological disorder: Implications for voluntary and functional movement. *Psychological Medicine, 47*(9), 1624–1636.

Benaderette, S., Zanotti Fregonara, P., Apartis, E., Nguyen, C., Trocello, J.-M., Remy, P., et al. (2006). Psychogenic parkinsonism: A combination of clinical, electrophysiological, and [(123)I]-FP-CIT SPECT scan explorations improves diagnostic accuracy. *Movement Disorders: Official Journal of the Movement Disorder Society, 21*(3), 310–317.

Blakemore, S. J., Goodbody, S. J., & Wolpert, D. M. (1998). Predicting the consequences of our own actions: The role of sensorimotor context estimation. *Journal of Neuroscience: The Official Journal of the Society for Neuroscience, 18*(18), 7511–7518.

Blakemore, S.-J., & Sirigu, A. (2003). Action prediction in the cerebellum and in the parietal lobe. *Experimental Brain Research, 153*(2), 239–245.

Burgmer, M., Konrad, C., Jansen, A., Kugel, H., Sommer, J., Heindel, W., et al. (2006). Abnormal brain activation during movement observation in patients with conversion paralysis. *NeuroImage, 29*(4), 1336–1343.

Carson, A., et al. (2016). Assessment of patients with functional neurologic disorders. *Handbook of Clinical Neurology, 139*, 169–188. PubMed-NCBI [Internet]. [cited 2018 Mar 7]. Available from: https://www.ncbi.nlm.nih.gov/pubmed/27719837.

Cojan, Y., Waber, L., Carruzzo, A., & Vuilleumier, P. (2009). Motor inhibition in hysterical conversion paralysis. *NeuroImage, 47*(3), 1026–1037.

Czarnecki, K., Jones, D. T., Burnett, M. S., Mullan, B., & Matsumoto, J. Y. (2011). SPECT perfusion patterns distinguish psychogenic from essential tremor. *Parkinsonism & Related Disorders, 17*(5), 328–332.

Edwards, M. J. (2017). Neurobiologic theories of functional neurologic disorders. *Handbook of Clinical Neurology, 139*, 131–137.

Elmalı, A. D., Gündüz, A., Poyraz, B. Ç., Kızıltan, M. E., & Ertan, S. (2017). A case illustrating how tremor of Wilson's disease may mimic functional tremor. *Acta Neurologica Belgica, 117*(1), 351–353.

Espay, A. J., Maloney, T., Vannest, J., Norris, M. M., Eliassen, J. C., Neefus, E., et al. (2018a). Impaired emotion processing in functional (psychogenic) tremor: A functional magnetic resonance imaging study. *NeuroImage Clinical, 17*, 179–187.

Espay, A. J., Maloney, T., Vannest, J., Norris, M. M., Eliassen, J. C., Neefus, E., et al. (2018b). Dysfunction in emotion processing underlies functional (psychogenic) dystonia. *Movement Disorders: Official Journal of the Movement Disorder Society, 33*(1), 136–145.

Gaig, C., Martí, M. J., Tolosa, E., Valldeoriola, F., Paredes, P., Lomeña, F. J., et al. (2006). 123I-Ioflupane SPECT in the diagnosis of suspected psychogenic Parkinsonism. *Movement Disorders: Official Journal of the Movement Disorder Society, 21*(11), 1994–1998.

Haggard, P. (2008). Human volition: Towards a neuroscience of will. *Nature Reviews. Neuroscience, 9*(12), 934–946.

Hassa, T., de Jel, E., Tuescher, O., Schmidt, R., & Schoenfeld, M. A. (2016). Functional networks of motor inhibition in conversion disorder patients and feigning subjects. *NeuroImage Clinical, 11*, 719–727.

Kägi, G., Bhatia, K. P., & Tolosa, E. (2010). The role of DAT-SPECT in movement disorders. *Journal of Neurology, Neurosurgery, and Psychiatry, 81*(1), 5–12.

Kranick, S. M., & Hallett, M. (2013). Neurology of volition. *Experimental Brain Research, 229*(3), 313–327.

Kwon, D.-Y., Kim, J., Yoon, H., & Park, M. H. (2018). Progressive myoclonic tremor mimicking functional tremor in Hirayama disease. *Acta Neurologica Belgica, 118*(3), 517–518.

Lang AE, et al. (1995). Psychogenic parkinsonism. PubMed-NCBI [Internet]. [cited 2018 Aug 29]. *Archives of Neurology, 52*(8), 802–810. Available from: https://www.ncbi.nlm.nih.gov/pubmed/?term=Lang+AE%2C+Koller+WC%2C+Fahn+S.+Psychogenic+parkinsonism.

Lau, H. C., Rogers, R. D., Haggard, P., & Passingham, R. E. (2004). Attention to intention. *Science, 303*(5661), 1208–1210.

Marshall, J. C., Halligan, P. W., Fink, G. R., Wade, D. T., & Frackowiak, R. S. (1997). The functional anatomy of a hysterical paralysis. *Cognition, 64*(1), B1–B8.

Maurer, C. W., LaFaver, K., Ameli, R., Epstein, S. A., Hallett, M., & Horovitz, S. G. (2016). Impaired self-agency in functional movement disorders: A resting-state fMRI study. *Neurology, 87*(6), 564–570.

Nahab, F. B., Kundu, P., Gallea, C., Kakareka, J., Pursley, R., Pohida, T., et al. (2011). The neural processes underlying self-agency. *Cerebral Cortex (New York, N.Y.: 1991), 21*(1), 48–55.

Nahab, F. B., Kundu, P., Maurer, C., Shen, Q., & Hallett, M. (2017). Impaired sense of agency in functional movement disorders: An fMRI study. *PLoS One, 12*(4), e0172502.

Nicholson, T. R., Aybek, S., Kempton, M. J., Daly, E. M., Murphy, D. G., David, A. S., et al. (2014). A structural MRI study of motor conversion disorder: Evidence of reduction in thalamic volume. *Journal of Neurology, Neurosurgery, and Psychiatry, 85*(2), 227–229.

Pareés, I., Saifee, T. A., Kojovic, M., Kassavetis, P., Rubio-Agusti, I., Sadnicka, A., et al. (2013). Functional (psychogenic) symptoms in Parkinson's disease. *Movement Disorders: Official Journal of the Movement Disorder Society, 28*(12), 1622–1627.

Perez, D. L., Matin, N., Barsky, A., Costumero-Ramos, V., Makaretz, S. J., Young, S. S., et al. (2017). Cingulo-insular structural alterations associated with psychogenic symptoms, childhood abuse and PTSD in functional neurological disorders. *Journal of Neurology, Neurosurgery, and Psychiatry, 88*(6), 491–497.

Perez, D. L., Matin, N., Williams, B., Tanev, K., Makris, N., LaFrance, W. C., et al. (2018). Cortical thickness alterations linked to somatoform and psychological dissociation in functional neurological disorders. *Human Brain Mapping, 39*(1), 428–439.

Perez, D. L., Williams, B., Matin, N., LaFrance, W. C., Costumero-Ramos, V., Fricchione, G. L., et al. (2017). Corticolimbic structural alterations linked to health status and trait anxiety in functional neurological disorder. *Journal of Neurology, Neurosurgery, and Psychiatry, 88*(12), 1052–1059.

Schrag, A. E., Mehta, A. R., Bhatia, K. P., Brown, R. J., Frackowiak, R. S. J., Trimble, M. R., et al. (2013). The functional neuroimaging correlates of psychogenic versus organic dystonia. *Brain: A Journal of Neurology, 136*(Pt. 3), 770–781.

Seifert, K. D., & Wiener, J. I. (2013). The impact of DaTscan on the diagnosis and management of movement disorders: A retrospective study. *American Journal of Neurodegenerative Disease, 2*(1), 29–34.

Sirigu, A., Daprati, E., Pradat-Diehl, P., Franck, N., & Jeannerod, M. (1999). Perception of self-generated movement following left parietal lesion. *Brain: A Journal of Neurology, 122*(Pt. 10), 1867–1874.

Spence, S. A., Crimlisk, H. L., Cope, H., Ron, M. A., & Grasby, P. M. (2000). Discrete neurophysiological correlates in prefrontal cortex during hysterical and feigned disorder of movement. *Lancet (London, England), 355*(9211), 1243–1244.

Stone, J., Carson, A., Duncan, R., Roberts, R., Warlow, C., Hibberd, C., et al. (2010). Who is referred to neurology clinics?—The diagnoses made in 3781 new patients. *Clinical Neurology and Neurosurgery, 112*(9), 747–751.

Tolosa, E., Coelho, M., & Gallardo, M. (2003). DAT imaging in drug-induced and psychogenic parkinsonism. *Movement Disorders: Official Journal of the Movement Disorder Society, 18*(Suppl. 7), S28–S33.

Umeh, C. C., Szabo, Z., Pontone, G. M., & Mari, Z. (2013). Dopamine transporter imaging in psychogenic parkinsonism and neurodegenerative parkinsonism with psychogenic overlay: A report of three cases. *Tremor and Other Hyperkinetic Movements (New York, N.Y.), 3.*

van Beilen, M., de Jong, B. M., Gieteling, E. W., Renken, R., & Leenders, K. L. (2011). Abnormal parietal function in conversion paresis. *PLoS One, 6*(10), e25918.

Voon, V., Brezing, C., Gallea, C., Ameli, R., Roelofs, K., LaFrance, W. C., et al. (2010). Emotional stimuli and motor conversion disorder. *Brain: A Journal of Neurology, 133*(Pt. 5), 1526–1536.

Voon, V., Brezing, C., Gallea, C., & Hallett, M. (2011). Aberrant supplementary motor complex and limbic activity during motor preparation in motor conversion disorder. *Movement Disorders: Official Journal of the Movement Disorder Society, 26*(13), 2396–2403.

Voon, V., Gallea, C., Hattori, N., Bruno, M., Ekanayake, V., & Hallett, M. (2010). The involuntary nature of conversion disorder. *Neurology, 74*(3), 223–228.

Vuilleumier, P., Chicherio, C., Assal, F., Schwartz, S., Slosman, D., & Landis, T. (2001). Functional neuroanatomical correlates of hysterical sensorimotor loss. *Brain: A Journal of Neurology, 124*(Pt. 6), 1077–1090.

CHAPTER SIX

Transcranial B-Mode Sonography in Movement Disorders

Rezzak Yilmaz*,1, Daniela Berg*,†
*Department of Neurology, Christian-Albrechts-University of Kiel, Kiel, Germany
†Department of Neurodegeneration, Hertie-Institute for Clinical Brain Research, University of Tuebingen,
Tuebingen, Germany
1Corresponding author: e-mail address: rezzak.yilmaz@uksh.de

Contents

Abstract

Applying a 2–4 MHz probe at the temporal bone window transcranial B-mode sonography (TCS) enables the depiction of the brain parenchyma through the intact skull. Meanwhile it has been applied for the diagnosis and the differential diagnosis of movement disorders for decades. In the first part of this chapter, we summarize the technical requirements and describe the ultrasound method for optimal TCS examination. Imaging planes and the relevant structures are explained in detail. In the second part of the chapter, we focus on the role of substantia nigra hyperechogenicity for the diagnosis of Parkinson's disease (PD) and prodromal PD. In this part, we also mention the role of TCS in atypical and secondary Parkinsonian syndromes and other movement disorders. Summarizing all these information we explain how TCS can be helpful for the differential diagnosis of movement disorders. The current data show that TCS is an easily applicable and economic imaging method which can be used as an additional tool for the diagnosis of PD with a high sensitivity (>85%), specificity (>80%) and inter-rater reliability (>84%) as well as for the differential diagnosis of movement disorders. Lately, TCS has

International Review of Neurobiology, Volume 143
ISSN 0074-7742
https://doi.org/10.1016/bs.irn.2018.10.008

also been utilized in further areas such as the detection of individuals at risk for PD or the determination of electrode localization in patients with deep brain stimulation. An insufficient temporal bone window especially in the elderly and the necessity of an experienced investigator are limitations of this method.

1. INTRODUCTION

In movement disorders, the first report on transcranial B-mode sonography (TCS) dates back to 1995. In this study Becker, Seufert, Bogdahn, Reichmann, and Reiners (1995) could show for the first time that the echogenicity of the substantia nigra (SN) is distinctly increased in patients with Parkinson's disease (PD). Since that date, the field of TCS has grown vastly and TCS became a fast and easy to apply supplementary method for the diagnosis and differential diagnosis of movement disorders as well as a powerful instrument for screening of individuals at risk for PD. In this chapter, we summarize the method of TCS and its application in movement disorders.

2. THE METHOD OF TRANSCRANIAL B-MODE SONOGRAPHY

2.1 Equipment, Settings and Examination

An optimal TCS examination requires a high-end ultrasound machine with a 2–4 MHz phased-array sector transducer which is also used for transcranial duplex sonography. The best visualization of the brain parenchyma in adults is in general achieved when the imaging depth is set to 14–15 cm and the dynamic range between 40 and 60 dB. Image brightness can be adjusted as needed (using the "Gain" or "Time Gain Compensation" button of the respective device). With these settings the best resolution is achieved in a distance of about 5–9 cm from the probe which is called the focal zone. Additionally, the focus can be manually adjusted on the screen to increase the local resolution of the relevant structure. Under optimal conditions, contemporary systems can achieve an axial (horizontal) resolution of up to 0.7 mm in the focal zone (Berg & Becker, 2002).

Ultrasound examination should be performed in a darkened room. Usually, the examiner sits behind the head of the patient who lies in supine position. The transducer is placed on the temporal bone in front of the upper part of the *auricle* parallel to the *orbitomeatal* line which passes through the outer

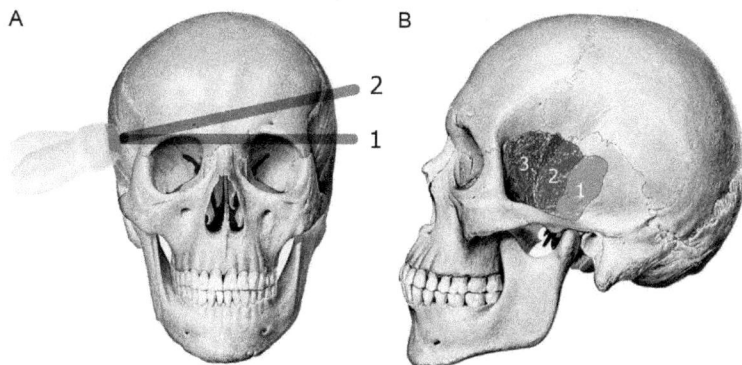

Fig. 1 The most frequently used sonographic planes and acoustic windows. (A) Depiction of the axial imaging planes. The mesencephalic plane (1) can be visualized by holding the transducer parallel to the orbitomeatal line. The third ventricle plane (2) can be seen by tilting the transducer 10–20 degree upward. (B) Depiction of the temporal bone window. The transducer should be moved carefully back and forth in order to find the best window for an optimal TCS examination.

canthus of the eye and the center of the *external auditory meatus* (Fig. 1A). In this position the axial image of the brain parenchyma can be obtained. The transducer should be carefully moved back and forth in order to find the best imaging window (or bone window) (Fig. 1B). In about 10–15% of the Caucasian population and especially in elderly females the examination may be unsuccessful or only partially successful due to an inadequate bone window. The rate of examination failure due to bone window insufficiency is even higher in the Asian population (Zhou et al., 2016).

2.2 Examination Planes and Visible Structures

The brain parenchyma can be visualized in several planes, although visualization of structures may vary depending on the quality of the bone window. For a routine clinical TCS examination, the most frequently used scanning planes are the mesencephalic and the third ventricle plane (Fig. 2).

2.2.1 Mesencephalic Plane

In a patient lying in supine position an axial image of the mesencephalic plane can be visualized by holding the transducer at the *orbitomeatal* line. In this plane the most important structure is the medially located and butterfly shaped *mesencephalon* at about 7 cm penetration depth, which appears hypoechogenic (dark) surrounded by the hyperechogenic (bright) basal

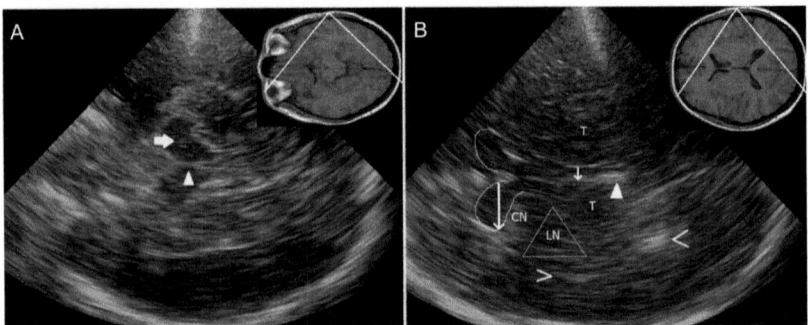

Fig. 2 Main axial imaging planes and structures for TCS in movement disorders. (A) Mesencephalic plane. In this plane the butterfly shaped mesencaphalon can be seen in the middle as a hypoechogenic structure, surrounded by the hyperechogenic basal cisterns. In the mesencephalon, the substantia nigra can be found as a lightly echogenic structure on the ipsilateral side (arrow). Another visible structure is the midline raphe which can be seen as an uninterrupted line in the middle of the mesencephalon (arrow head). (B) Third ventricle plane. In this plane the calcified pineal gland can be depicted in the middle of the image (arrow head). The borders of the third ventricle can be seen as two white parallel lines in front of the pineal gland. The width of the third ventricle can be measured as shown (short vertical arrow). On the contralateral side of the transducer, the anterior horn of the lateral ventricle can be depicted and measured as shown (long vertical arrow). Next to the anterior horn, the caudat nucleus is located. The lentiform nucleus has an isoechogenic texture and is located between the mildly echogenic caudate nucleus and the hypoechogenic thalamus. Some echogenicity of the insular cortex (arrow head showing right) and the calcified choroid plexus of the lateral ventricle (arrow head showing left) can also be found in this plane. CN, caudate nucleus; LN, lentiform nuclues; T, thalamus.

cisterns. In this plane, the most important structures are the midline *raphe* and the *substantia nigra* within the hypoechogenic *mesencephalon* (Fig. 2A).

Raphe: In healthy individuals, the *mesencephalic* part of the midline *raphe* is visualized as a continuous hyperechogenic line which is located in the middle of the *mesencephalon*. The midline *raphe* should be assessed regarding its continuity, i.e., an interruption or the absence of the hyperechogenic line is abnormal and has been related to depression (Drepper et al., 2017).

Substantia Nigra (SN): In the mesencephalic plane, the SN can be visualized as a hyperechogenic structure caudal of the *pedunculus cerebri*. It is assessed on the ipsilateral side of the transducer within the hypoechogenic *mesencephalon*. Since the SN is rather a small structure, the transducer should be carefully moved back and forth in order to find the image in which the SN is displayed at its largest size. After freezing and magnifying the image two- to threefold the so-called planimetric measurement should

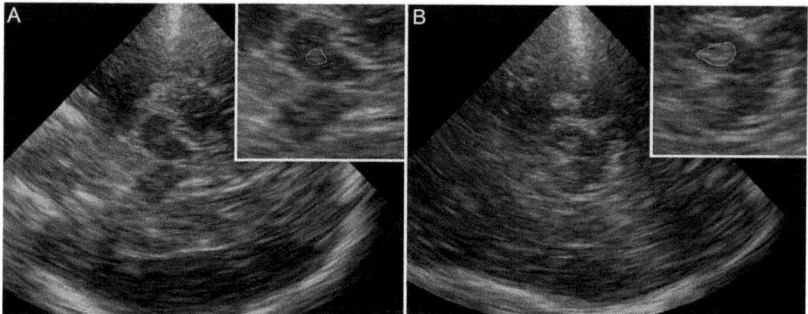

Fig. 3 Depiction of the substantia nigra (SN) in the mesencephalic plane. (A) Image of a healthy individual. After detecting the largest appearance of the ipsilateral SN, the image is frozen and zoomed. The planimetric measurement can then be performed by encircling the SN as shown. In this case, the area of the SN is 0.12 cm^2 which is lower than the cut-off value. (B) Image of a patient with Parkinson's disease. The increased echogenicity can be easily seen on both sides of the mesencephalon. In this case, the planimetric measurement of the ipsilateral SN is 0.39 cm^2, a value which is higher than the cut-off value, indicating substantia nigra hyperechogenicity.

be performed by encircling the area of the SN manually using the area measurement function of the machine (Fig. 3A). If the area of the SN is larger than the cut-off value, the term *substantia nigra* hyperechogenicity (SN+) is used (Fig. 3B). The cut-off value for SN+ should be calculated by taking the 90th percentile of SN values of at least 50 healthy individuals older than 50 years old. In some studies, the 75th percentile has been taken additionally as a cut-off value for detection of a "moderate SN+" which, however, is currently rarely done. The cut-off values may differ according to the ultrasound machine and the population assessed. For the cut-off values from different ultrasound systems, see Walter and Školoudík (2014).

One drawback of the planimetric measurement is that the two-dimensional size of the SN is taken into account but not its brightness. An option for also including SN brightness has been suggested by Skoloudik et al. (2014) who developed a quantitative digital image analysis method for measuring SN echogenicity. The analysis is performed by a computer program which captures the *mesencephalic* image, measures the brightness of the region of interest (in this case, SN) and calculates an index value by comparing the brightness of the SN with the rest of the captured image. Values higher than the 90th percentile of the healthy population are regarded as indicating SN+. Currently both SN measurements (planimetric and digital) are being clinically used. Research is also being conducted to achieve a three-dimensional sonographic display of the SN (Plate et al., 2012).

2.2.2 Third Ventricle Plane

The third ventricle plane is located rostral of the mesencephalic plane. This plane is reached by tilting the probe 10–25 degree upward from the axially scanned mesencephalic plane (Fig. 1A). In this plane, the third and the lateral ventricles, the *thalami* as well as the *basal ganglia* can be assessed (Fig. 2B). At the beginning of the examination the *pineal* gland can be spotted in the middle of the imaging plane as a calcified structure which can be used as a landmark. In front of the *pineal* gland the *ependyma* of the third ventricle appears as two parallel hyperechogenic lines. On both sides of the third ventricle the rather hypoechogenic egg-shaped *thalami* can be detected. Further to the nasal part, the frontal horn of the lateral ventricle can be seen on the contralateral side of the insonating probe. Both the third ventricle and the contralateral frontal horn are assessed by drawing a perpendicular line between the ventricle borders as shown in Fig. 2B. The cut-off values of the third ventricle are $<7/10\,mm$ and of the frontal horn $<17/20\,mm$ for individuals younger or older than 60 years, respectively. Adjacent to the frontal horn, the *caudate nucleus* is located, which is also visualized from the contralateral side. In between the *caudate nucleus* and *thalamus*, the *lentiform nucleus* (LN) is located, in healthy individuals hardly delimitable due to its isoechogenicity with the surrounding nuclei—although the *thalamus* is in general slightly more hypoechogenic.

The *basal ganglia* are assessed in terms of their echogenicity, i.e., the presence of hyperechogenicity in specific parts of the *basal ganglia* in relation to the surrounding brain parenchyma or calcified structures such as the *pineal gland, choroid plexus* or even skull (for instance in the evaluation of calcifications as seen in Fahr disease). This type of assessment is in general done in a semi-quantitative way, which does not provide an exact value but rather an approximate evaluation such as slight or moderately increased brightness. Of note, the *caudate nucleus* is almost always slightly hyperechogenic compared to the rest of the brain parenchyma (which also helps its identification). For the detailed video-demonstration of a TCS examination and the structures, see Yilmaz and Berg (2017).

2.2.3 Other Planes

Although not widely used, there are other imaging planes which may provide additional information. The (1) *cella media* plane can be reached by tilting the probe further upward from the third ventricle plane. In this plane the extension of the ventricles can be measured. The (2) cerebellar plane is reached from the mesencephalic plane by turning the top edge of the probe about 45 degrees posterior in order to visualize the *vermis* and, especially in case of atrophy the *arbor vitae* of the cerebellar hemispheres (Synofzik, Godau, Lindig, Schöls, & Berg, 2011). A (3) coronal visualization of the brain can also be achieved by turning the probe 90 degrees posterior to visualize the brain stem, the SN, the ventricles as well as the medial temporal lobe and the *basal ganglia* on the ipsilateral side in a coronal view (Yilmaz, Pilotto, et al., 2016).

3. TRANSCRANIAL B-MODE SONOGRAPHY IN MOVEMENT DISORDERS

3.1 Parkinson's Disease

Clinically, the most striking feature of TCS in its application in movement disorders is the visualization of the SN with its different sizes and echogenicities. It has been numerously shown that individuals with PD have an increased area of hyperechogenicity at the anatomical site of the SN compared to controls, a finding unique to TCS (Fig. 3B). A recently conducted meta-analysis of 4494 PD patients showed a prevalence of 84% (compared to 10% of healthy controls which is used as a cut-off value), which allows TCS to be employed as a reliable additional imaging tool in the diagnostic process of PD (Shafieesabet et al., 2017).

All sonographic assessments require expertise and depend on the examiner. Detection of SN+ for the diagnosis of PD is also not immune to examiner bias. Many studies blinded the examiner to patient's diagnosis to overcome this limitation. One interesting study investigated the performance of TCS after securing the complete blindness of the examiner by introducing the examiner to the darkened ultrasound room after the patients were covered completely with a blanket, leaving only holes for the scanning probe and for breathing. In this design, TCS differentiated PD patients from controls with a sensitivity and specificity of 86% and 83%, respectively, confirming the ability of TCS to contribute to diagnosis independent of the examiner bias (Prestel, Schweitzer, Hofer, Gasser, & Berg, 2006). The intra- and inter-rater reliability of TCS has also been extensively studied. In one of these studies, four blinded examiners demonstrated a substantial intra-rater (0.93–0.97) and inter-rater (0.84–0.89) reliability for SN measurements (van de Loo et al., 2010).

Finding of SN + in PD patients expanded the research to subgroups such as clinical PD subtypes or genetic PD. A study with 101 PD patients divided into subgroups according to the age of onset or clinical phenotype found that bilateral SN+ is more frequently found in patients with early-onset PD

(Walter, Dressler, Wolters, Wittstock, & Benecke, 2007). Bilateral SN+ and LN hyperechogenicity (LN+) were more common in patients with an akinetic-rigid type compared to a tremor-dominant (TD) subtype. This finding was subsequently supported by another study which reported larger SN sizes in non-TD patients compared to TD-PD (Lauckaite et al., 2014).

Genetic forms of PD have also been analyzed. It has been shown that PD patients with leucine-rich repeat kinase-2 gene mutations have similar TCS findings as individuals with idiopathic PD (Brockmann, Gröger, et al., 2011). This finding was confirmed by two other studies (Brüggemann, Hagenah, et al., 2011; Vilas et al., 2015) who found a higher prevalence of SN+ in PD patients with leucine-rich repeat kinase-2 gene mutations compared to controls. PD patients with glucocerebrosidase gene mutation also have an increased SN echogenicity similar to PD patients without any mutation (Brockmann, Srulijes, et al., 2011). Two independent studies confirmed the high prevalence of SN+ in PD patients with glucocerebrosidase gene mutation (Barrett et al., 2013; Kresojević et al., 2013).

With regard to the clinical subtypes of PD, more studies are needed. However, already at this stage it can be concluded that TCS features are similar in idiopathic PD and at least the two common genetic forms of leucine-rich repeat kinase-2 gene—and glucocerebrosidase gene—PD. Interestingly, almost in all genetic PD studies, mutation carriers without any PD symptom were also included in the study (Brüggemann, Hagenah, et al., 2011; Sierra et al., 2013; Vilas et al., 2015). These studies showed that asymptomatic carriers of leucine-rich repeat kinase 2 gene mutation also have an increased rate of SN+ compared to healthy controls. Percentage of SN+ in asymptomatic carriers of glucocerebrosidase gene mutation was likewise higher compared to individuals without any PD mutation (Kresojević et al., 2013), indicating that the existence of the SN+ may suggest the risk for future PD in asymptomatic mutation carriers (see prodromal PD).

Accumulating evidence from independent studies prompted further effort to understand the rational, consistency and significance of SN+ in PD. For example, it has been shown that the side of the larger echogenic area correlates with the clinically more affected contralateral side, indicating that SN+ size may be associated with clinical features or nigral pathology. Yet, reports with regard to association between SN and disease specific features are conflicting. In the first two studies no correlation between the size of the SN and age was found (Berg, Siefker, & Becker, 2001; Berg, Siefker, Ruprecht-Dörfler, & Becker, 2001), a finding that was supported in patients with rapid eye movement sleep behavior disorder (Iranzo et al., 2010), but contradicted another study (Behnke et al., 2007) which described an increasing prevalence of SN+ with age in elderly individuals. With regard to the association between SN size and motor disability measured by the United Parkinson's Disease Rating Scale or Hoehn and Yahr Scale, many studies (Iranzo et al., 2010; Jesus-Ribeiro et al., 2016; Lobsien et al., 2012 ; Spiegel et al., 2006) reported no such correlation, contrary to the positive findings of others (Behnke et al., 2007; Sanzaro, Iemolo, Duro, & Malferrari, 2014; Weise et al., 2009). A correlation between the SN size and disease duration or dopamine transporter uptake in functional imaging has also not been found in most of the studies (Lobsien et al., 2012; Spiegel et al., 2006).

Taking all studies into account, the current level of evidence favors the lack of association between SN size and other disease features. Nevertheless, more studies with longitudinal design are needed to elucidate this issue of debate.

3.1.1 The Significance of the SN Hyperechogenicity in Parkinson's disease

The finding of SN+ in patients with PD elicits the question: "What is the meaning of SN+ and to what extent does SN+ represent an underlying pathological process?" Within this context, the role of iron has been investigated considering that (1) iron accumulation appears highly echogenic by reflecting the ultrasound waves strongly and (2) iron is an important component in the pathophysiology of PD and nigral neurodegeneration. In normal conditions, iron in the SN is not toxic even in high concentrations as long as it is bound to storage proteins, which is primarily *neuromelanin* in the SN but also others like *ferritin*. The pathological mechanism of PD may be related to the accumulation of free (unbound) iron which may trigger or accelerate neurodegeneration via the Fenton reaction (Friedman, Arosio, Finazzi, Koziorowski, & Galazka-Friedman, 2011). With this background, the histopathological correlate for SN echogenicity has first been investigated by Berg, Grote, et al. (1999) who injected iron, *ferritin*, an iron chelator *desferroxamine* and *6-hydroxydopamine* into rat brains and followed by sonography. *6-Hydroxydopamine* is a destructive agent for dopaminergic cells which releases iron from *ferritin*. This study showed that injection of iron and *6-hydroxydopamine* positively correlates with SN echogenicity in a dose dependent manner. On the other hand, injection of the iron binders *ferritin*, or *desferroxamine* in addition to *6-hydroxydopamine* did not lead to an increase in SN echogenicity. In 2002, the same group analyzed postmortem brains of 20 individuals and showed that SN echogenicity correlates with the iron content of the SN but not with copper or calcium, which was also confirmed by histological analyses (Berg et al., 2002). Further a trend for negative correlation of SN echogenocity with *neuromelanin* was detected in 40 post-mortem brains (Zecca et al., 2005). Extending these results a correlation between SN echogenicity and *microglia* activation independent from the iron content was reported in 33 brains (Berg, Godau, Riederer, Gerlach, & Arzberger, 2010).

The outcome of these studies proves the close association between SN echogenicity and the known nigral pathology with increased amounts of non-protein bound (free) iron in PD. It seems that protein-bound iron does not (or to a lesser degree) cause increased echogenicity, which also may be the case in the *globus pallidus*, a structure which contains the highest level of iron in brain but looks isoechogenic on TCS. The lack of echogenicity of the SN in patients with restless legs syndrome, a syndrome associated with decreased levels of serum iron, further supports this argument (Godau, Schweitzer, Liepelt, Gerloff, & Berg, 2007). However, it is still unclear whether free iron is the initiator of the pathological mechanism leading to dopaminergic cell loss, or whether it is merely a by-product of the ongoing neurodegeneration. In that sense, the exact meaning of SN+ is also unclear. Interestingly, however, SN+ can also be detected in yet PD-wise healthy individuals, who in the long run have an increased risk of developing PD (see below). This observations and the evidence of (1) histopathological studies which suggest a relationship between non-bound iron and SN+ and (2) clinical studies, showing a lack of convincing evidence for a correlation between SN size and disease specific features, have led to the acceptance of SN+ as a marker of vulnerability for PD, which may indicate metabolic or immunologic instability of the area and the risk for nigral degeneration, rather than being a marker indicating the magnitude of the ongoing neuronal inflammation or the amount of remaining dopaminergic cells.

3.1.2 Prodromal Parkinson's Disease
By definition, taking the 90th percentile of the SN size as a cut-off value means that 10% of the population also have SN+ although they are healthy with regard to the classical PD symptoms. Interestingly, studies which focused on these individuals concluded that they may present some slight abnormalities typical in PD. For instance, in two different positron-emission

tomography studies it has been shown that healthy individuals with SN+ may have a decreased striatal [^{18}F]-dopa uptake compared to individuals with normoechogenicity (Berg, Becker, et al., 1999; Berg et al., 2002). Furthermore, individuals without any extrapyramidal disorder but with SN+ may perform slightly worse in motor tasks compared to those with normal SN echogenicity (Behnke et al., 2007; Berg, Siefker, Ruprecht-Dörfler, et al., 2001; Rupprecht et al., 2013). Substantia nigra hyperechogenicity has also been found more frequently in groups with an increased PD risk such as individuals with a positive family history for PD (Schweitzer et al., 2007) or asymptomatic PD-related mutation carriers (discussed above). Moreover, SN+ seems to be associated with other features of PD such as hyposmia or reduced uptake in presynaptic dopamine receptors as shown by functional SPECT imaging (Haehner et al., 2007; Sommer et al., 2004). Patients with a rapid eye movement sleep behavior disorder also show an increased rate of SN+ (Iranzo et al., 2010; Shin, Joo, Kim, Dhong, & Cho, 2013; Stockner et al., 2009). Lastly, individuals with SN+ may even have subtle cognitive deficits at group level, resembling the cognitive dysfunction seen in PD (Liepelt et al., 2008; Yilmaz, Behnke, et al., 2016).

The association of SN+ with other early motor or non-motor PD markers implies that SN+ is present not only in already diagnosed PD patients but also in individuals who will be diagnosed with PD in the future. In order to test this assumption, Berg et al. (2011) followed 1847 healthy elderly subjects older than 50 years in a longitudinal study in which participants were examined with TCS at baseline. After 3 years, 11 individuals were diagnosed with PD, 8 of them had SN+ at baseline (2 had normal echogenicity, 1 no bone window). This result corresponded to a relative risk of 17 for individuals older than 50 years with SN+ for the diagnosis of PD within 3 years. Follow-up for 5 years revealed a relative risk of 20. In another study, 43 individuals with rapid eye movement behavioral disorder were followed with dopaminergic transporter scan and TCS (Iranzo et al., 2010). After 2.5 years, eight of these individuals developed an alpha-synucleinopathy, seven of them had SN+ at baseline (one with normoechogenicity had multiple system atrophy).

These studies confirm that SN+ is present in PD patients even before motor impairments allow the clinical diagnosis. However, given that the current level of knowledge advocates no (or no easily detectable) change in the SN size with disease duration, SN+ should be regarded as a stable marker indicating risk for the disease. It should also be noted that the high prevalence (10%) of SN+ in the healthy population indicates that only a subgroup of those who display this marker will develop PD, which explains the low positive predictive value in prospective studies. In other words, the prevalence of PD in the elderly population is much lower (1–2%) than the prevalence of SN+ (10%), and therefore although SN+ can help to detect future PD patients, most of the healthy individuals with SN+ (the remaining 8–9%) will not be diagnosed with PD in their lifetime.

Thus a combination of SN+ with other PD markers such as hyposmia, rapid eye movement sleep behavior disorder or autonomic symptoms like constipation seems reasonable in narrowing the candidate group. Several studies have attempted to develop a frame for combining the different features for diagnosing PD successfully (Busse et al., 2012; Izawa, Miwa, Kajimoto, & Kondo, 2012). A more systematic approach on this issue was implemented by the "Movement Disorders Society Task Force for the Definition of PD". For the assessment of the probability of being in the prodromal phase of PD a Bayesian calculation method was created in which the positive (presence) or negative (absence) likelihood ratios of the risk and prodromal markers were multiplied with the overall probability of PD in the given age group (Berg et al., 2015). In the list of relevant risk markers SN+ is listed with a positive likelihood ratio of 4.7 which is higher than the likelihood ratio of many other risk and prodromal markers.

3.2 Dementia With Lewy Bodies and Atypical Parkinsonism

In contrast to PD, the literature covering TCS findings in other forms of Parkinsonism is limited. This is in part due to the relatively lower prevalence of these diseases and in part due to the typical findings in other structural imaging modalities make TCS less helpful which is different in PD, in which other routine structural imaging modalities do not reveal diagnostic clues.

In this section, we will focus on dementia with Lewy bodies, which according to the Movement Disorders Society Task Force for the Definition of PD is now regarded as part of the spectrum of idiopathic PD (Postuma et al., 2015), and the atypical parkinsonian syndromes: multiple system atrophy, progressive supranuclear palsy and corticobasal degeneration.

3.2.1 Dementia With Lewy Bodies

Three studies have been published on TCS findings in dementia with Lewy bodies so far. In the first study almost all of the 14 patients with dementia with Lewy bodies had bilateral equally large areas of SN+ compared to PD patients in whom there is in general more of side difference in size of

SN+ (Walter et al., 2006). Therefore the authors suggested calculating the asymmetry index by dividing the SN sizes of both sides. SN+ with an asymmetry index of ≤ 1.15 was suggested to indicate dementia with Lewy bodies in patients, in whom this diagnosis was clinically likely. Likewise in another study, none of the 22 patients with dementia with Lewy bodies had normal SN echogenicity confirming the very high prevalence of SN+ in dementia with Lewy bodies (Favaretto et al., 2016). Only in one pilot study a lower prevalence of 53% of SN+ in patients with dementia with Lewy bodies was described (Fernandes et al., 2011) which was still significantly higher compared to patients with Alzheimer's disease or vascular dementia (%11). This finding, however, may indicate that in individuals with dementia with Lewy bodies, in whom dementia is the predominating symptom, SN+ is less prevalent than in those in whom Parkinsonism prevails, which links this specific ultrasound finding to the movement disorder aspects rather than to the cognitive aspects of the disease.

In sum, SN+ seems to be a helpful diagnostic marker in individuals with dementia with Lewy bodies. Rather equally sized SN+ on both sides and a higher prevalence of SN+ in those in whom Parkinsonism is a predominating symptom seem to be characteristic, which needs to be confirmed in further studies.

3.2.2 Multiple System Atrophy

TCS findings in multiple system atrophy have been investigated in six different studies, so far. In these studies, the prevalence values for SN+ were close to those found in healthy controls with a mean prevalence of around 15% (6–25%) (Behnke, Berg, Naumann, & Becker, 2005; Fujita et al., 2016; Okawa et al., 2007; Walter et al., 2003). Other than PD, however, patients with multiple system atrophy seem to have a high rate of LN+ (67–75%) (Fig. 4). Ventricular enlargement was not detected in patients with multiple system atrophy. Summarizing the data from these studies, it can be concluded that multiple system atrophy is in general characterized by the presence of LN+ with normal SN and ventricle sizes. It should also be noted that out of the six studies, only three reported on subtypes of multiple system atrophy (multiple system atrophy-cerebellar type in Li et al. (2017) and multiple system atrophy-parkinsonism in Behnke et al. (2005) and Walter, Dressler, Probst, et al. (2007)) which revealed similar TCS findings in both types. As sample sizes are small and diagnosis during life time is difficult in the early stages, more studies would be helpful, especially on the subtypes.

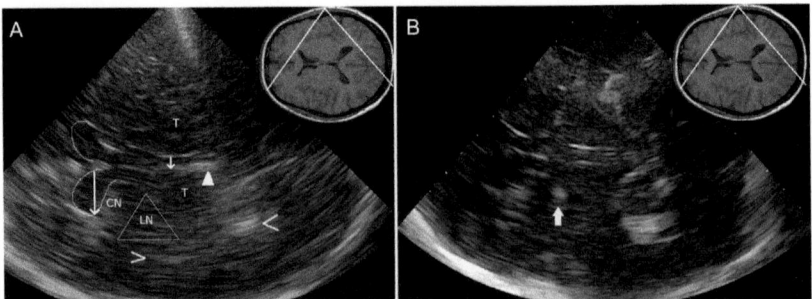

Fig. 4 Nucleus lentiformis hyperechogenicity. (A) The third ventricle plane of a healthy individual. The lentiform nucleus is seen isoechogenic (indicated by the triangle). CN, caudat nucleus; LN, lentiform nuclues; T, thalamus. Calcified pineal gland, (arrow head); ventricle borders, (vertical arrows); insular cortex, (arrow head showing right); calcified choroid plexus, (arrow head showing left). (B) Image of a patient with lentiform nucleus hyperechogenicity. An increased echogencity in the area of lentiform nucleus can been seen (arrow).

In these, other scanning planes may be of interest—e.g., the cerebellar plane, in which enlargement of the fourth ventricle and cerebellar atrophy can be visualized.

3.2.3 Progressive Supranuclear Palsy

In patients with progressive supranuclear palsy, TCS findings of several studies report a high prevalence of LN+ (around 73%, between 45% and 100%) (Behnke et al., 2005; Kostić et al., 2013; Sadowski, Serafin-Król, Szlachta, & Friedman, 2015; Walter, Dressler, Probst, et al., 2007; Walter, Dressler, Wolters, et al., 2007; Walter et al., 2003), which is far more frequent than in PD (around 20%, Mašková et al., 2016). An enlarged third ventricle has also been described frequently in patients with progressive supranuclear palsy (mean 79%, 56–100%) corresponding to the known midbrain atrophy. In contrast to PD, SN+ seems far less frequent in progressive supranuclear palsy with an average rate of 28% (20–36%). Seven of the nine studies published so far reported a SN+ rate of lower than 15% (two of them 0%), which is close to the prevalence of the healthy population (Ebentheuer, Canelo, Trautmann, & Trenkwalder, 2010; Fujita et al., 2016; Kostić et al., 2013; Okawa et al., 2007; Sadowski et al., 2015; Walter et al., 2004, 2003).

It is also noteworthy that very few studies have so far focused on the subtypes of progressive supranuclear palsy. The classical syndrome of progressive supranuclear palsy is Richardson's Syndrome with vertical gaze palsy, postural instability and frequent falls. Another subtype, namely, progressive supranuclear palsy-parkinsonism, has a milder phenotype with more PD-like symptoms and a better response to levodopa. A study examining these different subtypes reported that 14% of the patients with Richardson's Syndrome had SN+ compared to 73% of progressive supranuclear palsy-parkinsonism (Kostić et al., 2013). Nucleus lentiformis hyperechogenicity was found in 67% and 36% of patients with Richardson's Syndrome and of progressive supranuclear palsy-parkinsonism, respectively, with a wider third ventricle width in patients with Richardson's Syndrome. In another study almost all (six out of seven) of the patients with progressive supranuclear palsy-parkinsonism had SN+ and normal third ventricles compared to Richardson's Syndrome patients with an evident third ventricle enlargement and only one SN+ out of 27 (Ebentheuer et al., 2010).

Thus it can be postulated that a typical TCS finding in a patient with Richardson's Syndrome can be described as the presence of LN+ with an expanded third ventricle and a normoechogenic SN. Patients with of progressive supranuclear palsy-parkinsonism have TCS findings similar to PD. However, unlike the findings in PD neither the chronological relationship between the TCS markers and clinical findings nor the histopathological correlates of the TCS findings of progressive supranuclear palsy are investigated. More studies are needed to expand our understanding on this issue.

3.2.4 Corticobasal Degeneration

As far as we are aware of, only two studies have so far been published on TCS findings in patients with corticobasal degeneration. In the first study, seven of the eight patients with corticobasal degeneration had SN+ (Walter et al., 2004). Moreover five of the included six patients (two had no bone window on the third ventricle level) had hyperechogenic areas in the LN and caudate nucleus. No ventricle expansion was detected. In the second more recent study 67% of the 13 corticobasal degeneration patients had SN+ (Sadowski et al., 2015). However, they did not find LN+ in any of the patients contrasting the previous report. The results of these studies indicate a moderate to high frequency of SN+ in patients with corticobasal degeneration. Nonetheless, more studies are needed to determine the sonographic characteristics of corticobasal degeneration, which is clinically as well as probably etiologically quite heterogeneous.

3.3 Secondary Parkinsonism

Studies about secondary causes of Parkinsonism or other movement disorders are rather limited. In this part, we present TCS data of the most frequently. causes of secondary Parkinsonism.

3.3.1 Drug Induced Parkinsonism

Drug induced Parkinsonism is the second most common cause of secondary Parkinsonism and has been a subject of several TCS studies. Neuroleptics are the most common culprits causing drug induced parkinsonism. In a first study in this group 93 psychiatric patients were examined who were treated with neuroleptics (Berg, Jabs, Merschdorf, Beckmann, & Becker, 2001). They reported that the size of the SN+ is significantly larger in patients with neuroleptic induced parkinsonism compared to patients without. Moreover they also conducted a prospective study examining 11 patients with a diagnosis of acute psychosis, in whom a neuroleptic therapy was about to be started. Interestingly, baseline echogenicity correlated with the severity of extrapyramidal symptoms developed over time under neuroleptic therapy. This prospective study indicates that a vulnerability of the nigrostriatal system, visualized by TCS, may become clinically apparent with extrapyramidal symptoms, when the system is additionally blocked (in this case dopamine receptor blockage by neuroleptics). Another study reported an association between SN size and the existence and severity of extrapyramidal symptoms in 100 patients on neuroleptics (Jabs, Bartsch, & Pfuhlmann, 2003). Both of these studies indicate that patients with an increased SN size may be more susceptible to extrapyramidal side effects of neuroleptics. Apart from that, the TCS profile of patients with neuroleptic induced parkinsonism differs from PD patients. At group level, SN size is not as large in drug induced parkinsonism patients as in idiopathic PD (Olivares Romero, Arjona Padillo, Barrero Hernández, Martín González, & Gil Extremera, 2013).

In order to explain the association between the SN size and neuroleptic induced parkinsonism better, some studies performed subgroup analyses between patients who remained symptom-free (regarding extrapyramidal symptoms) after discontinuation of the drug and patients in whom parkinsonism did not resolve despite drug withdrawal. The latter group is defined as patients in whom the symptoms are triggered with a neuroleptic drug (subclinical drug exacerbated parkinsonism). With this regard 69 patients with drug induced parkinsonism were compared to PD patients and healthy controls. As a whole group, patients with drug induced parkinsonism did not have an increased SN echogenicity (Oh, Kwon, Kim, Park, & Berg, 2018). However, subgroup analyses revealed that patients with a complete recovery after discontinuation of the drug showed similar TCS features as healthy controls. On the contrary, patients with subclinical drug exacerbated parkinsonism had TCS features resembling PD patients. This outcome was also confirmed in a community-based prospective study cohort with 574 individuals (Mahlknecht et al., 2012). After a 3-year follow-up eight of them were diagnosed with drug induced parkinsonism, out of whom seven had normoechogenicity of the SN. The only patient with a baseline SN+ was the one in whom parkinsonism persisted in spite of drug discontinuation.

These results indicate that presence of SN+ in individuals with drug induced parkinsonism is not common which is in accordance with functional imaging studies. However, individuals with SN+ may be more susceptible to the extrapyramidal side-effects of neuroleptics, a finding which corroborates the hypothesis that SN+ constitutes a sign of vulnerability of the nigrostriatal system. In that sense TCS could be used a marker to predict the prognosis of drug induced parkinsonism in patients subjected to neuroleptic therapy.

3.3.2 Vascular Parkinsonism

Our knowledge about the value of TCS for the diagnosis of vascular parkinsonism is very limited. In one of the few studies regarding this issue 30 patients with vascular parkinsonism were examined using both B-mode and Doppler modalities of ultrasound. A slightly increased rate of SN+ (20%) as well as higher flow velocities in intracranial arteries compared to healthy controls (Tsai et al., 2007) was found. However, most of the vascular parkinsonism patients showed normal midbrain and basal ganglia echogenicity in another study. The authors commented on the possibility of overlapping vascular parkinsonism and PD (Venegas-Francke, 2010). The results suggest that combining B-mode with transcranial Doppler ultrasound may increase the diagnosis of vascular parkinsonism.

3.3.3 Hydrocephalus

As the ventricular system can be easily depicted using TCS, size of the third ventricle and the anterior horns of the lateral ventricle can be measured. Although no comparative study has been conducted, it can be said that in a patient with Parkinsonism, excessively increased ventricle sizes with an absence of SN+ may indicate a normal pressure hydrocephalus rather than a PD (Fig. 5).

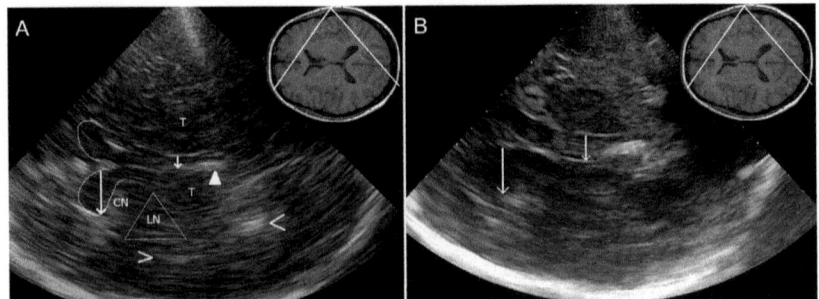

Fig. 5 Detection of enlarged ventricles on TCS. The cut-off values for ventricle sizes differ with regard to age. The values for the third ventricle: <7/10 mm for </>60 years, and for the frontal horns <17/20 mm for </>60 years. (A) The third ventricle plane of a healthy individual. Ventricle sizes are normal (vertical arrows). CN, caudat nucleus; LN, lentiform nuclues; T, thalamus. The calcified pineal gland, (arrow head); insular cortex, (arrow head showing right); calcified choroid plexus, (arrow head showing left). (B) An image of a patient with increased ventricle sizes (arrows).

3.4 Other Movement Disorders

3.4.1 Essential Tremor

Essential tremor (ET) is the most frequent movement disorder and one of the most important diseases for the differential diagnosis of PD. Therefore the critical question regarding the diagnosis of essential tremor on TCS is the presence of SN+. Sonographic findings in essential tremor have been studied extensively.

In all TCS studies investigating patients with essential tremor low values for the prevalence SN+ is documented ranging between 0% and 18% (Alonso-Cánovas et al., 2014; Bártová et al., 2014; Budisic et al., 2009; Chitsaz et al., 2013; Doepp et al., 2008; Kim et al., 2012; Luo, Zhang, Sheng, Fang, & Liu, 2012; Okawa et al., 2007; Richter et al., 2017; Stockner et al., 2007). Some speculated that since healthy individuals with SN+ have an increased rate of future PD, essential tremor patients with SN+ may also have an increased risk for future PD compared to normoechogenic patients (Budisic et al., 2009). The highest prevalence of SN+ for essential tremor has been reported by one group who reported values of 31% and 32% in two studies (Laučkaite et al., 2014, 2012). They also reported a LN+ prevalence of 17.2%. The reason for this deviation from other studies may be related to the measurement technique.

The SN size and prodromal PD markers in patients with essential tremor may be correlated which supports the hypothesis of SN+ being a risk marker for future PD (Kim et al., 2012). This assumption was later tested in a prospective study with 54 of 70 patients with essential tremor (Sprenger et al., 2016). The authors could show that seven of the nine patients who were diagnosed with PD after 6 years of follow-up had SN+ at baseline TCS, which indicates that essential tremor patients with SN+ have a high likelihood to develop PD in the future.

3.4.2 Idiopathic Dystonia

The investigation of TCS in dystonia started with 86 patients with several types of dystonias and it was found that 75% of patients with cervical and 83% of those with upper limb dystonia and one-third of the patients with facial dystonia had LN+ on the contralateral side of the symptoms, a finding which was less frequent in controls (12%) or in patients with secondary dystonia (0%) (Naumann, Becker, Toyka, Supprian, & Reiners, 1996). Another small scale study with early with cervical dystonia supported these results with a rate of 70% LN+ which was contralateral to the side of the torticollis (Becker et al., 1997). In these patients, structural magnetic resonance imaging (MRI) revealed no abnormalities; however, a tendency for reduced tracer uptake in the region of hyperechogenicity was shown on functional neuroimaging. Later, it was reported that 57% of 84 patients with dystonia (66% cervical dystonia, 23% Blepharospasm) have LN+ compared to the surprisingly high prevalence (50%) of LN+ in controls (Hagenah et al., 2011). Percentages of SN, *caudate* or *thalamic nuclei* hyperechogenicities were also similar. Here, differences in the ultrasound technique applied may account for the differences to all other studies published in this field. In another study, 12 of the 14 patients with spasmodic dysphonia displayed LN+, which was reported only in 1 of the 12 controls (Walter, Blitzer, Benecke, Grossmann, & Dressler, 2014). *Substantia nigra* sizes were not significantly different. A recent study examined 80 patients with focal dystonia (cervical dystonia, $n = 30$; blepharospasm, $n = 30$; oromandibular dystonia, $n = 10$) and found that LN+ was present in 51% of the patients with dystonia compared to 12% of controls (Zhang et al., 2016). Similarly, SN sizes were not different. Detailed subgrouping revealed a rate of LN+ in 73% of patients with cervical dystonia, in 33% of blepharospasm and in 40% of oromandibular dystonia. In another interesting study SN+ was found in 63% of the dopa-responsive dystonia patients, a value closer to PD patients (87%) than other dystonia types (20%). Another study confirmed this finding with LN+ in 54.5% of dopa-responsive dystonia patients compared with 7.3%, 43.3%, 4.5% of patients with PD, focal dystonia and controls, respectively (Svetel et al., 2017).

These studies show that TCS results vary to a certain extent in the field of dystonia. This is probably due to the heterogeneous clinical profile of the study patients and the possibly different underlying pathomechanisms. First postmortem studies indicate an elevation of the tissue copper content of the lentiform nucleus in dystonia (Becker et al., 1999) and changes in copper metabolism (Berg et al., 2000). These findings need to be substantiated with further studies. So far, studies indicate an increased frequency of LN+, especially in cervical dystonia patients up to 75%. This effect seems to be weaker in blepharospasm and oromandibular dystonia. It can also be said that SN+ does not differentiate between patients with dystonia and controls with an exception of dopa-responsive dystonia, in which a higher rate of SN+ has been reported.

3.4.3 Huntington's Disease

The first report on TCS findings in 45 patients with Huntington's disease described hyperechogenicities in at least one brain region in 40% of the patients compared to 12% of the controls (Postert, Lack, Kuhn, & Jergas, 1999). At least 6% of the patients with Huntington's disease had LN+, 13% had hyperechogenicity of the *caudate nucleus* and 27% had SN+. Also, a correlation between clinical severity or number of CAG repeats and echogenic findings was described. In a later report 41% of the patients had hyperechogenicity of the SN, 21% of the *caudate nucleus* and 17% of the LN (Krogias et al., 2011). The width of the third ventricle was significantly larger in Huntington's disease patients than in controls. More significant differences between Huntington's disease patients and controls were reported in another study (Lambeck et al., 2015) which examined a subgroup of Huntington's disease patients with hypokinesia. As opposed to the previous studies they found a strikingly higher prevalence of SN+ (93.3%), LN+ (53.3%) and *caudatus* hyperechogenicity (80%) compared to controls. In contrast, in a small group with hyperkinetic Huntington's disease, SN+ echogenicity was found only in few individuals. As increased tissue iron content is known to be part of the neurodegenerative process in the *caudate nucleus* in Huntington's disease, it is well perceivable, that hyperechogenicity of this nucleus can be detected. Further TCS studies focusing more on this brain structure are needed to evaluate, whether the atrophy of the *caudate nucleus* may be helpful in the diagnosis of Huntington's disease.

3.4.4 Accumulation Syndromes

Increased tissue iron, copper and manganese content as well as increased calcium content seem to lead to alterations of the brain structure which reflect ultrasound in a strong way resulting in hyperechogenicity. For instance, accumulation of calcium in Fahr disease (Toscano et al., 2011) or increased tissue manganese content caused by substance abuse or welding-related can be visualized in the *basal ganglia* using TCS (Skowronska, Dziezyc, & Członkowska, 2014; Walter, Dressler, Lindemann, Slachevsky, & Miranda, 2008). These increased tissue accumulations primarily manifest themselves as LN+. Here, we focus on "Neurodegeneration with Brain Iron Accumulation" and Wilson's disease.

3.4.4.1 Neurodegeneration With Brain Iron Accumulation

The first findings of "neurodegeneration with brain iron accumulation" in TCS were reported as a case report with LN+ and normal SN echogenicity (Brüggemann, Wuerfel, et al., 2011). Further, a study with 7 patients with genetically proven "neurodegeneration with brain iron accumulation" (including pediatric patients) and 13 controls showed that the patient group had a significantly increased size of SN echogenicity (Liman, Wellmer, Rostasy, Bähr, & Kermer, 2012). Interestingly, they failed to show any difference in the *basal ganglia* using TCS. On the contrary, bilateral LN+ and SN+ were detected in all of five patients with "panthotenat kinase-associated neurodegeneration" in accordance with low signals of the *globus pallidus* on T2-weighted MRI (Kostić et al., 2012). Also, 12 of 13 patients with "mitochondrial membrane protein associated neurodegeneration" were found to display LN+ and none SN+ (Skowronska, Walter, Kmiec, & Czlonkowska, 2013). The detected LN+ in these studies seems to correspond to the "Eye of the tiger" sign in T2-weighted MRI.

3.4.4.2 Wilson's Disease

Wilson's disease is characterized by the accumulation of copper in tissues like liver, cornea and brain. Consistent with other heavy metals, the main site of accumulation in the brain is the *basal ganglia*. In the first TCS study it has been shown that all (18) patients with neurological symptoms (neuro-Wilson) display LN+, 89% of them bilaterally (Walter et al., 2005). Additionally, the size of the area of LN hyperechogenicity was shown to correlate with disease severity. *Substantia nigra* hyperechogenicity was found in 25% of the patients. Interestingly, the *basal ganglia* abnormality depicted as LN+ was confirmed only in 12 of the 19 patients in MRI, indicating, that TCS may be more sensitive in detecting increased tissue concentrations of copper. In another study with 54 patients with Wilson's disease, 42% of patients with neuro-Wilson showed SN+ compared to patients with hepatic-Wilson (7%) and controls (8%) (Svetel et al., 2012). Hyperechogenicity of the LN was detected in 82% patients with neuro-Wilson and in 25% with hepatic-Wilson compared to 7% of controls. Similar to the earlier study, most of the LN+ was seen bilaterally. However, the echogenicity of SN was found to be similar in patients with Wilson's disease as in controls in another study (Mašková et al., 2016). The prevalence of LN+ was as usual markedly higher in Wilson's disease patients compared to PD and controls. No correlation was detected between echogenicity and disease duration or severity. Moreover, six Wilson's disease patients who had no NL abnormality on MRI showed increased LN echogenicity. Correlation between the LN+ and disease features was also tested in a 7-Tesla MRI study (Dusek et al., 2018). The authors reported that LN+ is correlated with neither the disease features, nor with changing R2 values of the basal ganglia, supporting the finding of Walter et al. (2005) that patients with neuro-Wilson may display a normoechogenic SN but LN+, which may be detected even before abnormalities are seen on MRI. This issue needs to be confirmed in prospective studies.

3.5 Differential Diagnosis of Parkinsonism

The diagnosis of most movement disorders depends on clinical examination given that no laboratory or imaging method is sensitive and specific enough for an undisputed diagnosis. Besides, clinical diagnosis may be challenging in patients with discrete or vague symptoms since there is a considerable overlap between the phenotype and presentation of movement disorders. Therefore it is not uncommon to follow patients for months or even years before the clinical picture becomes clear and the final diagnosis can be made.

It has been numerously shown that TCS can detect an alteration of the SN which is highly predictive for PD in individuals with respective movement abnormalities. A recent meta-analysis of 31 studies including 1926 PD patients and 2460 healthy controls revealed that TCS can differentiate PD patients from healthy controls with a sensitivity and specificity of 83% and 87%, respectively (Li, He, Liu, & Chen, 2016). Sensitivity and specificity of TCS in relation to dopamine transporter SPECT data, which is currently the best diagnostic tool to identify presynaptic neurodegeneration as seen in PD and other neurodegenerative forms of Parkinsonism, yielded values of 86% and 93% for the detection of PD (Doepp et al., 2008). However, the main clinical challenge is the differentiation of PD not from healthy individuals but from other movement disorders, which may mimic PD. One is the differentiation from essential tremor. Here, several studies reported that TCS can differentiate PD from essential tremor patients with sensitivity and specificity values higher than 90% and 80%, respectively (Budisic et al., 2009; Doepp et al., 2008). Yet, a lower value for specificity of (55.2%) was also reported. Collecting the data of all 15 studies with 642 essential tremor patients revealed that, the presence of SN+ differentiates PD from essential tremor with 78% (69–85%) sensitivity and 85% (77–91%) specificity (Shafieesabet et al., 2017). The odds of having PD versus essential tremor in the presence of SN+ increases approximately five times, while the odds decreased by 74% with an absence of SN+. As previously discussed, patients with essential tremor diagnosis and SN+ also have an increased risk for future PD. As essential tremor patients are more likely to develop PD than controls, prevalence of SN+ is higher in this group of individuals compared to controls accounting for the relatively low sensitivity.

Another crucial point is the differentiation of PD from atypical parkinsonism. Taking the sonographic differences into account, Gaenslen et al. (2008) reported that baseline TCS findings can differentiate PD patients from atypical parkinsonism with 91% sensitivity and 82% specificity (and with a positive predictive value of 93%) in a prospective study with 60 patients with unclear symptoms. According to another prospective study of 9 months duration, the sensitivity and specificity for detecting atypical parkinsonism from PD were 100% and 85% for TCS, respectively (Hellwig et al., 2014). A successful separation was also reported from other studies (Behnke et al., 2005; Okawa et al., 2007; Walter et al., 2003). On the contrary, Fujita et al. (2016) found a relatively lower sensitivity (50%) with a high specificity (93.8%) for the role of TCS among early-stage patients with PD compared to atypical parkinsonism. The data of all these studies were gathered in an extensive review and a meta-analysis which included a total of 71 TCS studies with a number of 5730 participants. In this study, Shafieesabet et al. (2017) have found that the pooled prevalence rate of SN+ was 28% (20–36%) in atypical parkinsonism. They further reported that, taking only the presence of SN+ into account, TCS can discriminate PD from atypical parkinsonism with 75% (60–86%) sensitivity and 69% (55–81%) specificity. Furthermore they reported that SN+

increases the odds of having PD versus atypical parkinsonism almost 2.5-fold whereas the odds are decreased by 64% with a normoechogenic SN.

With regard to the within-comparison of atypical parkinsonism, the combination of SN normoechogenicity and LN+ separates progressive supranuclear palsy from corticobasal degeneration with 100% sensitivity and positive predictive value (Sadowski et al., 2015). In another study, the presence of SN+ and a normal third ventricle (not LN+) indicated corticobasal degeneration with a sensitivity of 100% and a specificity of 83% (Walter et al., 2004). Additionally, hyperechogenicities of SN and LN differentiated early-onset PD patients from Wilson's disease with a sensitivity of 93.8%, 95.5% and specificity of 90.9%, 93.8%, respectively (Mašková et al., 2016).

3.6 Transcranial B-Mode Sonography and Deep Brain Stimulation

Besides its value in diagnosis and differential diagnosis, TCS has gained increasing attention in the localization of the electrodes in deep brain stimulation. As the position of the inserted electrodes is one of the most important factors for success in deep brain stimulation, a postoperative imaging is always part of the protocol. The detection of electrode positioning during or after surgery in patients with deep brain stimulation has been investigated in several studies. In a study with eight PD patients who underwent *subthalamic nucleus* deep brain stimulation, the position of the electrodes could be detected using TCS by taking the anatomical proximity of SN and *subthalamic nucleus* into account (Moringlane, Fuss, & Becker, 2005). Interestingly, perioperative electrode assessment seems also feasible in patients with *globus pallidus interna* deep brain stimulation (Walter et al., 2009). The authors argue that even arterial structures in the vicinity of the electrodes can be visualized using the duplex mode of the ultrasound, which can help refining the electrode placement and avoid hemorrhages. Further, the same group followed 34 patients with *ventral intermedius, globus pallidus interna* or *subthalamic nucleus* deep brain stimulation who were assessed postoperatively with TCS. In two patients, displacement of the electrodes was detected by TCS which was confirmed by conventional imaging methods (Walter et al., 2011). Moreover, it has been shown that patients with an optimal lead localization (<3 mm to the target) detected by the TCS showed a better motor improvement in comparison to patients with a suboptimal lead localization (3–5 mm) after a follow-up period of 12 months.

Detection of electrode localization using TCS has been further investigated with another TCS method, a recently developed image fusion technology. In this method, the system matches the previously obtained MRI images of the patient with ultrasound images by tracking the position and the orientation of the ultrasound transducer, which enables the examiner to visualize both imaging modalities on top of each other in real-time during the ultrasound examination (Fig. 6). Using the MRI-TCS fusion, Walter et al. (2016) integrated preoperative MRI images with postoperative ultrasound in 15 patients with deep brain stimulation (6 *subthalamic nucleus*, 5 *globus pallidus interna*, 4 *ventral intermedius*) and detected the electrode displacement of successfully. These findings indicate that perioperative or postoperative lead localization using TCS is possible in patients that undergo deep brain stimulation.

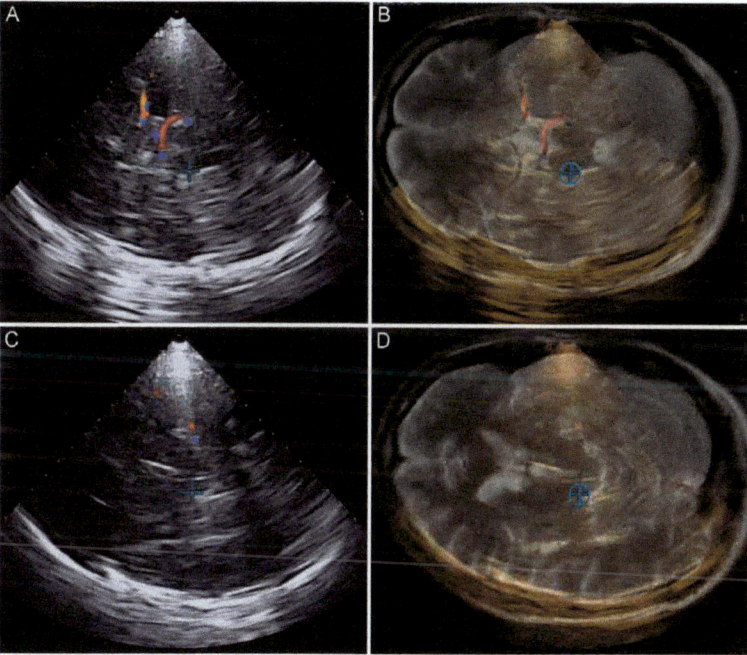

Fig. 6 TCS-MRI fusion technique. For application of this method, the previously obtained MRI images of a patient are uploaded to the system before the ultrasound examination. The system can detect the position of the ultrasound transducer in 3D space and performs a real-time matching with the MRI images. Transcranial sonography at the (A) mesencephalic and (C) the third ventricle planes. (B and D) With the help of TCS-MRI fusion technique, the exact corresponding MRI image can been seen during the ultrasound examination.

4. LIMITATIONS OF TRANSCRANIAL B-MODE SONOGRAPHY

The main limitation of the TCS is the lack of bone window in about 10% of the caucasian population. This rate seems to be increasing with advancing age and may reach to up to 25% especially in elderly women. It has been speculated that osteoporosis may play a role in the deterioration of scanning conditions, possibly by changes in the bone structure which may lead to an increased dispersion of the ultrasound beam. Additionally, reports show that in the Asian population a higher rate of insufficient bone window is found. Unfortunately this cannot be known before the patient lies down to have the examination. Moreover, TCS examination may be difficult in individuals with small or semi-sufficient temporal bone window. In those patients, the success of the examination is highly dependent on the examiner since the detection and measurement of the structures necessitates a high expertise. This limitation can be overcome with an increased number of skilled examiners.

5. CONCLUSION

Transcranial B-mode sonography is a time- and cost-efficient, practical and a patient-friendly imaging tool which can be used safely as a supplementary method in the clinical routine. It has already been recommended by the "European Federation of Neurological Societies and Movement Disorders Society" for the early detection of PD, differential diagnosis of PD from atypical parkinsonism and secondary syndromes, as well as the detection of individuals at risk for future PD (Berardelli et al., 2013). The inadequacy of the bone window and the necessity of examiner expertise may limit the value of TCS in some patients.

REFERENCES

Alonso-Cánovas, A., López-Sendón, J. L., Buisán, J., DeFelipe-Mimbrera, A., Guillán, M., García-Barragán, N., et al. (2014). Sonography for diagnosis of parkinson disease—From theory to practice: A study on 300 participants. *Journal of Ultrasound in Medicine, 33*(12), 2069–2074. https://doi.org/10.7863/ultra.33.12.2069.
Barrett, M. J., Hagenah, J., Dhawan, V., Peng, S., Stanley, K., Raymond, D., et al. (2013). Transcranial sonography and functional imaging in glucocerebrosidase mutation Parkinson disease. *Parkinsonism & Related Disorders, 19*(2), 186–191. https://doi.org/10.1016/j.parkreldis.2012.09.007.

Bártová, P., Kraft, O., Bernátek, J., Havel, M., Ressner, P., Langová, K., et al. (2014). Transcranial sonography and 123I-FP-CIT single photon emission computed tomography in movement disorders. *Ultrasound in Medicine and Biology, 40*(10), 2365–2371. https://doi.org/10.1016/j.ultrasmedbio.2014.05.014.

Becker, G., Berg, D., Rausch, W. D., Lange, H. K., Riederer, P., & Reiners, K. (1999). Increased tissue copper and manganese content in the lentiform nucleus in primary adult-onset dystonia. *Annals of Neurology, 46*(2), 260–263. Retrieved from http://www.ncbi.nlm.nih.gov/pubmed/10443894.

Becker, G., Naumann, M., Scheubeck, M., Hofmann, E., Deimling, M., Lindner, A., et al. (1997). Comparison of transcranial sonography, magnetic resonance imaging, and single photon emission computed tomography findings in idiopathic spasmodic torticollis. *Movement Disorders, 12*(1), 79–88. https://doi.org/10.1002/mds.870120114.

Becker, G., Seufert, J., Bogdahn, U., Reichmann, H., & Reiners, K. (1995). Degeneration of substantia nigra in chronic Parkinson's disease visualized by transcranial color-coded real-time sonography. *Neurology, 45*(1), 182–184.

Behnke, S., Berg, D., Naumann, M., & Becker, G. (2005). Differentiation of Parkinson's disease and atypical parkinsonian syndromes by transcranial ultrasound. *Journal of Neurology, Neurosurgery, and Psychiatry, 76*(3), 423–425. https://doi.org/10.1136/jnnp.2004.049221.

Behnke, S., Double, K. L., Duma, S., Broe, G. A., Guenther, V., Becker, G., et al. (2007). Substantia nigra echomorphology in the healthy very old: Correlation with motor slowing. *NeuroImage, 34*(3), 1054–1059. https://doi.org/10.1016/j.neuroimage.2006.10.010.

Berardelli, A., Wenning, G. K., Antonini, A., Berg, D., Bloem, B. R., Bonifati, V., et al. (2013). EFNS/MDS-ES recommendations for the diagnosis of Parkinson's disease. *European Journal of Neurology, 20*(1), 16–34. https://doi.org/10.1111/ene.12022.

Berg, D., & Becker, G. (2002). Perspectives of B-mode transcranial ultrasound. *NeuroImage, 15*(3), 463–473. https://doi.org/10.1006/nimg.2001.1014.

Berg, D., Becker, G., Zeiler, B., Tucha, O., Hofmann, E., Preier, M., et al. (1999). Vulnerability of the nigrostriatal system as detected by transcranial ultrasound. *Neurology, 53*(5), 1026–1031. Retrieved from http://www.ncbi.nlm.nih.gov/pubmed/10496262.

Berg, D., Godau, J., Riederer, P., Gerlach, M., & Arzberger, T. (2010). Microglia activation is related to substantia nigra echogenicity. *Journal of Neural Transmission, 117*(11), 1287–1292. https://doi.org/10.1007/s00702-010-0504-6.

Berg, D., Grote, C., Rausch, W. D., Mäurer, M., Wesemann, W., Riederer, P., et al. (1999). Iron accumulation in the substantia nigra in rats visualized by ultrasound. *Ultrasound in Medicine & Biology, 25*(6), 901–904. Retrieved from http://www.ncbi.nlm.nih.gov/pubmed/10461717.

Berg, D., Jabs, B., Merschdorf, U., Beckmann, H., & Becker, G. (2001). Echogenicity of substantia nigra determined by transcranial ultrasound correlates with severity of parkinsonian symptoms induced by neuroleptic therapy. *Biological Psychiatry, 50*(6), 463–467. Retrieved from http://www.ncbi.nlm.nih.gov/pubmed/11566164.

Berg, D., Postuma, R. B., Adler, C. H., Bloem, B. R., Chan, P., Dubois, B., et al. (2015). MDS research criteria for prodromal Parkinson's disease. *Movement Disorders, 30*(12), 1600–1611. https://doi.org/10.1002/mds.26431.

Berg, D., Roggendorf, W., Schröder, U., Klein, R., Tatschner, T., Benz, P., et al. (2002). Echogenicity of the substantia nigra: Association with increased iron content and marker for susceptibility to nigrostriatal injury. *Archives of Neurology, 59*(6), 999–1005. Retrieved from http://www.ncbi.nlm.nih.gov/pubmed/12056937.

Berg, D., Seppi, K., Behnke, S., Liepelt, I., Schweitzer, K., Stockner, H., et al. (2011). Enlarged substantia nigra hyperechogenicity and risk for Parkinson disease. *Archives of Neurology, 68*(7), 932. https://doi.org/10.1001/archneurol.2011.141.

Berg, D., Siefker, C., & Becker, G. (2001). Echogenicity of the substantia nigra in Parkinson's disease and its relation to clinical findings. *Journal of Neurology, 248*(8), 684–689. Retrieved from http://www.ncbi.nlm.nih.gov/pubmed/11569897.

Berg, D., Siefker, C., Ruprecht-Dörfler, P., & Becker, G. (2001). Relationship of substantia nigra echogenicity and motor function in elderly subjects. *Neurology, 56*(1), 13–17. Retrieved from http://www.ncbi.nlm.nih.gov/pubmed/11148229.

Berg, D., Weishaupt, A., Francis, M. J., Miura, N., Yang, X. L., Goodyer, I. D., et al. (2000). Changes of copper-transporting proteins and ceruloplasmin in the lentiform nuclei in primary adult-onset dystonia. *Annals of Neurology, 47*(6), 827–830. Retrieved from http://www.ncbi.nlm.nih.gov/pubmed/10852553.

Brockmann, K., Gröger, A., Di Santo, A., Liepelt, I., Schulte, C., Klose, U., et al. (2011). Clinical and brain imaging characteristics in leucine-rich repeat kinase 2-associated PD and asymptomatic mutation carriers. *Movement Disorders, 26*(13), 2335–2342. https://doi.org/10.1002/mds.23991.

Brockmann, K., Srulijes, K., Hauser, A.-K., Schulte, C., Csoti, I., Gasser, T., et al. (2011). GBA-associated PD presents with nonmotor characteristics. *Neurology, 77*(3), 276–280. https://doi.org/10.1212/WNL.0b013e318225ab77.

Brüggemann, N., Hagenah, J., Stanley, K., Klein, C., Wang, C., Raymond, D., et al. (2011). Substantia nigra hyperechogenicity with LRRK2 G2019S mutations. *Movement Disorders, 26*(5), 885–888. https://doi.org/10.1002/mds.23644.

Brüggemann, N., Wuerfel, J., Petersen, D., Klein, C., Hagenah, J., & Schneider, S. A. (2011). Idiopathic NBIA-clinical spectrum and transcranial sonography findings. *European Journal of Neurology, 18*(6), 2010–2011. https://doi.org/10.1111/j.1468-1331.2010.03298.x.

Budisic, M., Trkanjec, Z., Bosnjak, J., Lovrencic-Huzjan, A., Vukovic, V., & Demarin, V. (2009). Distinguishing Parkinson's disease and essential tremor with transcranial sonography. *Acta Neurologica Scandinavica, 119*(1), 17–21. https://doi.org/10.1111/j.1600-0404.2008.01056.x.

Busse, K., Heilmann, R., Kleinschmidt, S., Abu-Mugheisib, M., Höppner, J., Wunderlich, C., et al. (2012). Value of combined midbrain sonography, olfactory and motor function assessment in the differential diagnosis of early Parkinson's disease. *Journal of Neurology, Neurosurgery & Psychiatry, 83*(4), 441–447. https://doi.org/10.1136/jnnp-2011-301719.

Chitsaz, A., Mehrbod, N., Saadatnia, M., Fereidan-Esfahani, M., Akbari, M., & Abtahi, S.-H. (2013). Transcranial sonography on Parkinson's disease and essential tremor. *Journal of Research in Medical Sciences: The Official Journal of Isfahan University of Medical Sciences, 18*(Suppl. 1), S28–S31.

Doepp, F., Plotkin, M., Siegel, L., Kivi, A., Gruber, D., Lobsien, E., et al. (2008). Brain parenchyma sonography and 123I-FP-CIT SPECT in Parkinson's disease and essential tremor. *Movement Disorders, 23*(3), 405–410. https://doi.org/10.1002/mds.21861.

Drepper, C., Geißler, J., Pastura, G., Yilmaz, R., Berg, D., Romanos, M., et al. (2017). Transcranial sonography in psychiatry as a potential tool in diagnosis and research. *World Journal of Biological Psychiatry, 30*, 1–13. https://doi.org/10.1080/15622975.2017.1386325.

Dusek, P., Skoloudik, D., Maskova, J., Huelnhagen, T., Bruha, R., Zahorakova, D., et al. (2018). Brain iron accumulation in Wilson's disease: A longitudinal imaging case study during anticopper treatment using 7.0T MRI and transcranial sonography. *Journal of Magnetic Resonance Imaging, 47*(1), 282–285. https://doi.org/10.1002/jmri.25702.

Ebentheuer, J., Canelo, M., Trautmann, E., & Trenkwalder, C. (2010). Substantia nigra echogenicity in progressive supranuclear palsy. *Movement Disorders, 25*(6), 773–777. https://doi.org/10.1002/mds.22981.

Favaretto, S., Walter, U., Baracchini, C., Pompanin, S., Bussè, C., Zorzi, G., et al. (2016). Accuracy of transcranial brain parenchyma sonography in the diagnosis of dementia with Lewy bodies. *European Journal of Neurology, 23*(8), 1322–1328. https://doi.org/10.1111/ene.13028.

Fernandes, R. C. L., Rosso, A. L. Z., Vincent, M. B., Silva, K. S., Bonan, C., Araújo, N. C., et al. (2011). Transcranial sonography as a diagnostic tool for Parkinson's disease: A pilot study in the city of Rio de Janeiro, Brazil. *Arquivos de Neuro-Psiquiatria, 69*(6), 892–895. Retrieved from http://www.ncbi.nlm.nih.gov/pubmed/22297874.

Friedman, A., Arosio, P., Finazzi, D., Koziorowski, D., & Galazka-Friedman, J. (2011). Ferritin as an important player in neurodegeneration. *Parkinsonism & Related Disorders, 17*(6), 423–430. https://doi.org/10.1016/j.parkreldis.2011.03.016.

Fujita, H., Suzuki, K., Numao, A., Watanabe, Y., Uchiyama, T., Miyamoto, T., et al. (2016). Usefulness of cardiac MIBG scintigraphy, olfactory testing and substantia nigra hyperechogenicity as additional diagnostic markers for distinguishing between Parkinson's disease and atypical parkinsonian syndromes. *PLoS One, 11*(11), e0165869. https://doi.org/10.1371/journal.pone.0165869.

Gaenslen, A., Unmuth, B., Godau, J., Liepelt, I., Di Santo, A., Schweitzer, K. J., et al. (2008). The specificity and sensitivity of transcranial ultrasound in the differential diagnosis of Parkinson's disease: A prospective blinded study. *Lancet Neurology, 7*(5), 417–424. https://doi.org/10.1016/S1474-4422(08)70067-X.

Godau, J., Schweitzer, K. J., Liepelt, I., Gerloff, C., & Berg, D. (2007). Substantia nigra hypoechogenicity: Definition and findings in restless legs syndrome. *Movement Disorders: Official Journal of the Movement Disorder Society, 22*(2), 187–192. https://doi.org/10.1002/mds.21230.

Haehner, A., Hummel, T., Hummel, C., Sommer, U., Junghanns, S., & Reichmann, H. (2007). Olfactory loss may be a first sign of idiopathic Parkinson's disease. *Movement Disorders, 22*(6), 839–842. https://doi.org/10.1002/mds.21413.

Hagenah, J., König, I. R., Kötter, C., Seidel, G., Klein, C., & Brüggemann, N. (2011). Basal ganglia hyperechogenicity does not distinguish between patients with primary dystonia and healthy individuals. *Journal of Neurology, 258*(4), 590–595. https://doi.org/10.1007/s00415-010-5795-x.

Hellwig, S., Reinhard, M., Amtage, F., Guschlbauer, B., Buchert, R., Tüscher, O., et al. (2014). Transcranial sonography and [18F]fluorodeoxyglucose positron emission tomography for the differential diagnosis of parkinsonism: A head-to-head comparison. *European Journal of Neurology, 21*(6), 860–866. https://doi.org/10.1111/ene.12394.

Iranzo, A., Lomeña, F., Stockner, H., Valldeoriola, F., Vilaseca, I., Salamero, M., et al. (2010). Decreased striatal dopamine transporter uptake and substantia nigra hyperechogenicity as risk markers of synucleinopathy in patients with idiopathic rapid-eye-movement sleep behaviour disorder: A prospective study. *Lancet Neurology, 9*(11), 1070–1077. https://doi.org/10.1016/S1474-4422(10)70216-7.

Izawa, M. O., Miwa, H., Kajimoto, Y., & Kondo, T. (2012). Combination of transcranial sonography, olfactory testing, and MIBG myocardial scintigraphy as a diagnostic indicator for Parkinson's disease. *European Journal of Neurology, 19*(3), 411–416. https://doi.org/10.1111/j.1468-1331.2011.03533.x.

Jabs, B. E., Bartsch, A. J., & Pfuhlmann, B. (2003). Susceptibility to neuroleptic-induced parkinsonism—Age and increased substantia nigra echogenicity as putative risk factors. *European Psychiatry: The Journal of the Association of European Psychiatrists, 18*(4), 177–181. Retrieved from http://www.ncbi.nlm.nih.gov/pubmed/12814851.

Jesus-Ribeiro, J., Sargento-Freitas, J., Sousa, M., Silva, F., Freire, A., & Januário, C. (2016). Substantia nigra hyperechogenicity does not correlate with motor features in Parkinson's disease. *Journal of the Neurological Sciences, 364*, 9–11. https://doi.org/10.1016/j.jns.2016.03.002.

Kim, J. S., Oh, Y. S., Kim, Y. I., Koo, J. S., Yang, D. W., & Lee, K. S. (2012). Transcranial sonography (TCS) in Parkinson's disease (PD) and essential tremor (ET) in relation with putative premotor symptoms of PD. *Archives of Gerontology and Geriatrics, 54*(3), 436–439. https://doi.org/10.1016/j.archger.2012.01.001.

Kostić, V. S., Mijajlović, M., Smajlović, D., Lukić, M. J., Tomić, A., & Svetel, M. (2013). Transcranial brain sonography findings in two main variants of progressive supranuclear palsy. *European Journal of Neurology, 20*(3), 552–557. https://doi.org/10.1111/ene.12034.

Kostić, V. S., Svetel, M., Mijajlović, M., Pavlović, A., Ječmenica-Lukić, M., & Kozić, D. (2012). Transcranial sonography in pantothenate kinase-associated neurodegeneration. *Journal of Neurology, 259*(5), 959–964. https://doi.org/10.1007/s00415-011-6294-4.

Kresojević, N., Mijajlović, M., Perić, S., Pavlović, A., Svetel, M., Janković, M., et al. (2013). Transcranial sonography in patients with Parkinson's disease with glucocerebrosidase mutations. *Parkinsonism & Related Disorders, 19*(4), 431–435. https://doi.org/10.1016/j.parkreldis.2012.12.006.

Krogias, C., Strassburger, K., Eyding, J., Gold, R., Norra, C., Juckel, G., et al. (2011). Depression in patients with Huntington disease correlates with alterations of the brain stem raphe depicted by transcranial sonography. *Journal of Psychiatry and Neuroscience, 36*(3), 187–194. https://doi.org/10.1503/jpn.100067.

Lambeck, J., Niesen, W. D., Matthias, R., Weiller, C., Matthias, D., & Birgit, Z. (2015). Substantia nigra hyperechogenicity in hypokinetic Huntington's disease patients. *Journal of Neurology, 262*(3), 711–717. https://doi.org/10.1007/s00415-014-7587-1.

Laučkaite, K., Rastenyte, D., Šurkiene, D., Vaidelyte, B., Dambrauskaite, G., Sakalauskas, A., et al. (2014). Ultrasonographic (TCS) and clinical findings in overlapping phenotype of essential tremor and Parkinson's disease (ET-PD). *BMC Neurology, 14*, 54. https://doi.org/10.1186/1471-2377-14-54.

Laučkaite, K., Rastenyte, D., Šurkiene, D., Vaitkus, A., Sakalauskas, A., Lukoševičius, A., et al. (2012). Specificity of transcranial sonography in parkinson spectrum disorders in comparison to degenerative cognitive syndromes. *BMC Neurology, 12*. https://doi.org/10.1186/1471-2377-12-12.

Li, D. H., He, Y. C., Liu, J., & Chen, S. D. (2016). Diagnostic accuracy of transcranial sonography of the substantia nigra in Parkinson's disease: A systematic review and meta-analysis. *Scientific Reports, 6*, 20863. https://doi.org/10.1038/srep20863.

Li, X., Xue, S., Jia, S., Zhou, Z., Qiao, Y., Hou, C., et al. (2017). Transcranial sonography in idiopathic REM sleep behavior disorder and multiple system atrophy. *Psychiatry and Clinical Neurosciences, 71*(4), 238–246. https://doi.org/10.1111/pcn.12483.

Liepelt, I., Wendt, A., Schweitzer, K. J., Wolf, B., Godau, J., Gaenslen, A., et al. (2008). Substantia nigra hyperechogenicity assessed by transcranial sonography is related to neuropsychological impairment in the elderly population. *Journal of Neural Transmission, 115*(7), 993–999. https://doi.org/10.1007/s00702-008-0043-6 [Vienna, Austria: 1996].

Liman, J., Wellmer, A., Rostasy, K., Bähr, M., & Kermer, P. (2012). Transcranial ultrasound in neurodegeneration with brain iron accumulation (NBIA). *European Journal of Paediatric Neurology, 16*(2), 175–178. https://doi.org/10.1016/j.ejpn.2011.07.009.

Lobsien, E., Schreiner, S., Plotkin, M., Kupsch, A., Schreiber, S. J., & Doepp, F. (2012). No correlation of substantia nigra echogenicity and nigrostriatal degradation in Parkinson's disease. *Movement Disorders: Official Journal of the Movement Disorder Society, 27*(3), 450–453. https://doi.org/10.1002/mds.24070.

Luo, W. F., Zhang, Y. C., Sheng, Y. J., Fang, J. C., & Liu, C. F. (2012). Transcranial sonography on Parkinson's disease and essential tremor in a Chinese population. *Neurological Sciences, 33*(5), 1005–1009. https://doi.org/10.1007/s10072-011-0876-x.

Mahlknecht, P., Stockner, H., Kiechl, S., Willeit, J., Rastner, V., Gasperi, A., et al. (2012). Is transcranial sonography useful to distinguish drug-induced parkinsonism from

Parkinson's disease? *Movement Disorders: Official Journal of the Movement Disorder Society*, *27*(9), 1194–1196. https://doi.org/10.1002/mds.25071.

Mašková, J., Školoudík, D., Burgetová, A., Fiala, O., Brůha, R., Záhoráková, D., et al. (2016). Comparison of transcranial sonography-magnetic resonance fusion imaging in Wilson's and early-onset Parkinson's diseases. *Parkinsonism & Related Disorders*, *28*, 87–93. https://doi.org/10.1016/j.parkreldis.2016.04.031.

Moringlane, J. R., Fuss, G., & Becker, G. (2005). Peroperative transcranial sonography for electrode placement into the targeted subthalamic nucleus of patients with Parkinson disease: Technical note. *Surgical Neurology*, *63*(1), 66–69. https://doi.org/10.1016/j.surneu.2004.01.029.

Naumann, M., Becker, G., Toyka, K. V., Supprian, T., & Reiners, K. (1996). Lenticular nucleus lesion in idiopathic dystonia detected by transcranial sonography. *Neurology*, *47*(5), 1284–1290. https://doi.org/10.1212/WNL.47.5.1284.

Oh, Y. S., Kwon, D. Y., Kim, J. S., Park, M. H., & Berg, D. (2018). Transcranial sonographic findings may predict prognosis of gastroprokinetic drug-induced parkinsonism. *Parkinsonism and Related Disorders*, *46*, 36–40. https://doi.org/10.1016/j.parkreldis.2017.10.011.

Okawa, M., Miwa, H., Kajimoto, Y., Hama, K., Morita, S., Nakanishi, I., et al. (2007). Transcranial sonography of the substantia nigra in Japanese patients with Parkinson's disease or atypical parkinsonism: Clinical potential and limitations. *Internal Medicine*, *46*(18), 1527–1531. https://doi.org/10.2169/internalmedicine.46.0271.

Olivares Romero, J., Arjona Padillo, A., Barrero Hernández, F. J., Martín González, M., & Gil Extremera, B. (2013). Utility of transcranial sonography in the diagnosis of drug-induced parkinsonism: A prospective study. *European Journal of Neurology*, *20*(11), 1451–1458. https://doi.org/10.1111/ene.12131.

Plate, A., Ahmadi, S.-A., Pauly, O., Klein, T., Navab, N., & Bötzel, K. (2012). Three-dimensional sonographic examination of the midbrain for computer-aided diagnosis of movement disorders. *Ultrasound in Medicine & Biology*, *38*(12), 2041–2050. https://doi.org/10.1016/j.ultrasmedbio.2012.07.017.

Postert, J., Lack, B., Kuhn, W., & Jergas, M. (1999). Basal ganglia alterations and brain atrophy in Huntington's disease depicted by transcranial real time sonography. *Journal of Neurology, Neurosurgery, and Psychiatry*, *67*(4), 457–462.

Postuma, R. B., Berg, D., Stern, M., Poewe, W., Olanow, C. W., Oertel, W., et al. (2015). MDS clinical diagnostic criteria for Parkinson's disease. *Movement Disorders*, *30*(12), 1591–1601. https://doi.org/10.1002/mds.26424.

Prestel, J., Schweitzer, K. J., Hofer, A., Gasser, T., & Berg, D. (2006). Predictive value of transcranial sonography in the diagnosis of Parkinson's disease. *Movement Disorders*, *21*(10), 1763–1765. https://doi.org/10.1002/mds.21054.

Richter, D., Woitalla, D., Muhlack, S., Gold, R., Tönges, L., & Krogias, C. (2017). Coronal transcranial sonography and M-mode tremor frequency determination in Parkinson's disease and essential tremor. *Journal of Neuroimaging*, *27*(5), 524–530. https://doi.org/10.1111/jon.12441.

Rupprecht, S., Walther, B., Gudziol, H., Steenbeck, J., Freesmeyer, M., Witte, O. W., et al. (2013). Clinical markers of early nigrostriatal neurodegeneration in idiopathic rapid eye movement sleep behavior disorder. *Sleep Medicine*, *14*(11), 1064–1070. https://doi.org/10.1016/j.sleep.2013.06.008.

Sadowski, K., Serafin-Król, M., Szlachta, K., & Friedman, A. (2015). Basal ganglia echogenicity in tauopathies. *Journal of Neural Transmission*, *122*(6), 863–865. https://doi.org/10.1007/s00702-014-1310-3.

Sanzaro, E., Iemolo, F., Duro, G., & Malferrari, G. (2014). A new assessment tool for Parkinson disease. *Journal of Ultrasound in Medicine*, *33*(9), 1635–1640. https://doi.org/10.7863/ultra.33.9.1635.

Schweitzer, K. J., Behnke, S., Liepelt, I., Wolf, B., Grosser, C., Godau, J., et al. (2007). Cross-sectional study discloses a positive family history for Parkinson's disease and male gender as epidemiological risk factors for substantia nigra hyperechogenicity. *Journal of Neural Transmission, 114*(9), 1167–1171 [Vienna, Austria: 1996] https://doi.org/10.1007/s00702-007-0725-5.

Shafieesabet, A., Fereshtehnejad, S. M., Shafieesabet, A., Delbari, A., Baradaran, H. R., Postuma, R. B., et al. (2017). Hyperechogenicity of substantia nigra for differential diagnosis of Parkinson's disease: A meta-analysis. *Parkinsonism and Related Disorders, 42*, 1–11. https://doi.org/10.1016/j.parkreldis.2017.06.006.

Shin, H. Y., Joo, E. Y., Kim, S. T., Dhong, H. J., & Cho, J. W. (2013). Comparison study of olfactory function and substantia nigra hyperechogenicity in idiopathic REM sleep behavior disorder, Parkinson's disease and normal control. *Neurological Sciences, 34*(6), 935–940. https://doi.org/10.1007/s10072-012-1164-0.

Sierra, M., Sanchez-Juan, P., Martinez-Rodriguez, M. I., Gonzalez-Aramburu, I., Garcia-Gorostiaga, I., Quirce, M. R., et al. (2013). Olfaction and imaging biomarkers in premotor LRRK2 G2019S-associated Parkinson disease. *Neurology, 80*(7), 621–626. https://doi.org/10.1212/WNL.0b013e31828250d6.

Skoloudik, D., Jelinkova, M., Blahuta, J., Cermak, P., Soukup, T., Bartova, P., et al. (2014). Transcranial sonography of the substantia Nigra: Digital image analysis. *American Journal of Neuroradiology, 35*(12), 2273–2278. https://doi.org/10.3174/ajnr.A4049.

Skowronska, M., Dziezyc, K., & Członkowska, A. (2014). Transcranial sonography in manganese-induced parkinsonism caused by drug abuse. *Clinical Neuroradiology, 24*(4), 385–387. https://doi.org/10.1007/s00062-013-0256-4.

Skowronska, M., Walter, U., Kmiec, T., & Czlonkowska, A. (2013). Transcranial sonography in mitochondrial membrane protein-associated neurodegeneration. *Parkinsonism & Related Disorders, 19*(11), 1061–1063. https://doi.org/10.1016/j.parkreldis.2013.06.020.

Sommer, U., Hummel, T., Cormann, K., Mueller, A., Frasnelli, J., Kropp, J., et al. (2004). Detection of presymptomatic Parkinson's disease: Combining smell tests, transcranial sonography, and SPECT. *Movement Disorders: Official Journal of the Movement Disorder Society, 19*(10), 1196–1202. https://doi.org/10.1002/mds.20141.

Spiegel, J., Hellwig, D., Möllers, M.-O., Behnke, S., Jost, W., Fassbender, K., et al. (2006). Transcranial sonography and [123I]FP-CIT SPECT disclose complementary aspects of Parkinson's disease. *Brain, 129*(5), 1188–1193. https://doi.org/10.1093/brain/awl042.

Sprenger, F. S., Wurster, I., Seppi, K., Stockner, H., Scherfler, C., Sojer, M., et al. (2016). Substantia nigra hyperechogenicity and Parkinson's disease risk in patients with essential tremor. *Movement Disorders, 31*(4), 579–583. https://doi.org/10.1002/mds.26515.

Stockner, H., Iranzo, A., Seppi, K., Serradell, M., Gschliesser, V., Sojer, M., et al. (2009). Midbrain hyperechogenicity in idiopathic REM sleep behavior disorder. *Movement Disorders, 24*(13), 1906–1909. https://doi.org/10.1002/mds.22483.

Stockner, H., Sojer, M., Seppi, K., Mueller, J., Wenning, G. K., Schmidauer, C., et al. (2007). Midbrain sonography in patients with essential tremor. *Movement Disorders, 22*(3), 414–417. https://doi.org/10.1002/mds.21344.

Svetel, M., Mijajlović, M., Tomić, A., Kresojević, N., Pekmezović, T., & Kostić, V. S. (2012). Transcranial sonography in Wilson's disease. *Parkinsonism and Related Disorders, 18*(3), 234–238. https://doi.org/10.1016/j.parkreldis.2011.10.007.

Svetel, M., Tomić, A., Mijajlović, M., Dobričić, V., Novaković, I., Pekmezović, T., et al. (2017). Transcranial sonography in dopa-responsive dystonia. *European Journal of Neurology, 24*(1), 161–166. https://doi.org/10.1111/ene.13172.

Synofzik, M., Godau, J., Lindig, T., Schöls, L., & Berg, D. (2011). Transcranial sonography reveals cerebellar, nigral, and forebrain abnormalities in Friedreich's ataxia. *Neurodegenerative Diseases, 8*(6), 470–475. https://doi.org/10.1159/000327751.

Toscano, M., Canevelli, M., Giacomelli, E., Zuco, C., Di Piero, V., Lenzi, L., et al. (2011). Transcranial sonography of basal ganglia calcifications in Fahr disease. *Journal of Ultrasound in Medicine, 30*(7), 1032–1033.

Tsai, C. F., Wu, R. M., Huang, Y. W., Chen, L. L., Yip, P. K., & Jeng, J. S. (2007). Transcranial color-coded sonography helps differentiation between idiopathic Parkinson's disease and vascular parkinsonism. *Journal of Neurology, 254*(4), 501–507. https://doi.org/10.1007/s00415-006-0403-9.

van de Loo, S., Walter, U., Behnke, S., Hagenah, J., Lorenz, M., Sitzer, M., et al. (2010). Reproducibility and diagnostic accuracy of substantia nigra sonography for the diagnosis of Parkinson's disease. *Journal of Neurology, Neurosurgery, and Psychiatry, 81*(10), 1087–1092. https://doi.org/10.1136/jnnp.2009.196352.

Venegas-Francke, P. (2010). Transcranial sonography in the discrimination of Parkinson's disease versus vascular parkinsonism. *International Review of Neurobiology,* Vol. 90, 147–156. https://doi.org/10.1016/S0074-7742(10)90010-X.

Vilas, D., Ispierto, L., Álvarez, R., Pont-Sunyer, C., Martí, M. J., Valldeoriola, F., et al. (2015). Clinical and imaging markers in premotor LRRK2 G2019S mutation carriers. *Parkinsonism & Related Disorders, 21*(10), 1170–1176. https://doi.org/10.1016/j.parkreldis.2015.08.007.

Walter, U., Blitzer, A., Benecke, R., Grossmann, A., & Dressler, D. (2014). Sonographic detection of basal ganglia abnormalities in spasmodic dysphonia. *European Journal of Neurology, 21*(2), 349–352. https://doi.org/10.1111/ene.12151.

Walter, U., Dressler, D., Lindemann, C., Slachevsky, A., & Miranda, M. (2008). Transcranial sonography findings in welding-related parkinsonism in comparison to Parkinson's disease. *Movement Disorders, 23*(1), 141–145. https://doi.org/10.1002/mds.21795.

Walter, U., Dressler, D., Probst, T., Wolters, A., Abu-Mugheisib, M., Wittstock, M., et al. (2007). Transcranial brain sonography findings in discriminating between parkinsonism and idiopathic Parkinson disease. *Archives of Neurology, 64*(11), 1635. https://doi.org/10.1001/archneur.64.11.1635.

Walter, U., Dressler, D., Wolters, A., Probst, T., Grossmann, A., & Benecke, R. (2004). Sonographic discrimination of corticobasal degeneration vs progressive supranuclear palsy. *Neurology, 63*(3), 504–509. Retrieved from http://www.ncbi.nlm.nih.gov/pubmed/15304582.

Walter, U., Dressler, D., Wolters, A., Wittstock, M., & Benecke, R. (2007). Transcranial brain sonography findings in clinical subgroups of idiopathic Parkinson's disease. *Movement Disorders, 22*(1), 48–54. https://doi.org/10.1002/mds.21197.

Walter, U., Dressler, D., Wolters, A., Wittstock, M., Greim, B., & Benecke, R. (2006). Sonographic discrimination of dementia with Lewy bodies and Parkinson's disease with dementia. *Journal of Neurology, 253*(4), 448–454. https://doi.org/10.1007/s00415-005-0023-9.

Walter, U., Kirsch, M., Wittstock, M., Müller, J. U., Benecke, R., & Wolters, A. (2011). Transcranial sonographic localization of deep brain stimulation electrodes is safe, reliable and predicts clinical outcome. *Ultrasound in Medicine and Biology, 37*(9), 1382–1391. https://doi.org/10.1016/j.ultrasmedbio.2011.05.017.

Walter, U., Krolikowski, K., Tarnacka, B., Benecke, R., Czlonkowska, A., & Dressler, D. (2005). Sonographic detection of basal ganglia lesions in asymptomatic and symptomatic Wilson disease. *Neurology, 64*(10), 1726–1732. https://doi.org/10.1212/01.WNL.0000016184/.46465.D9.

Walter, U., Müller, J.-U., Rösche, J., Kirsch, M., Grossmann, A., Benecke, R., et al. (2016). Magnetic resonance-transcranial ultrasound fusion imaging: A novel tool for brain electrode location. *Movement Disorders: Official Journal of the Movement Disorder Society, 31*(3), 302–309. https://doi.org/10.1002/mds.26425.

Walter, U., Niehaus, L., Probst, T., Benecke, R., Meyer, B. U., & Dressler, D. (2003). Brain parenchyma sonography discriminates Parkinson's disease and atypical parkinsonian syndromes. *Neurology, 60*(1), 74–77. Retrieved from http://www.ncbi.nlm.nih.gov/pubmed/12525721.

Walter, U., & Školoudík, D. (2014). Transcranial sonography (TCS) of brain parenchyma in movement disorders: Quality standards, diagnostic applications and novel technologies. *Ultraschall in Der Medizin—European Journal of Ultrasound, 35*(4), 322–331. https://doi.org/10.1055/s-0033-1356415.

Walter, U., Wolters, A., Wittstock, M., Benecke, R., Schroeder, H. W., & Müller, J. U. (2009). Deep brain stimulation in dystonia: Sonographic monitoring of electrode placement into the globus pallidus internus. *Movement Disorders, 24*(10), 1538–1541. https://doi.org/10.1002/mds.22663.

Weise, D., Lorenz, R., Schliesser, M., Schirbel, A., Reiners, K., & Classen, J. (2009). Substantia nigra echogenicity: A structural correlate of functional impairment of the dopaminergic striatal projection in Parkinson's disease. *Movement Disorders, 24*(11), 1669–1675. https://doi.org/10.1002/mds.22665.

Yilmaz, R., Behnke, S., Liepelt-Scarfone, I., Roeben, B., Pausch, C., Runkel, A., et al. (2016). Substantia nigra hyperechogenicity is related to decline in verbal memory in healthy elderly adults. *European Journal of Neurology, 23*(5), 973–978. https://doi.org/10.1111/ene.12974.

Yilmaz, R., & Berg, D. (2017). Evaluating a patient with transcranial sonography to look for echogenicity. *Movement Disorders Clinical Practice, 4*(6), 907. https://doi.org/10.1002/mdc3.12503.

Yilmaz, R., Pilotto, A., Roeben, B., Preiche, O., Suenkel, U., Heinzel, S., et al. (2016). Structural ultrasound of the medial temporal lobe in Alzheimer's disease | Struktureller Ultraschall des medialen Temporallappens bei Alzheimer-Demenz. *Ultraschall in Der Medizin, 38*, 294–300. https://doi.org/10.1055/s-0042-107150.

Zecca, L., Berg, D., Arzberger, T., Ruprecht, P., Rausch, W. D., Musicco, M., et al. (2005). In vivo detection of iron and neuromelanin by transcranial sonography: A new approach for early detection of substantia nigra damage. *Movement Disorders, 20*(10), 1278–1285. https://doi.org/10.1002/mds.20550.

Zhang, Y., Zhang, Y. C., Sheng, Y. J., Chen, X. F., Wang, C. S., Ma, Q., et al. (2016). Sonographic alteration of basal ganglia in different forms of primary focal dystonia: A cross-sectional study. *Chinese Medical Journal, 129*(8), 942–945. https://doi.org/10.4103/0366-6999.179792.

Zhou, H. Y., Sun, Q., Tan, Y. Y., Hu, Y. Y., Zhan, W. W., Li, D. H., et al. (2016). Substantia nigra echogenicity correlated with clinical features of Parkinson's disease. *Parkinsonism & Related Disorders, 24*, 28–33. https://doi.org/10.1016/j.parkreldis.2016.01.021.

CHAPTER SEVEN

Imaging Transplantation in Movement Disorders

Edoardo Rosario de Natale, Heather Wilson, Gennaro Pagano, Marios Politis[1]

Neurodegeneration Imaging Group, Maurice Wohl Clinical Neuroscience Institute, Institute of Psychiatry, Psychology and Neuroscience (IoPPN), King's College London, London, United Kingdom
[1]Corresponding author: e-mail address: marios.politis@kcl.ac.uk

Contents

Abstract

Cell replacement therapy with graft transplantation has been tested as a disease-modifying treatment in neurodegenerative diseases characterized by the damage of a predominant cell type, such as substantia nigra dopaminergic neurons in Parkinson's disease (PD) or striatal medium spiny projection neurons in Huntington's disease (HD). The results of these trials are mixed with success in preclinical and pilot open-label trials,

International Review of Neurobiology, Volume 143
ISSN 0074-7742
https://doi.org/10.1016/bs.irn.2018.10.002

which were not consistently reproduced in randomized controlled trials. Positron emission tomography (PET) and single photon emission computed tomography (SPECT) molecular imaging and functional magnetic resonance imaging allow the graft survival, and its relationship with the host tissues to be studied *in vivo*. In PD, PET with [^{18}F]DOPA showed that graft survival does not necessarily correlate with the clinical improvement and PD patients with worse outcome had lower binding in the ventral striatum and a high serotonin ([^{11}C]DASB PET) to dopamine ([^{18}F]DOPA PET) ratio in the grafted neurons. In HD, PET with [^{11}C]PK11195 showed the graft survival and the clinical responses may be related to the reactive activation of the host inflammatory/immune system. Findings from these studies have been used to refine study protocols and patient selection in current clinical trials, which includes identifying suitable candidates for transplantation using imaging markers and employing multiple and/or novel PET tracers to better assess graft functions and inflammatory responses to grafts.

1. INTRODUCTION

Progressive damage and death of specific neuronal populations, such as dopaminergic neurons in the substantia nigra and striatal medium spiny projection neurons, are associated with the development of Parkinson's disease (PD) and Huntington's disease (HD), respectively. Despite advances in our understanding of the pathogenesis of these neurodegenerative diseases, treatment approaches are merely symptomatic, and no disease-modifying treatments are currently available. Cell replacement therapy is a potentially attractive strategy that aims to restore functional neuronal circuits, modifying, at least in theory, the course of the disease. Cell replacement therapy is based on the use of stem cells, a cellular population with a high self-renewal capacity, which can mature into different cellular lineages (Lindvall & Kokaia, 2006). The history of cell replacement therapy in human subjects affected by movement disorders started about 30 years ago (Lindvall et al., 1988). Since then, enthusiasm about this therapeutic approach has been mixed over the years as promising and disappointing results were published. However, preclinical research has progressed in parallel with advancements in the comprehension of the biological properties of stem cells *in vitro* and in animal models. There is hope that these insights could increase the efficacy and efficiency of cell replacement therapy.

Stem cells can be obtained through direct transplantation of embryonic mesenchymal tissue into the adult brain. Experiments carried out during the last decades in animal disease models have demonstrated the feasibility and safety of this technique, and that stem cells can survive, integrate, differentiate, and recover physiological functions, with subsequent clinical improvement (Bjorklund & Stenevi, 1979; Deckel, Robinson, Coyle, & Sanberg, 1983; Grasbon-Frodl, Nakao, Lindvall, & Brundin, 1997; Perlow et al., 1979; Pundt, Kondoh, & Low, 1994; Stromberg, Bygdeman, Goldstein, Seiger, & Olson, 1986; Watts, Brasted, & Dunnett, 2000). This encouraged researchers to test this technique in human subjects with neurodegenerative diseases.

Neuroimaging techniques allow structural and functional changes within the brain to be studied at a molecular level *in vivo* (Politis, 2010; Politis & Lindvall, 2012; Politis & Piccini, 2010, 2012). Positron emission tomography (PET) and single photon emission computed tomography (SPECT) are non-invasive imaging techniques which have been employed across the spectrum of neurodegenerative disorders to study a variety of biological functions (Niccolini & Politis, 2014; Pagano, Niccolini, & Politis, 2016; Politis, 2014; Wilson, De Micco, Niccolini, & Politis, 2017). Magnetic resonance imaging (MRI) has a high spatial resolution and enables structural and functional connectivity changes to be tracked over time. These neuroimaging techniques have played a critical role in elucidating the pathophysiology of movement disorders and have been implemented as outcome measures for human-based trials with fetal graft transplantation.

In this chapter, we review molecular and functional neuroimaging studies on cell replacement therapy performed in patients with PD, HD, and other movement disorders including primary dystonia, multiple system atrophy and degenerative ataxias, and progressive supranuclear palsy.

2. TRANSPLANTATION IN PARKINSON'S DISEASE

Bradykinesia, rigidity, tremor, and postural instability constitute the cardinal motor symptoms of PD and are mainly associated with the loss of dopaminergic neurons in the substantia nigra, which causes a disconnection of the nigro-striatal pathways. The main objective from cell transplantation in PD is the reconstitution of a surviving and functional substantia nigra, in terms of dopamine production and release, as well as of physiological reconnection with distant brain regions. The embryonic ventral mesencephalic tegmentum contains a large population of dopaminergic precursory cells. When these cells are implanted in the striatum of the 6-hydroxydopamine (6-OHDA) immunosuppressed rat model of PD, they survive, produce dopamine, restore functional connections with other brain areas, and revert the motor symptoms (Bjorklund & Stenevi, 1979; Brundin et al., 1986; Perlow et al., 1979; Stromberg et al., 1986). The success of this methodology led various European and U.S.-based specialized centers to attempt transplantation of fetal mesencephalic tissue in a limited number of advanced PD patients with motor complications. In these pilot studies, fetal mesencephalic tissue transplantation showed a good feasibility and a reasonable short-term safety (Lindvall et al., 1990, 1989, 1988; Madrazo et al., 1990; Sawle et al., 1992). The clinical complications of fetal mesencephalic tissue transplantation were infrequent, and mainly consisted of post-operatory confusion. Rare cases however led to cortical hemorrhage along the needle tract (Brundin et al., 2000; Freeman et al., 1995).

These first pioneering trials were conducted using a variety of different methodologies, and therefore it was difficult to compare the results from each study. In order to guarantee the highest possible consistency between centers a specific committee, named the Core Assessment Program for Intra-cerebral Transplantation (also known as CAPIT) (Langston et al., 1992), developed a guideline program for intracerebral transplantation in PD. The guidelines covered inclusion criteria for the selection of patients, the side and size of the graft transplantation, immunosuppressive therapy, and outcome measures. All subsequent open-label studies and randomized controlled trial on PD patients adhered to these guidelines. An overview of the main studies conducted on PD patients is summarized in Table 1.

2.1 Molecular Imaging of the Dopaminergic System
2.1.1 Fluoro-Dopa
$[^{18}F]DOPA$ is a fluorinated analogous of endogenous dopamine that is absorbed by dopaminergic neurons in the brain. Within dopaminergic neurons, $[^{18}F]DOPA$ is converted, by the enzyme amino acid decarboxylase (AADC), to $[^{18}F]$-labeled dopamine and subsequently concentrated and retained in dopaminergic terminals. It has been demonstrated that the levels of $[^{18}F]DOPA$ uptake in the striatum reflect the state of dopaminergic terminals both anatomically and functionally (Leenders et al., 1990) and are correlated with the *post mortem* quantification of substantia nigra neurons (Snow et al., 1993). Therefore, $[^{18}F]DOPA$ PET imaging has been used to evaluate, *in vivo*, the survival of the grafts and their functional recovery over time (Fig. 1).

In the initial stages of the disease, PD patients typically show a 50% reduction of $[^{18}F]DOPA$ uptake in the putamen contralateral to the clinically most affected side (Morrish, Sawle, & Brooks, 1996). Reduction in $[^{18}F]DOPA$ uptake correlates with longer disease duration and greater

Table 1 Summary of the Open-Label and Randomized Controlled Studies With Transplantation of Fetal Graft Tissue in Parkinson's Disease

Reference	Patients	Procedure	Imaging Modality	Main Findings
Lindvall et al. (1988)	2 PD	Unilateral transplantation in the caudate and putamen	[^{18}F]DOPA PET	Clinical improvement in one of the two patients not mirrored by increases in striatal tracer uptake
Lindvall et al. (1990)	1 PD	Unilateral transplantation in the putamen	[^{18}F]DOPA PET	Clinical improvement with restoration of dopamine synthesis in the grafted putamen
Sawle et al. (1992)	2 PD	Unilateral transplantation in the putamen	[^{18}F]DOPA PET	Clinical improvement at 1 year with evidence of increased dopamine synthesis in the grafted tissue
Freed et al. (1992)	7 PD	Two received unilateral grafts in caudate and putamen, five received bilateral grafts in the putamen	[^{18}F]DOPA PET	Five out of seven patients improved in motor scales and showed sustained increase of dopamine synthesis
Widner et al. (1992)	2 MPTP-induced PD	Bilateral graft in the caudate and putamen	[^{18}F]DOPA PET	Substantial and sustained improvement in motor function with increase of tracer uptake after 1 year from surgery
Spencer et al. (1992)	4 PD	Unilateral grafting in the caudate	[^{18}F]DOPA PET	Three cases showed improvement on both motor clinical and imaging outcomes
Hoffer et al. (1992)	2 PD	Unilateral putamen and caudate transplantation	[^{18}F]DOPA PET [^{11}C]Nomifensine PET	Clinical sustained improvement in one patient, clinical transitory improvement in the other, mirrored by similar PET outcomes

Continued

Table 1 Summary of the Open-Label and Randomized Controlled Studies With Transplantation of Fetal Graft Tissue in Parkinson's Disease—cont'd

Reference	Patients	Procedure	Imaging Modality	Main Findings
Peschanski et al. (1994)	2 PD	Unilateral intrastriatal transplantation	[^{18}F]DOPA PET	Increase in daily activity autonomy, with gradual increase of tracer uptake at the site of grafting
Freeman et al. (1995)	4 PD	Bilateral transplantation of the putamen	[^{18}F]DOPA PET	Clinically significant improvement, with bilateral evidence of restoration of dopaminergic synthesis
Remy et al. (1995)	5 PD	Unilateral transplantation in the putamen	[^{18}F]DOPA PET	Increase in the tracer uptake in the putamen that was proportional to the extent of daily time spent on "ON" phase, and with functional motor scores
Wenning et al. (1997)	6 PD	Unilateral graft in the putamen (4 PD) and unilateral putaminal and caudate transplantation (2 PD)	[^{18}F]DOPA PET	Clinical improvement evident in four out of six PD patients, with evidence of increase of dopamine synthesis in the grafted putamen, but not in the caudate
Levivier et al. (1997)	3 PD	Bilateral grafting of the putamen	[^{18}F]DOPA PET	Clinical improvement after 3–6 months after surgery, characterized by improvement on UPDRS score and increase of "ON" periods, accompanied by increase of [^{18}F]DOPA uptake in the grafted areas

Hauser et al. (1999)	6 PD	Bilateral grafting in the putamen	[18F]DOPA PET	Clinical improvement of patients and improvement on motor scores. Imaging evidence of dopaminergic synthesis by the graft from 6 months
Hagell et al (1999)	5 PD	Initial unilateral striatal transplantation followed by transplantation in the putamen (4/5) or putamen and caudate (1/5) contralaterally to the initial surgical procedure	[18F]DOPA PET	Two patients showed additional significant clinical improvement, one improved moderately, two worsened
Piccini et al. (1999)	1 PD	Unilateral transplantation in the putamen. Follow-up of 10 years. Drug challenge with methamphetamine	[11C]Raclopride PET	Sustained and marked clinical benefit. Imaging evidence of restoration of normal striatal pool physiology of D_2 receptors, both basal and drug induced
Ross et al. (1999)	24 PD, 14 HD	Bilateral graft in the putamen (PD patients) and in the putamen and caudate (HD patients)	[1H]MRS	Evidence of increased amounts of N-acetyl aspartate in grafts from both PD and HD subjects, compared with fetal transplants
Bluml et al. (1999)	2 PD	Bilateral graft in the putamen	fMRI	Increase of activation of the grafted putamen in response to contralateral motor tasks in both patients
Piccini et al. (2000)	4 PD	Bilateral grafting in the caudate nuclei and putamen. Joystick movement task as motor paradigm	[18F]DOPA PET $H_2^{15}O$ PET	Immediate increase in dopaminergic synthesis activity in the grafted tissue, with delayed clinical improvement. Functional and delayed increase in the supplementary motor area and dorsal prefrontal cortex

Continued

Table 1 Summary of the Open-Label and Randomized Controlled Studies With Transplantation of Fetal Graft Tissue in Parkinson's Disease—cont'd

Reference	Patients	Procedure	Imaging Modality	Main Findings
Brundin et al. (2000)	5 PD	Bilateral putamen and caudate transplantation with additional administration of tirizalad mesylate	[^{18}F]DOPA PET	After 1–2 years from surgery, increase in tracer uptake in both putamen and caudate, bilaterally. Improvement in the UPDRS motor score in "OFF" phase
Freed et al. (2001)	40 PD	Randomized controlled trial. Bilateral implantation in the putamen vs sham surgery. Follow-up of 1 year	[^{18}F]DOPA PET	No clinical improvement in the whole transplanted group. In patients younger than 60 years UPDRS score decreased. Increase of [^{18}F]DOPA uptake in 17 out of 20 transplantation recipients. 15% developed GIDs
Mendez et al. (2002)	3 PD	Transplantation of grafts in bilateral striatal and nigral areas	[^{18}F]DOPA PET	After 1 year, evidence of increase of dopaminergic synthesis in both areas, with clinical motor improvement
Ma et al. (2002)	17 PD	Bilateral transplantation in the caudate and putamen. 5 PD developed GIDs, 12 did not	[^{18}F]DOPA PET	[^{18}F]DOPA uptake was increased in specific areas of the putamen of transplant recipients who developed GIDs
Olanow et al. (2003)	34 PD	Randomized Controlled Trial. Bilateral implantation in the putamen vs sham surgery. Follow-up of 2 years	[^{18}F]DOPA PET	No significant treatment effect between arms. Patients with milder baseline disease tended to improve more than patients with more severe disease at baseline. Significant increase of [^{18}F]DOPA intake in transplanted grafts. 56% developed GIDs

Cochen et al. (2003)	6 PD	Bilateral grafting in the putamen, in two successive times	[18F]DOPA PET [76Br]CBT PET	Patients showed clinical benefit. No change in the [76Br]CBT PET findings despite a postoperatively increase of [18F]DOPA uptake in the grafted tissue
Huang et al. (2003)	2 PD	Bilateral grafting in patients who developed GIDs	[11C]Raclopride PET	Administration of L-DOPA caused prolonged displacement of tracer from striatal D_2 receptors
Piccini et al. (2005)	9 PD	Implantation of the right putamen (2 PD), of the bilateral putamen (2 PD) and bilateral putamen and caudate nuclei (5 PD)	[11C]Raclopride PET [18F]DOPA PET	Patients who attained the best clinical and functional outcome after transplantation were those with no preliminary evidence of dopaminergic denervation outside the grafted areas. There was a correlation between the level of [18F]DOPA uptake and the level of displacement of [11C]Raclopride from striatal areas after methamphetamine challenge. Withdrawal of immunosuppression resulted in an increase of dyskinesia scores
Pogarell et al. (2006)	2 PD	Bilateral transplantation in the putamen and caudate. Follow-up to 8 years	[123I]IPT SPECT	Increase of striatal DAT availability at three ears that was sustained over the 8 years' follow-up. Moderate clinical improvement, but presence of GIDs

Continued

Table 1 Summary of the Open-Label and Randomized Controlled Studies With Transplantation of Fetal Graft Tissue in Parkinson's Disease—cont'd

Reference	Patients	Procedure	Imaging Modality	Main Findings
Politis et al. (2010)	2 PD	Bilateral transplantation in the putamen (1 PD) and in the putamen and caudate (1 PD). Follow-up after 13 and 16 years after transplantation	[18F]DOPA PET [11C]DASB PET [11C]Raclopride PET	Patients with motor improvement after surgery but subsequent development of severe GIDs. Marked increase of serotonergic terminals in the grafted striatum. Introduction of a 5-HT$_{1A}$ antagonist reverted the symptoms
Politis et al. (2011)	1 PD	Bilateral transplantation in the caudate and putamen. 14 Years' follow-up	[123I]FP-CIT SPECT [18F]DOPA PET [11C]DASB PET	Moderate clinical improvement despite the onset of severe GIDs. Increased serotonin/dopamine transporter ratio in the grafted tissue
Politis et al. (2012)	3 PD	Bilateral transplantation in the putamen and, for two patients, also in the caudate. Follow-up 13–16 years	[18F]DOPA PET [11C]DASB PET	After long follow-up, significant clinical improvement compared to preoperative scored. Dopaminergic innervation still restored to normal levels in the basal ganglia and preserved in extrastriatal areas. Decrease of [11C]DASB uptake in the raphe and in cortical areas
Kefalopoulou et al. (2014)	2 PD	Bilateral transplantation in the putamen (1 PD) and in the putamen and caudate (1 PD). Follow-up after 15 and 18 years	[18F]DOPA PET	Sustained clinical improvement after surgery, with corresponding increase in dopaminergic synthesis activity from the grafts

Abbreviations: *PD*, Parkinson's disease; *PET*, positron emission tomography; *SPECT*, single photon emission computed tomography; *HD*, Huntington's disease; *GID*, graft-induced dyskinesia; *MPTP*, 1-methyl-4-phenyl-1,2,3,6-tetrahydropyridine; *UPDRS*, unified Parkinson's disease rating scale; *5-HT$_{1A}$*, serotonin 1A; *DAT*, dopamine transporter; *L-DOPA*, levodopa; *D$_2$ receptors*, dopamine type 2 receptors; *fMRI*, functional magnetic resonance imaging.

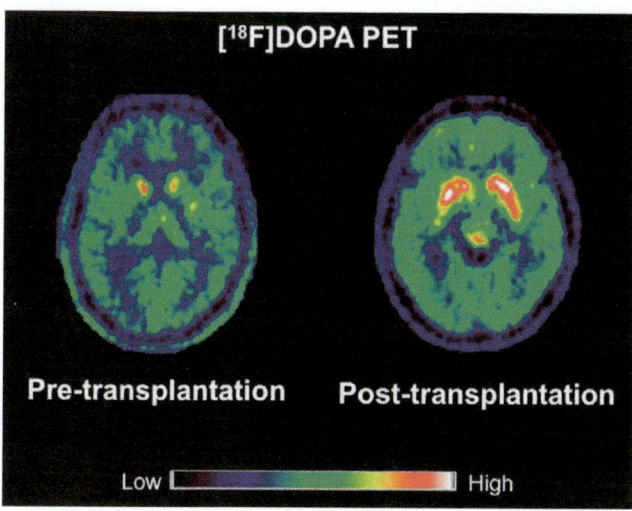

Fig. 1 PET images of a Parkinson's disease patient scanned with [¹⁸F]DOPA before (*left*) and after (*right*) grafting with fetal mesencephalic stem cells.

severity of rigidity and bradykinesia (Brooks et al., 1990; Broussolle et al., 1999; Morrish et al., 1996; Vingerhoets, Schulzer, Calne, & Snow, 1997). The reduction of striatal [¹⁸F]DOPA uptake progresses over time in parallel with the worsening of the clinical picture (Morrish, Rakshi, Bailey, Sawle, & Brooks, 1998).

In PD patients who received cell transplantation, the changes in [¹⁸F]DOPA uptake have been quantified in comparison with their baseline levels or with levels obtained from age- and gender-matched healthy controls. The first estimate permits to measure the net quantification of the modification obtained in a subject by the surgical intervention, while the second allows a more reliable comparison of this outcome among different PET studies.

In one of the first PET transplantation studies, [¹⁸F]DOPA PET was performed at baseline and at 5-month follow-up in a patient with a 13-year history of PD who underwent unilateral left putaminal transplantation of ventral mesencephalic cells. The clinical improvement was associated with a 130% increase of [¹⁸F]DOPA uptake in the left putamen, compared to the same exam performed at the baseline (Lindvall et al., 1990). The right-sided motor improvement, along with the selective increase of the dopaminergic pool in the left implanted putamen, demonstrated the direct effect of grafting in the amelioration of the clinical picture of this patient (Lindvall et al., 1990). Similarly, two unilateral recipients of putaminal grafts, experienced after 1 year from transplantation, an increase of [¹⁸F]DOPA uptake in the grafted putamen (Sawle et al., 1992). In addition, the non-grafted putamen and caudate showed a progressive reduction of [^{18}F]DOPA uptake. This has been interpreted as a sign of natural disease progression in the non-operated side and as an evidence of the direct role of the grafted tissue in modifying the dopaminergic function (Sawle et al., 1992). Some studies have shown that the first signs of graft function, detectable with PET imaging, may be delayed by up to 1 year (Freed et al., 1992). This finding is consistent with the assumption that it takes time for the cells to survive in the host tissue, mature, and establish functional connections. As evidence of this, it has been found that the increase in [^{18}F]DOPA uptake progresses steadily up to 3–4 years after transplantation (Brundin et al., 2000; Freed et al., 1992; Hagell et al., 1999). Therefore, this may be a surrogate imaging marker of the progressive maturation and gain of function of the grafted tissue.

Grafting has also been proven to be beneficial in MPTP-induced parkinsonism. In a study from Lund, Sweden, bilateral fetal grafts were transplanted in two patients with severe MPTP-induced parkinsonism (Widner et al., 1992). Both patients improved significantly, in comparison to their baseline motor performances, and at 24 months there was a retention of striatal [^{18}F] DOPA at about twice the preoperative values. This result is relevant in the view of the poor clinical condition that these patients experienced at baseline. The investigators chose to implant mesencephalic tissue bilaterally and to increase the total amount of grafted tissue; this might have had a role in the favorable clinical and imaging outcome (Widner et al., 1992).

The clinical significance of PET imaging findings was investigated in an interesting French study conducted on five PD patients. It was demonstrated that the increase in [^{18}F]DOPA *Ki* uptake after transplantation correlated significantly with the percentage of daily time spent by the patients in the "ON" time as well as with fine movement dexterity contralaterally to the grafted putamen (Remy et al., 1995). In addition, the patients who attained post-operative *Ki* within the normal range of healthy controls were those who had the best clinical results. Interestingly, these results were obtained only for the putaminal grafts. Four of the five patients who were grafted also in the caudate, did not experience any change in [^{18}F]DOPA uptake in this region (Remy et al., 1995).

Grafting of the caudate nucleus has been tried also in other open-label studies but, similarly to the aforementioned study, the outcome in this case has been less consistent. Spencer and colleagues performed unilateral caudate transplantation in four severe PD patients. Only one patient underwent pre-

and post-transplantation [^{18}F]DOPA PET scans demonstrated that caudate levels of dopaminergic activity were doubled bilaterally to levels similar to controls. This was mirrored by an improvement in motor daily activities and by the discontinuation of his dopamine agonist regimen and the significant reduction in levodopa (L-DOPA) therapy (Spencer et al., 1992). These results were similar to reports from five PD patients who were bilaterally transplanted in the putamen and caudate. In these patients, a 24% [^{18}F] DOPA increase in the caudate was apparent after 6 months (Brundin et al., 2000). However, a different report on two PD patients failed to find such improvements in the caudate (Wenning et al., 1997).

A great deal of information on the long-term biological effects of grafting has come from a fruitful collaboration between centers in Sweden, Germany, and United Kingdom, named the Lund/London/Marburg collaboration which, over the years, has gathered a remarkable bulk of data on transplanted PD patients (Hagell et al., 2002; Piccini et al., 2005; Wenning et al., 1997). In six PD patients, of which four received putaminal and two received putaminal and caudate unilateral grafts (Wenning et al., 1997), [^{18}F]DOPA uptake in the grafted putamen increased from 42% to 91%, 1 year following transplantation, whereas, in the contralateral putamen, the uptake decreased by a mean 28%, ranging from 2% to 57%. One patient was followed up at 72 months and demonstrated that [^{18}F] DOPA uptake in the putamen was sustained within normal values (Wenning et al., 1997). [^{18}F]DOPA uptake within the normal range was further supported by the fact that this patient was able to withdraw from dopaminergic therapy with good motor response. A group of five PD patients from the cohort studied by Wenning and colleagues received transplantation also in the contralateral putamen and were followed up for 18–24 months. In these patients, [^{18}F]DOPA uptake in the newly grafted putamen increased by 85%, similarly to what was obtained in the previous transplantation, whereas the contralateral putamen plateaued to the levels they had reached after 3–4 years. Furthermore, clinical scores improved markedly after the second transplant (Hagell et al., 1999). This work was important as it set grounds for the viability and efficacy of sequential grafting in the same patient. Some studies within the U.S.-based cohort, which underwent bilateral putaminal grafting, have yielded similar results. In one study, four PD patients were grafted bilaterally into the postcommissural putamen, immunosuppressed for 6 months and then studied with clinical and PET assessments. All patients showed clinical amelioration of motor and disability scores, as well as a reduction of "OFF" time and an increase in the time spent in the "ON "state. [^{18}F]DOPA PET detected a significant increase of 53% in the right putamen and a minor increase of 33% in the left (Freeman et al., 1995). These patients underwent a follow-up [^{18}F]DOPA PET after an additional 6 months, and demonstrated further increase of dopaminergic activity in the putamen, which correlated with the clinical improvements in the total unified Parkinson's disease rating scale (UPDRS) "OFF" scores and with the reduction of motor complications, quantified as a lower percentage of the time spent with dyskinesia (Hauser et al., 1999).

Two PD patients of this cohort underwent a very long-term clinical and PET imaging follow-up at 13 years from the transplantation. These patients still clinically benefited from grafting, not taking any dopaminergic medication and their levels of [^{18}F]DOPA were still within the range of normality in the striatum. Interestingly, they showed [^{18}F]DOPA levels within normality also in extrastriatal areas such as the amygdala, thalamus, hypothalamus, cingulate, insular, prefrontal cortex, and locus coeruleus (Politis et al., 2012). A further clinical follow-up at 16 years from the transplantation showed that, in these subjects, motor performances were still better than their preoperative baseline and they were still free from taking dopaminergic supplementation (Kefalopoulou et al., 2014). Although these results are available only for two patients, they provide clinical and neuroimaging evidence that ventral mesencephalic neuronal transplantation can provide long-term relief in recipients.

Although these studies were un-blinded, open-label, and heterogeneous in methodology, the promising results prompted the set-up of two large double-blind, placebo-controlled, randomized controlled trials (Freed et al., 2001; Olanow et al., 2003).

The first of the two randomized controlled trials, Freed et al. (2001) was performed in the Universities of Colorado and Columbia in the United States and involved 40 advanced PD patients who were randomly allocated to receiving either graft transplantation in the bilateral putamina with embryonic mesencephalic tissue obtained from 7 to 8 week fetuses, or sham surgery. [^{18}F]DOPA PET was performed before, and 12 months after receiving surgery, and measure of the success of the surgery was set as the difference between baseline and follow-up levels of putamen uptake. No immunosuppression was used in this trial. [^{18}F]DOPA PET imaging revealed a significant 40% increase in the putaminal uptake after 12 months from surgery in transplant recipients, without any difference between younger and older patients. A patient underwent a second [^{18}F]DOPA PET scan after 2 years from transplantation and showed that the [^{18}F]DOPA uptake was doubled bilaterally compared to baseline. Despite the imaging results, the trial failed in meeting the clinical endpoints, as the global rating score of self-perceived improvement was not different between arms despite a significant 18% improvement in the UPDRS score in the transplanted arm (Freed et al., 2001). However, when stratifying groups according to age (cut-off 60 years), younger patients improved significantly better than older patients (Freed et al., 2001). In a further analysis of the imaging trial data, it was demonstrated that 12 months after surgery, [^{18}F]DOPA PET uptake levels obtained by transplantation recipients rose within 2 standard deviations of normal values in 8 out of 19 patients, without any age difference, with the highest effect of the engraftment localized in the posterior putamen. Moreover, regression analysis between PET variables and "OFF" state UPDRS score, controlling for age, revealed a significant correlation between clinical and imaging measures that were retained only in the younger group, although this group comprised also of the sham arm (Nakamura et al., 2001). The cohort of this study was retrospectively re-assessed 2 and 4 years after surgery (Ma et al., 2010), and patients showed an improvement in UPDRS scores which was correlated with [^{18}F]FDOPA uptake levels in the putamen.

The second randomized controlled trial took place in Tampa, United States and was conducted by Olanow and collaborators (Olanow et al., 2003). Here, 34 PD patients received either ventral mesencephalic tissue obtained from one or four donors per side, or sham surgery. The follow-up was at 24 months and the primary endpoint was the change in the UPDRS motor score in "OFF" state from baseline to follow-up. Changes in putaminal [^{18}F]DOPA uptake were set as secondary endpoint. Patients received a 6-month immunosuppression with cyclosporine. Patients belonging to the treatment group showed a bilateral increase in [^{18}F]DOPA uptake, evident at 12-month follow-up and persisted, unchanged, at 24 months. However, the primary endpoint of the study was not met, even after stratification for age. When patients were stratified according to baseline UPDRS scores, it was demonstrated that patients with less severe disease attained better results from the grafting, if compared to patients with higher UPDRS score, one-donor grafts, and sham surgery recipients. Interestingly, patients improved clinically when they were still under cyclosporine immunosuppression, but deteriorated after, implying that the stopping of immunosuppression might have affected the outcome of the trial (Olanow et al., 2003).

These disappointing results from the two randomized controlled trials raised methodological and scientific considerations that may be useful for the set-up of future trials. First, preliminary data, although promising, was obtained from un-blinded, open-label studies which were performed according to heterogeneous methodologies, endpoints, and immunotherapy regimens, and the results varied greatly even within single studies. The lack of a standardized protocol may therefore have affected the results of the randomized controlled trials (Barker, Barrett, Mason, & Bjorklund, 2013). Second, the combined clinical and imaging deterioration experienced by the sham group in both studies rules out a possible placebo effect. Third, the data from the trial by Olanow indicated that the severity of PD may be a contributing factor to the success of the surgery. Retrospective studies examining this phenomenon have shown that the PD patients who were most liable to benefit from this procedure were the ones in whom the baseline [18F] DOPA PET scan alterations were limited to the dorsal striatum (Ma et al., 2010; Piccini et al., 2005). This characteristic might be taken into account in the selection of participants for future trials. Moreover, despite signs from [18F]DOPA PET of immediate graft survival and function, longer follow-up assessment could have allowed better results to be obtained, as embryonic cells in a highly deranged system like the putamen of advanced PD patients may take longer to grow and build functionally relevant connections with neighboring cells. This assumption is suggested by *post mortem* data (Freed et al., 2001) and by the initial age difference in imaging and clinical outcomes (Ma et al., 2010). Finally, there is the possibility that the initial increase of [18F]DOPA uptake may not reflect entirely what is visible at a clinical level. It is known that connections within the dopaminergic nigro-striatal system are complex and the growth and connections of the newly transplanted cells may need to reach a threshold before becoming clinically functional (Lao-Kaim, Piccini, & Tai, 2017; Ma, Peng, Dhawan, & Eidelberg, 2011). This may partly explain the apparent gap between clinical and imaging data that have been reported in some patients.

The initial enthusiasm brought about by the first open-label studies was dampened not only by the disappointing results from the two randomized controlled trials but also by the presence of a motor complication, characterized by involuntary and persistent dyskinetic or dystonic movements, which was experienced by a relevant percentage of transplantation recipients (Freed et al., 2001; Hagell et al., 2002; Olanow et al., 2003; Piccini et al., 2005). Early reports from the open-label studies described the presence of dyskinesias immediately after transplantation (Brundin et al., 2000; Freed et al., 1992; Hagell et al., 1999; Levivier et al., 1997; Peschanski et al., 1994; Wenning et al., 1997). These dyskinesias were associated with an increase of [^{18}F]DOPA uptake and were responsive to adjustments of dopaminergic medication. It was hypothesized that the development of dyskinesias was the effect of immature connections formed by the graft tissue in the host environment (Lindvall & Hagell, 2000).

However, longer follow-up studies and the two randomized controlled trials pointed to an apparently different, late-onset, troublesome, dyskinetic, and dystonic movement disorder which affected grafted patients in "OFF" state medication and persisted also if the dopaminergic therapy was withdrawn, that was called graft-induced dyskinesia (GID) (Hagell & Cenci, 2005; Hagell et al., 2002). The onset of GID was generally between the first and the second year after transplantation, but it could ensue as early as 5 months after surgery (Freed et al., 2001; Hagell & Cenci, 2005; Hagell et al., 2002; Olanow et al., 2003; Piccini et al., 2005; Politis et al., 2011, 2012, 2010). GIDs were not severe in intensity but were disabling and in some cases required surgery (Politis et al., 2012). During the two large randomized controlled trials conducted by Freed and Olanow, the incidence of GID was of 15% and 56%, respectively (Freed et al., 2001; Olanow et al., 2003). Imaging of GIDs will be discussed in Section 2.6.

[^{18}F]DOPA PET has been employed to assess the viability of alternative ways of transplantation, aimed at optimizing the dose of injected cells, or at increasing the efficacy of the transplant, through the targeting of more nuclei. The viability of low-quantity bilateral putamen and caudate grafting with the use of the lazaroid tirilazad mesylate was evaluated in five PD patients, who were followed for up to 20 months (Brundin et al., 2000). Lazaroid tirilazad mesylate is a substance that promotes the survival of embryonic human dopaminergic neurons and may be useful to prevent early post-operative graft cell death. Six months after transplantation, [^{18}F]DOPA uptake level was increased by 53% in the putamen and by 24% in the caudate when compared with preoperative measures, showing the efficacy of the neuroprotective treatment with lazaroids; however, these levels did not change significantly at later time points. The steadiness of uptake levels of [^{18}F]DOPA was contrasted by a progressive improvement in clinical motor scores and by a 50% reduction in L-DOPA dosage, with one subject discontinuing dopaminergic therapy (Brundin et al., 2000). More recently, a pilot study has been performed on three PD patients, who underwent bilateral dopaminergic graft in both the striatal and nigral nuclei; this study reported a significant increase in [^{18}F]DOPA uptake levels in all but two transplanted sites, interpreted as a good survival rate of the grafts (Mendez et al., 2002). However, further studies on this particular surgical method have not been replicated.

The suspension of immunosuppressive therapy following transplantation might influence the long-term survival and clinical efficacy of the grafts, because of detrimental events such as, for example, microglial activation and inflammation. In the randomized controlled trial conducted by Olanow and colleagues, it was noted that the suspension of immunosuppression could have had a role in the long-term insufficient improvement of the transplantation recipients (Olanow et al., 2003). The effect of suspension of immunosuppressive therapy on graft survival and functionality was studied in nine transplanted PD patients, who underwent immunosuppression for an average of 23 months, and were tested with [^{18}F]DOPA PET after an average of 21 months from immunosuppression withdrawal (Piccini et al., 2005). It was found that the levels of [^{18}F]DOPA uptake (expressed as *Ki* in percentage of normal mean of healthy controls) did not change, thus demonstrating that stopping immunosuppressive therapy did not affect graft survival and function. Furthermore, motor severity, as measured by UPDRS, did not change after immunosuppression withdrawal (Piccini et al., 2005). However, the investigators found that, after immunosuppression withdrawal, dyskinesia scores increased, which may have been the effect of either the growth of the graft or the worsening of a low-grade inflammation around the graft, caused by the suspension of immunosuppression.

Neuropathological data are available for a few transplantation recipients, who participated to the open-label studies, and died from PD-unrelated conditions. The case of a 59-year-old man who received unilateral putaminal grafting with good clinical outcome and evidence of about a 70% recovery of [^{18}F]FDOPA striatal uptake has been described. At neuropathological examination, all graft sites contained neurons reactive to tyrosine hydroxylase, a marker of dopaminergic neurons, and demonstrated sprouting of terminals to neighboring sites in the striatum (Kordower et al., 1995, 1996). These findings were confirmed by successive cases (Kordower et al., 1998) and have provided the first evidence of a neuropathological correlate of clinical and imaging findings on graft transplantation. Interesting data come from a patient from one of the first randomized controlled trials (Freed et al., 2001) who displayed, after 18-month follow-up, bilateral doubled values of [^{18}F]DOPA putaminal uptake compared with baseline. This patient died from causes unrelated to the study procedures. Neuropathological examination revealed that there was a remarkable outgrowth of dopaminergic neuronal fibers, implying that a significant re-innervation of the putamen was present. Interestingly, there was a striking difference in the number of surviving graft neurons between sides, which was not detectable on PET imaging study (Freed et al., 2001). A neuropathological report is also available from one patient who participated to the randomized controlled trial conducted by Olanow et al. (2003), who died from graft-non-related causes, after discontinuation of immunosuppressive therapy and consequent clinical deterioration. Neuropathological examination revealed the presence of prominent microglial activation, reinforcing the hypothesis of a role of immunosuppression in the long-term outcome (Olanow et al., 2003). Taken together, these studies suggest that a good clinical response can associated with the survival of at least 100,000 dopaminergic neurons in the putamen, which re-innervate about 30–50% of putaminal volume. It has been hypothesized that such levels of graft survival and fiber outgrowth probably corresponds to a recovery of putaminal [^{18}F]FDOPA PET uptake to 50% of normal values (Hagell & Brundin, 2001).

Recent autopsies have also reported the formation of Lewy body pathology in grafted tissue. Two cases grafted 12 and 16 years before their death, who showed good response to clinical and [^{18}F]DOPA PET imaging, showed *post mortem* α-synuclein and ubiquitin-positive Lewy body inclusions within the grafted tissue (Li et al., 2008). Similar results were reported in two case reports, for two different patients who died after 14 years from grafting, who presented Lewy body pathology in the grafted tissue, positive staining for α-synuclein and ubiquitin, and demonstrated low levels of dopamine transporters. These patients responded well to the grafting but were lately reported to have experienced an impairment of their motor condition (Kordower, Chu, Hauser, Freeman, & Olanow, 2008; Kordower, Chu, Hauser, Olanow, & Freeman, 2008). This finding may not have important consequences in the view of the long-term viability of this technique (Cooper et al., 2009; Li, 2012) but could have tremendous implications in the understanding of the pathophysiology of PD. These reports, however, confirm that the magnitude of increase of putaminal [^{18}F]DOPA uptake does not always match the clinical improvement after transplantation.

2.1.2 Dopamine Transporter

An alternative way to assess the presynaptic dopaminergic system in grafted patients is the study of the dopamine transporter. The dopamine transporter is a symporter, located in the presynaptic end of the dopaminergic termination, which clears the synapses from extracellular dopamine. Several PET and SPECT radiotracers which target this enzyme have been developed and all present high affinity for dopamine transporter *in vivo* with some affinity also for the serotonergic and the noradrenergic transporters. Dopamine transporters are downregulated in PD; therefore, the uptake levels of radioligands in the putamen of parkinsonian patients are consistently decreased compared with healthy controls (Chou et al., 2004; Eshuis et al., 2009; Filippi et al., 2005; Guttman et al., 1997; Lee et al., 2000). Given the selective localization of dopamine transporters in the dopaminergic end terminals, the estimated quantity of this enzyme can represent a better indicator of the local dopaminergic re-innervation than $[^{18}F]$DOPA PET. Preclinical PET studies in transplanted rats have indeed shown that the quantification of dopamine transporter could be useful even in determining the amount of surviving graft, as there was a correlation between the degree of striatal binding of the tracer $[^{11}C]$RTI-121 and the amount of surviving tissue measured *post mortem* (Sullivan, Pohl, & Blunt, 1998).

In PD patients who underwent mesencephalic cell transplantation, PET studies with radioligands selective for dopamine transporter, such as $[^{11}C]$ Nomifensine and $[^{76}Br]$CBT, have investigated dopaminergic re-innervation *in vivo*. Using PET with $[^{11}C]$Nomifensine, it was demonstrated, in two patients, that a significant increase in the putaminal binding was detectable 6 months after unilateral transplantation. One of the patients underwent a clinical deterioration over the successive 12 months, that was reflected in a decline in $[^{18}F]$DOPA PET uptake levels and a decrease of $[^{11}C]$Nomifensine binding in the putamen (Hoffer et al., 1992). $[^{76}Br]$ CBT PET imaging was used to estimate the degree of dopaminergic re-innervation in six PD patients who were grafted bilaterally, in two sessions, but did not find any change in $[^{76}Br]$CBT uptake 24 months after grafting, despite a significant increase of $[^{18}F]$DOPA uptake levels (Cochen et al., 2003). This suggests that the clinical benefit induced by the surgical procedure could be related to an increased dopaminergic activity rather than to the restored dopaminergic innervation in the striatum.

Studies with a longer follow-up period, after transplantation, have allowed the long-term clinical benefits of restoring dopaminergic terminal density in the host striatum to be evaluated. Dopamine transporter availability was investigated using $[^{123}I]$PT-CIT SPECT imaging in two bilaterally transplanted patients followed for up to 8 years (Pogarell et al., 2006). A marked bilateral increase in $[^{123}I]$PT-CIT uptake was observed, which peaked after 2 years and was persistent even at the 8-year follow-up scan.

In addition, increased $[^{123}I]PT$-CIT uptake was associated with increased striatal $[^{18}F]DOPA$ uptake as well as improved clinical outcomes measured with the UPDRS scale (Pogarell et al., 2006). At a 14-year follow-up, one of the patients still showed a normal $[^{123}I]FP$-CIT SPECT scan in the right caudate and putamen, with a slight reduction in the contralateral caudate and a significant reduction in the contralateral putamen (Politis et al., 2011). The finding of a satisfactory and a time persistent restoration of dopaminergic nerve endings is in apparent contradiction with neuropathological reports of 2 PD patients who died 14 years after transplantation, for which little expression of dopamine transporter was found in the grafts (Kordower, Chu, Hauser, Freeman, et al., 2008; Kordower, Chu, Hauser, Olanow, et al., 2008). However, the small sample size of these studies limits these findings and studies on larger cohorts are needed to draw conclusions on this issue.

2.1.3 Dopaminergic Receptors

The study of post-synaptic dopaminergic receptors is important to understand the dynamic behavior of grafted tissue in the release of dopamine. $[^{11}C]$Raclopride is a benzamide PET radioligand with affinity for the dopamine type 2 (D_2) receptor that competes with dopamine for its binding. When an external substance (L-DOPA, or metamphetamine) is administered, this induces an acute release of dopamine in the synaptic cleft, that displaces Raclopride binding from the receptors. The amount of displacement of the radioligand can be measured and is an indicator of the degree of dopamine released. It has been demonstrated that a 10% reduction in the D_2 receptor binding potential seen on $[^{11}C]$Raclopride PET corresponds to about 500% increase of dopamine in the synaptic space (Breier et al., 1997). In PD patients, this value is significantly reduced due to an inability of the damaged system to release dopamine after external pharmacological stimulation (Piccini, Pavese, & Brooks, 2003). The ability of grafted tissue to store and release dopamine was studied with a $[^{11}C]$ Raclopride PET challenge in one PD patient, from the Lund/London/Marburg cohort, 10 years after receiving unilateral ventral mesencephalic transplantation. He had a sustained clinical response from the transplantation, and underwent two $[^{11}C]$Raclopride PET scans, one after the injection of saline solution and a second after a metamphetamine challenge (Piccini et al., 1999). Metamphetamine was chosen because it induces a short-latency endogenous release of dopamine (Breier et al., 1997; Tsukada et al., 1999). After the saline injection, the levels of $[^{11}C]$Raclopride binding in the grafted side were unchanged, whereas the non-grafted striatum showed a receptor upregulation, suggesting that grafting had restored a normal dopamine release. After metamphetamine administration, the grafted tissue expressed a marked 26.6% reduction of D_2 receptor occupancy by $[^{11}C]$ Raclopride, indicating an increased release of dopamine. This, together with the finding of normal $[^{18}F]DOPA$ levels, has been interpreted as a sign of the restoration of physiological mechanisms of dopamine production, storage, and release in the grafted tissue, 10 years after transplantation (Piccini et al., 1999). Additionally, a correlation was observed between the degree of striatal increase of $[^{18}F]DOPA$ uptake and the methamphetamine-induced change in $[^{11}C]$Raclopride release in nine patients (Piccini et al., 2005). These results have also been confirmed in later studies (Politis et al., 2011, 2010). In particular, three PD patients who underwent $[^{11}C]$ Raclopride PET according to a metamphetamine challenge 13–16 years after transplantation, continued to display rates of D_2 receptor displacement similar to those obtained by normal controls (Politis et al., 2012). This latter finding indicates that grafts are able to induce an enduring restoration of physiological dopamine function in the host striatum.

2.2 Molecular Imaging of the Serotonergic System

A dysfunction of the serotonergic system has been well documented in PD and has been linked with the presence of both motor and non-motor symptoms (Politis & Loane, 2011; Politis & Niccolini, 2015). The serotonin transporter clears the synaptic cleft from extracellular serotonin and its density, as assessed by the PET radioligand $[^{11}C]DASB$, is a good measure of the state of the serotonergic system in the brain. PD patients show a major, widespread reduction of $[^{11}C]DASB$ binding that progresses over time (Pagano, Niccolini, Fusar-Poli, & Politis, 2017). It has been noted that, despite a significant amelioration in motor symptoms, transplanted PD patients do not improve in terms of non-motor symptoms. $[^{11}C]DASB$ PET imaging has been used in three long-standing graft recipients, 13–16 years after transplantation, who displayed stabilization of their motor symptoms but showed a great deal of non-motor symptoms (Politis et al., 2012). The patients showed significant reductions in $[^{11}C]DASB$ to lower levels than non-grafted PD patients, in the raphe nucleus, as well as in nuclei that receive serotonergic projections from the raphe, such as the midbrain, forebrain, and limbic areas. This was indicative of the progressive deterioration of the serotonergic system in long-lasting PD. It has been suggested that this continuous decline in serotonergic activity might underlie the persistence of non-motor symptoms in these patients (Politis et al., 2012).

2.3 Molecular Imaging of Cerebral Blood Flow

Regional cerebral blood flow (rCBF) is a valid measure of local neuronal activity. This can be achieved by means of PET imaging with $H_2[^{15}O]$ that detects changes in the rCBF in response to motor tasks, such as limb movements. Consequently, the level of local connectivity can be inferred, in healthy and pathological conditions. In PD, it has been demonstrated that, during the performance of voluntary movements, there was a reduction in the activation of the supplementary motor area and of the dorsolateral prefrontal cortex, compared with healthy controls, as a measure of the functional disconnection between striatal and prefrontal areas, that, in turn, reflects the degree of akinesia experienced by these patients (Jahanshahi et al., 1995; Rascol et al., 1992). It was demonstrated that dopaminergic and stereotactic surgical therapy were able to restore this connection (Ceballos-Baumann et al., 1999; Fukuda et al., 2001; Rascol et al., 1994). Researchers have tried to test the effects of graft transplantation in the recovery of these connections. Four graft recipients underwent a series of $H_2[^{15}O]$ PET scans during paced joystick movements with their left hand in freely selected directions. After transplantation, patients showed an increase in the activation of the supplementary motor area and dorsolateral prefrontal cortex regions that reached statistical significance at 18.3 months follow-up, meaning that graft transplantation was able to restore cortical activation during movement. By contrast, a discrepancy was observed with $[^{18}F]$ DOPA striatal uptake levels, which increased at earlier time points compared to $H_2[^{15}O]$ PET. This result may be interpreted by postulating that $[^{18}F]$ DOPA PET modifications are related to the local reconstitution of the dopaminergic pool at the presynaptic level, whereas $H_2[^{15}O]$ PET detects the establishment of effective afferent and efferent graft-cortical connections (Piccini et al., 2000). The timeline of functional improvement seen by $H_2[^{15}O]$ PET is in line with the clinical experience, and it can be hypothesized that the clinical improvement of graft recipients is related to the re-establishment of functional reconnection of dopaminergic terminals to distant areas.

2.4 Magnetic Resonance Spectroscopy

In addition to PET and SPECT, other imaging techniques have been employed to evaluate the survival and function of grafted tissue into host PD patients. Proton magnetic resonance spectroscopy ($[^1H]MRS$) is an MRI technique with less spatial and temporal resolution than PET and

SPECT, but that allows the direct visualization of metabolites and small molecules, which reflect the presence and function of specific neuronal populations. N-acetylaspartate (NAA) is a specific adult marker for neurons and axons (Jacobson, 1959) but is lacking in fetal tissues. Therefore, its presence in grafted tissue would be a marker of survival and maturation of transplanted tissue in PD recipients. Twenty-four PD patients underwent bilateral transplantation of fetal tissue in the putamen, and an [^{1}H]MRS exam approximately 1 year after transplantation. They showed increased levels of NAA around the grafted tissue compared to fetal data, which acted as controls, indicative of the survival and maturation of the grafted tissue into adult neuronal cells (Ross et al., 1999). This suggests a potential role of this imaging technique in the follow-up assessment of graft survival and maturation.

2.5 Functional Magnetic Resonance Imaging

Functional MRI (fMRI) detects localized changes in the levels of blood oxygenation in brain regions activated by specific tasks; simultaneous regional activation or deactivation in response to a task paradigm is interpreted as the reflection of an enhanced or decreased connectivity between distant brain areas. In PD, it has been demonstrated that motor sequence tasks induce a reduction in the activation of the supplementary motor area compared with healthy controls, in accordance to the aforementioned studies using $H_2[^{15}O]$ PET (Haslinger et al., 2001; Rowe et al., 2002). Moreover, the extent of supplementary motor area activation correlates with the severity of motor alteration (Tessa et al., 2013). Functional MRI techniques have been used to investigate whether grafted tissue in the putamen of PD patients established new connections with cortical regions (Bluml, Kopyov, Jacques, & Ross, 1999). Two PD patients were studied, 2 and 3 years after receiving bilateral transplantation, using task-based fMRI with a motor paradigm consisting in shifting and rotating a lever in random directions as fast as possible. They detected an increase in the activation of the grafted putamen when patients performed the motor task with the contralateral arm, and no activation when this task was performed with the ipsilateral arm, suggesting a selective functional rewiring of the grafted tissue (Bluml et al., 1999).

2.6 Imaging of Graft-Induced Dyskinesia

PET imaging studies in transplanted PD patients with GIDs have been central in verifying the possible hypotheses about the pathogenesis of this long-term motor complication (Fig. 2). In the first large randomized

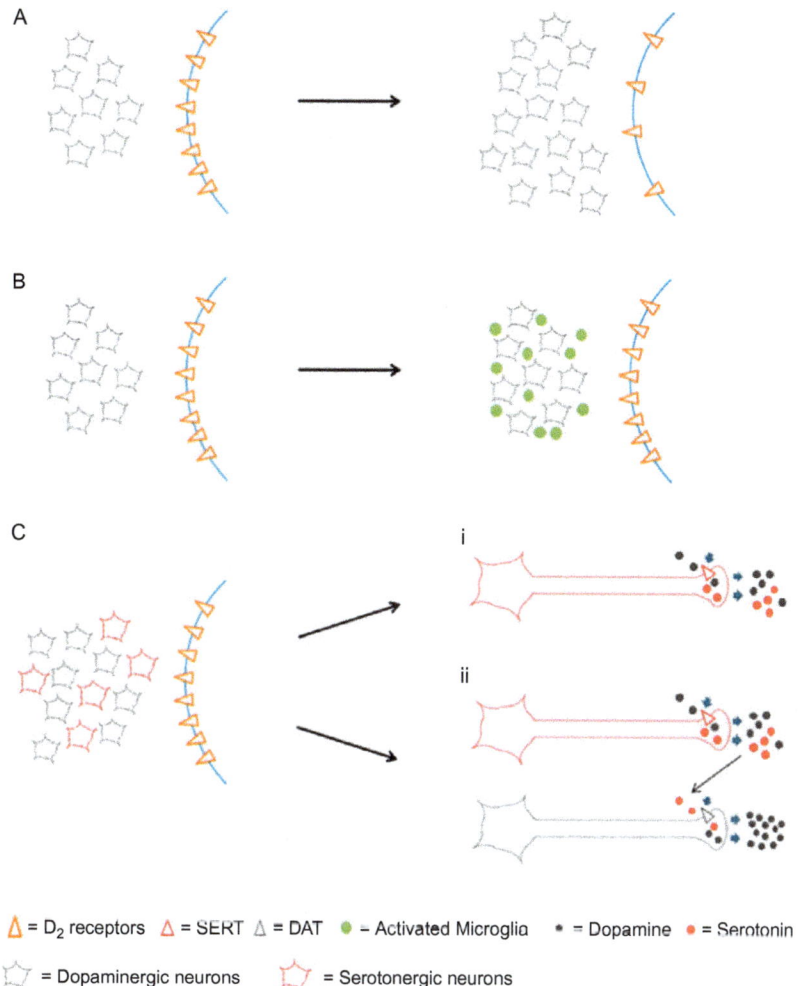

$\triangle$ = D$_2$ receptors $\triangle$ = SERT $\triangle$ = DAT   ● = Activated Microglia   • = Dopamine   • = Serotonin

⬠ = Dopaminergic neurons   ⬠ = Serotonergic neurons

Fig. 2 Hypothesized mechanisms underlying graft-induced dyskinesias. (A) The grafted tissue contains functioning dopaminergic neurons (*black*) that restore dopaminergic metabolism both on the presynaptic and the post-synaptic end (*left*). With time, the grafted tissue may overgrow, leading to increased production of endogenous dopamine that causes a downregulation of D$_2$ receptors (*orange triangles*) on the membrane of striatal neurons (*right*). (B) The functioning grafted tissue in the striatum may be attacked by activated microglia (*green circles*), before or after the discontinuation of immunosuppressive therapy, and may hinder the function of the transplanted tissue. (C) The grafted tissue may contain a proportion of serotonergic neurons (*red*), which alter the normal physiology of the dopaminergic terminals through two steps: (i) alongside releasing serotonin (*red circles*) in the synaptic cleft, they may take up dopamine through the serotonergic transporter (SERT) on the presynaptic terminal (*red triangle*) and release it as a "false neurotransmitter" and (ii) the circulating serotonin in the synaptic cleft may activate the dopamine transporter (DAT; *black triangle*). This results in an increased release of dopamine (*black circles*) from the dopaminergic terminal.

controlled trial, it was suggested that an overgrowth of dopaminergic terminals inside the grafted tissue was responsible for this disorder (Freed et al.,
2001). In support to this hypothesis, in a successive study, the five dyskinetic
patients from the trial were studied with [^{18}F]DOPA PET, up to 24 months
after transplantation (Ma et al., 2002). The authors reported that transplant
recipients with GIDs showed a significant increase in [^{18}F]DOPA uptake in
the left putamen, irrespective from the side of transplantation. When a
voxel-by-voxel approach was used, it was seen that the areas of tracer
over-uptake were localized in the dorsal (which showed preoperative
impairment of dopaminergic innervation) and ventral putamen (which conversely was normal at baseline), suggesting that local imbalances of dopaminergic innervation in specific areas of the left putamen could be responsible
for the appearance of GIDs (Ma et al., 2002). Interestingly, three of the five
patients studied, showed dyskinesias in their head and neck, which are
somatotopically represented in the ventral part of the putamen
(Ma et al., 2002).

The study of the density of the post-synaptic D$_2$ receptors has provided
additional data about the amount of local dopamine release in the grafted
tissue. According to the [^{11}C]Raclopride paradigm, the induction of dopamine release in the synaptic extracellular space can displace the radiotracer
binding from the D$_2$ receptor, in a way that is proportional to the amount of
released dopamine. Consequently, if patients with GIDs suffer from an
excessive release of dopamine from the transplanted tissue, the pharmacological challenge would induce an excessive reduction of [^{11}C]Raclopride
uptake. Huang and colleagues tested two patients with GIDs with [^{11}C]
Raclopride PET before and after an acute administration of L-DOPA. They
found that acute administration of L-DOPA caused an excessive reduction
in [^{11}C]Raclopride binding in the grafted putamen, which was therefore
indicative of an excessive dopaminergic drive from the grafted tissue and
seemed to confirm this hypothesis (Huang et al., 2003). However, successive
works did not replicate this result. [^{11}C]Raclopride PET was tested in nine
grafted patients with GIDs from the Lund/London/Marburg cohort
according to a metamphetamine challenge. The investigators did not notice
any pathological putaminal D$_2$ receptor displacement of [^{11}C]Raclopride
caused by either methamphetamine or saline solution, finding no indirect
evidence of an increased dopamine release from grafted tissue. Furthermore,
they did not find any correlation between [^{11}C]Raclopride uptake levels
pre- and post-metamphetamine challenge and dyskinesia severity scores
(Piccini et al., 2005). This data were confirmed by other studies which fell

short in detecting any correlation between the severity of dyskinesias and the extent of $[^{18}F]$DOPA putaminal uptake (Hagell et al., 2002). Overall, these latter studies challenge the hypothesis of an excessive dopaminergic drive from the grafted tissue pointing rather to an incomplete or aberrant dopaminergic re-innervation of the striatum (Hagell & Cenci, 2005; Olanow et al., 2003).

Other hypotheses have been formulated to justify the appearance of GIDs. The relative delay in the onset of this motor complication has been put in relation with the discontinuation of immunosuppressive therapy. Particularly, the data from one of the two randomized controlled trials indicated that the onset of GIDs was in all cases successive to the suppression of immunotherapy, suggesting that the presence of local inflammatory response after withdrawal of these medications may play a role (Olanow et al., 2003). Piccini and colleagues showed that the dyskinesia scores in a cohort of six grafted patients with GIDs impaired markedly after discontinuation of immunosuppression (Piccini et al., 2005). However, three of the six patients studied developed GIDs before immunosuppression withdrawal. PET studies in ungrafted PD patients, using the first-generation PET tracer $[^{11}C]$ PK11195, which binds the translocator protein (TSPO) expressed on activated microglial, have indicated the presence of widespread cerebral microglial activation (Gerhard et al., 2006; Ouchi et al., 2005). Furthermore, neuropathological findings from patients who deteriorated clinically after discontinuing immunosuppression therapy indicated the presence of prominent microglial activation around the grafted tissue (Kordower et al., 1998; Olanow et al., 2003). Despite these observations, no studies have been performed to date with imaging methods aimed at verifying the levels of inflammation in grafted patients at early stages (when the degree of inflammation is likely to be at its highest) and according to the onset of GIDs.

More recently, the hypothesis for the role of graft cellular composition has gained ground to explain the pathogenesis of GIDs. It is known that the composition of cell suspension derived from mesencephalic fetal tissue is not homogeneous but is composed of non–dopaminergic neurons and non-neuronal cells that may influence the function of the tissue after transplantation (Isacson, Bjorklund, & Schumacher, 2003). Among these populations, a relevant component of human grafts is constituted by serotonergic neurons, as found in neuropathological studies (Mendez et al., 2008). Serotoninergic neurons are able to take up exogenous L-DOPA, convert it to dopamine and store and release it; this process can be relevant when there is a shortage of dopaminergic neurons (Carta, Carlsson, Kirik, & Bjorklund,

2007). Two patients from the Lund/London/Marburg cohort who underwent ventral mesencephalic cells transplantation were studied with [^{11}C] DASB PET imaging (Politis et al., 2010). They had an initial motor recovery but later development of GIDs. [^{11}C]DASB binding in the striatum of these two subjects was remarkably increased bilaterally (by 123% and 252%, respectively, in the caudate and by 77% and 150%, respectively, in the putamen). These values were superior compared to advanced PD patients without transplantation and, in the case of caudate uptake, superior also to normal values (Fig. 3). Moreover, the [^{11}C]DASB/[^{18}F]DOPA binding ratio was excessively increased in the two PD patients with GIDs compared to normal controls. The subsequent administration of the serotonin 1A (5-HT_{1A}) agonist buspirone, which inhibits serotonergic neurotransmission by activating inhibitory autoreceptors, improved dramatically the dyskinetic symptoms in these patients (Politis et al., 2010). Based on these results, it was hypothesized that an excessive serotonergic drive in the graft may be a crucial player in the generation of GIDs. The serotoninergic terminals may take up dopamine through the serotonin transporter and release it as a "false neurotransmitter." Alternatively, or in addition, the excessive release of serotonin may influence negatively the capacity of the dopamine transporter in the dopaminergic terminals to take up dopamine and, thus, to release it in the synaptic cleft (Politis, 2010; Politis et al., 2010). However, despite the rate of serotonergic excessive activity was significantly different between the two patients studied, the severity of their GIDs was not. Therefore, a pathological increase only in the serotonergic activity was not sufficient to explain the development of GIDs. To test the hypothesis of a

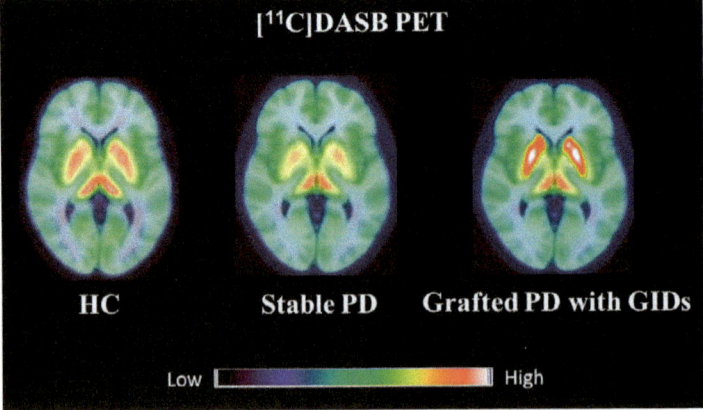

Fig. 3 PET images with [^{11}C]DASB of a healthy control (HC) (*left*), a Parkinson's disease (PD) patient without graft-induced dyskinesias (GIDs) (*center*) and with GIDs (*right*).

dysregulation in the relationship between serotonergic and dopaminergic innervation in the graft, a patient with GIDs was further studied to measure the relationship between the availability of the serotonin transporter, with $[^{11}C]DASB$ PET, and of the dopamine transporter, with $[^{18}F]DOPA$ PET and $[^{123}I]FP$-CIT SPECT (Politis et al., 2011). It was found that the ratio, in the grafted striatum, between the serotonin transporter activity and dopaminergic presynaptic activity was significantly increased (Politis et al., 2011). This finding yielded further support to the hypothesis of an aberrant serotonergic-to-dopaminergic ratio in the generation of GIDs (Fig. 2). A study on the 6-OHDA animal model confirmed this hypothesis. In the grafted mouse model, it was demonstrated that the presence of dopaminergic or serotonergic neurons alone was not able to induce GIDs; however, when mice were exposed to grafted tissue containing both serotonergic and dopaminergic neurons they developed GIDs. Furthermore, the administration of $5\text{-}HT_{1A}$ agonists reverted successfully the dyskinetic symptoms (Shin, Garcia, Winkler, Bjorklund, & Carta, 2012).

To date, this remains the most convincing hypothesis for the pathogenesis of GIDs in transplanted PD patients. Similar mechanisms for the generation of dyskinesias have been identified also in ungrafted PD patients who develop levodopa-induced dyskinesias (Pagano et al., 2017; Politis et al., 2014; Roussakis, Politis, Towey, & Piccini, 2016), reinforcing the role of a dysregulation of the serotonergic over the dopaminergic system in the generation of these motor complications in PD. Although selective studies on transplantation recipients who did not develop GIDs and confirmatory neuropathological studies are lacking, these findings have potentially critical implications in the future methodology of grafting. Techniques aimed at isolating dopaminergic neurons from the serotonergic neurons in the cell suspension may minimize the risk of long-term development of GIDs; moreover, treatment with $5\text{-}HT_{1A}$ agonists may be used in grafted PD patients to reduce the severity of GIDs (Politis & Piccini, 2012).

2.7 Conclusions

Neuroimaging has played a central role in studying the outcome and success of graft transplantation in PD. Although it allows the survival and functionality of grafted tissues to be defined, neuroimaging data do not always reflect the clinical outcome experienced by the patients. PET imaging can be used to select patients who would most benefit from this surgical procedure, as well as to monitor the survival and functionality of the grafted tissue. Moreover, by use of PET imaging, it has been possible to unravel the pathophysiology underlying the presence of motor complications successive to grafting and to suggest possible therapeutic measures. More studies, with more accurate selection of participants, design and definition of outcomes, are warranted to improve our knowledge about the outcome of graft transplantation in PD patients.

3. TRANSPLANTATION IN HUNTINGTON'S DISEASE

The main pathophysiological alteration underlying HD, an inherited neurodegenerative disorder caused by a trinucleotide CAG repeat expansion in the huntingtin gene, is the selective neurodegeneration of the medium spiny neurons in the striatum. These GABAergic neurons mediate inhibitory signals along the cortico-striato-pallidal circuitry and present, on their surface, the dopamine D_1 and D_2 receptors expressed in the *direct* and *indirect* dopaminergic pathway, respectively. The loss of these neurons causes a complex array of motor, cognitive, and behavioral symptoms that start insidiously and progress relentlessly for 2 or 3 decades, ultimately leading to death. Currently, no therapy has demonstrated any capability to modify the disease course. The rather circumscribed population of degenerated striatal neurons in HD makes this disease a potentially ideal target site for cell transplantation therapy.

The rationale underlying cellular transplantation for HD resides in the potential of selected stem cells to differentiate into medium spiny neuron cell lineage and, once transplanted in the host brain, to replace the function of the degenerated neurons (Precious, Zietlow, Dunnett, Kelly, & Rosser, 2017). The human ganglionic eminence is a region that contains stem cells that eventually mature to become medium spiny neurons (Deacon, Pakzaban, & Isacson, 1994; Olsson, Campbell, Wictorin, & Bjorklund, 1995). Initial preclinical studies have used cell populations coming from the lateral part of the ganglionic eminence, with satisfying results in terms of graft outgrowth and function (Grasbon-Frodl et al., 1997). Other studies found that human grafts containing a mixture of lateral and medial ganglionic eminence grafts displayed the highest chance to well integrate in the host brain, differentiate into large and functional striatal tissue (Pundt et al., 1994; Watts et al., 2000), and ameliorate cognitive and motor symptoms in animal models of HD (McLeod et al., 2013; Schackel, Pauly, Piroth, Nikkhah, & Dobrossy, 2013; Yhnell, Dunnett, & Brooks, 2016). Therefore, this represents the "gold standard" source for grafts in human HD studies. The open-label trials on graft transplantation conducted in HD are summarized in Table 2.

Table 2 Summary of the Open-Label Studies With Transplantation of Fetal Graft Tissue in Huntington's Disease

References	Patients	Procedure	Imaging Modality	Main Findings
Ross et al. (1999)	14 HD 24 PD	Bilateral graft in the putamen and caudate (HD patients) and in the putamen (PD patients)	[^{1}H]MRS	Evidence of increased amounts of N-acetyl aspartate in grafts from both HD and PD subjects, compared with fetal transplants
Bachoud-Levi et al. (2000)	5 HD	Transplantation in the left striatum and, after 1 year, in the right striatum	[^{18}F]FDG PET	Increased metabolic activity in the striatum of three out of five patients, who maintained steady clinically over the 2-year follow-up period. Two patients did not show any metabolic modification and worsened clinically
Hauser et al. (2002)	7 HD	Bilateral transplantation in the striatum	[^{18}F]FDG PET [^{11}C]Raclopride PET [^{11}C]SCH23390 PET	Stabilization of motor scores assessed with UHDRS. Stabilization of [^{11}C]SCH23390 uptake in the grafted tissue, with decline of [^{11}C]Raclopride uptake in the same regions
Rosser et al. (2002)	4 HD	Unilateral transplantation in the striatum	[^{11}C]Raclopride PET	During the 6 months transplantation period, there were no clinical signs of progression of the disease
Gaura et al. (2004)	5 HD	Transplantation in the left striatum and, after 1 year, in the right striatum. 4-Year follow-up	[^{18}F]FDG PET	Three patients improved and exhibited sustained increase of metabolic activity in the striatum over the entire course of follow-up that was accompanied by a stabilization of the clinical picture. Two patients who did not improve initially continued to progress both clinically and radiologically
Furtado et al. (2005)	7 HD	Bilateral transplantation in the striatum. 2-Year follow-up	[^{18}F]FDG PET [^{11}C]Raclopride PET [^{11}C]SCH23390 PET	Patients showed widespread reduction of glucose metabolism over the course of the follow-up. This was accompanied by a loss of D_2 receptors, but not of D_1 receptors. There was no clinical benefit from the procedure

Continued

Table 2 Summary of the Open-Label Studies With Transplantation of Fetal Graft Tissue in Huntington's Disease—cont'd

References	Patients	Procedure	Imaging Modality	Main Findings
Bachoud-Levi et al. (2006)	5 HD	Transplantation in the left striatum and, after 1 year, in the right striatum. Follow-up of 6 years	$[^{18}F]FDG$ PET	Three patient exhibited clinical improvement that stabilized after 2 years and started progressing again after 4–6 years. Imaging showed heterogeneous levels of hypometabolism diffusely, with the exception of the graft sites and of the frontal cortex. The two patients who initially worsened continued to progress
Reuter et al. (2008)	2 HD	Bilateral transplantation in the striatum	$[^{11}C]$Raclopride PET	One patient showed an increase of D_2 receptors in the striatum that was sustained over a 5-years follow-up. The other patient did not show any improvement
Krebs et al. (2011)	10 HD	Bilateral intrastriatal transplantation	$[^{18}F]FDG$ PET	No evidence of increase in glucose metabolism in the grafted tissue. Five patients out of 10 developed antibodies against HLA type I and II
Pagamini et al. (2014)	10 HD	Bilateral transplantation in putamen and caudate nuclei. Follow-up of a median 11.2 years	$[^{18}F]FDG$ PET $[^{123}I]IDZM$ SPECT	Reduction of the progression in motor and cognitive scores. Increase in glucose metabolism in the grafted tissue that decreased slightly starting from the fourth year after surgery. Increase in striatal D_2 concentration after surgery

Abbreviations: *PD*, Parkinson's disease; *PET*, positron emission tomography; *SPECT*, single photon emission computed tomography; *HD*, Huntington's disease; *UPDRS*, unified Huntington's disease rating scale; D_2 *receptors*, dopamine type 2 receptors; D_1 *receptors*, dopamine type 1 receptors; *HLA*, human leukocyte antigen.

3.1 Molecular Imaging of the Dopaminergic System

The *in vivo* quantification of dopaminergic receptor density in medium spiny neurons provides information on whether the grafted neuroblasts have evolved into physiologically functional striatal neuronal cells. Several PET and SPECT tracers exist to selectively study the D_1 (e.g., $[^{11}C]$ SCH23390 PET) and D_2 (e.g., $[^{11}C]$Raclopride PET and $[^{123}I]$IBZM SPECT) receptor density and have found extensive application in the study of HD pathophysiology. In ungrafted HD patients, the availability of D_1 and D_2 receptors in putamen and caudate nuclei has been found to be significantly reduced by up to 75% in symptomatic stages, with an annual progression rate of decline of 5% for D_1 receptors and 3% for D_2 receptors (Andrews et al., 1999; Antonini et al., 1996; Backman, Robins-Wahlin, Lundin, Ginovart, & Farde, 1997; Ciarmiello et al., 2006; Herben-Dekker et al., 2014; Squitieri et al., 2009). It has been estimated that, to obtain clinically relevant consequences, a loss of post-synaptic dopaminergic receptors in the striatum of at least 50% may be needed (Weeks, Piccini, Harding, & Brooks, 1996). One theoretical limitation for the use of these tracers in HD is that, due to behavioral symptoms, many patients take neuroleptics and other drugs that activate dopaminergic receptors. For this reason, to perform PET and SPECT imaging with these tracers, patients need to have a short-term washout period of dopaminergic medication, which may not be sustained by all patients.

Seven HD subjects underwent bilateral striatal transplantation in two-stage procedures with 8–9-week-old fetal tissue. They were evaluated in their regional glucose metabolism using $[^{18}F]$FDG PET, as well as the density of post-synaptic D_1 and D_2 receptors using $[^{11}C]$SCH23390 and $[^{11}C]$ Raclopride PET, respectively, at an average of 12 months after surgery (Hauser et al., 2002). These patients remained clinically stable at follow-up; $[^{11}C]$SCH23390 striatal levels remained at preoperative levels, whereas $[^{11}C]$Raclopride binding declined (Hauser et al., 2002). After 2 years from transplantation, the density of post-synaptic D_1 receptors remained stable at preoperative levels, whereas striatal D_2 receptor density declined by 14–24%, despite the presence of stable global metabolic levels (Furtado et al., 2005). These results, although obtained from a small sample, suggest that the reconstitution, suggested that in grafted HD patients, the progression of reconstitution of D_1 and D_2 receptors in the striatum may follow different timeframes.

In another study conducted in the United Kingdom, two HD subjects underwent bilateral striatal transplantation of 9–10-week-old fetal tissue and were followed up for up to 30 months by clinical and imaging ($[^{11}C]$ Raclopride PET) assessments (Reuter et al., 2008). In one subject, $[^{11}C]$ Raclopride binding in the striatum increased by 23% after 6 months and then declined steadily in the following months, similarly to non-grafted HD patients, although at 30 months, striatal D_2 density remained higher than preoperative data and functional independence and depression scores were better than baseline. In contrast, the other patient declined in both clinical and imaging measures similarly to non-grafted HD patients, suggesting that, in this case, the graft did not survive (Reuter et al., 2008).

In the NEST-UK study, a multicenter collaboration study aimed at assessing the long-term feasibility of neural transplantation in mild HD patients, five subjects who received bilateral sequential striatal graft transplantation and have been followed for about 10 years (Rosser et al., 2002). After surgery, the patients continued to decline clinically; additionally, the follow-up $[^{11}C]$Raclopride PET scans showed a rate of decline of the striatal density of D_2 receptors similar to what was observed in control, ungrafted HD patients (Barker, Mason, et al., 2013). The poor outcome of this study may have been influenced by the fact that, for safety reasons and to avoid any risk of graft overgrowth in the host tissue, this trial was designed as to transplant a low number of cells. Contrasting results were, however, obtained by another open-label trial (Paganini et al., 2014) in which 10 HD patients received similarly small amounts of graft tissue and were tested with $[^{123}I]$IBZM SPECT imaging. Those who tolerated washout and received follow-up scans, showed a significant bilateral increase of striatal D_2 receptor binding, that was reached at 12 months and remained stable at 18 and 24 months; these results were paired by a similar improvement in regional glucose metabolism (Paganini et al., 2014). However, at 18- and 24-month time points, data were gathered from only one patient who could stand a washout period.

3.2 Molecular Imaging of Glucose Metabolism

The regional cerebral cellular integrity in HD can be measured indirectly according to its metabolic level. The PET tracer $[^{18}F]$Fluoro-deoxy-glucose ($[^{18}F]$FDG) is an analog of glucose, which is avidly taken up by neuronal cells in a degree that is proportional to their integrity. All manifest HD patients, display high degrees of $[^{18}F]$FDG PET hypometabolism in various striatal and cortical areas, which correlates with the degree of motor and cognitive symptoms and progresses in the striatum at a rate of about 7% per year

(Hayden et al., 1987; Kremer et al., 1999; Kuwert et al., 1990; Young et al., 1986). It is expected, therefore, that cellular replacement therapy may restore normal metabolic levels and be beneficial in modifying the clinical picture.

An open-label, pilot study on five manifest HD patients was designed according to the guidelines on methodology of graft transplantation and clinical follow-up on HD patients (Quinn et al., 1996). The patients received bilateral caudate and putamen transplantation surgery in two sessions 12 months apart, and were followed up for an additional 12 months after the final surgery, with clinical and neuroimaging measures obtained by serial [^{18}F]FDG PET scans (Bachoud-Levi et al., 2000). Three HD patients showed stability or a slight improvement in neuropsychological and motor tests. In those patients, [^{18}F]FDG PET demonstrated increased metabolic activity in small patchy striatal areas, which co-registered with regions of hypointensity on T1-weighted MRI and represented metabolically active grafted tissue. By contrast, two patients did not respond clinically to transplantation and their glucose metabolic activity in the striatum progressively decreased, at a rate similar to ungrafted manifest HD control patients. The motor improvement in HD patients who responded clinically to the transplantation was delayed rather than immediate, as was reported in PD, and is consistent with a mechanism of maturation of the grafted neurons that must take place before yielding clinical effects. A further, successive voxel-based analysis of these data showed that the three patients who stabilized or slightly improved clinically had proportional levels of decrease in cortical hypometabolic voxels that were more represented in the frontal lobes. By contrast, [^{18}F]FDG PET data in the two patients who did not respond clinically to the grafting, were consistent with a widespread spread of cortical hypometabolism. In addition, the changes in glucose metabolism of cortical and striatal areas were paralleled in all responders. This was interpreted as the effect of the reconstruction of the physiological cortico–striato–thalamo-cortical loop, brought about by the graft tissue (Gaura et al., 2004).

When the three responder HD patients were followed up for 6 years after grafting, [^{18}F]FDG average uptake levels were found to be declined by only 7%, which is the average degree of metabolic decline that is normally expected over only 1 year in HD (Kremer et al., 1999). In addition, voxel-based analysis revealed that the number of hypometabolic voxels was still lower or unchanged compared to the preoperative state. This was accompanied by regional unchanged striatal and cortical levels of glucose consumption when compared with the earlier time point. Clinically, patients experienced a slight, non-significant, deterioration in motor and cognitive performances consistent with imaging data (Bachoud-Levi et al., 2006). The study of this transplanted HD cohort has demonstrated that grafting was able to provide enduring improvements on both clinical and imaging outcomes, thereby modifying the course of the disease. However, the positive outcomes were only slightly favorable as the change was not definitive and grafting did not prevent for the disease to progress. In another study, 10 HD patients received bilateral striatal transplantation of whole ganglionic eminence cells. After a 36-month follow-up, [^{18}F]FDG PET failed to detect any sign of survival or activity of the grafts in any patient. The imaging data were mirrored by the lack of clinical improvement in all patients but one, who displayed a decrease of Unified Huntington's disease rating scale (UHDRS) score (Krebs et al., 2011).

Opposite findings were obtained by a more recent open-label study based in Italy, that studied brain glucose metabolism in 10 manifest HD patients grafted bilaterally, 2–6 months apart, and followed up by means of [^{18}F]FDG PET, for an average of 4 years. At 24 months, all patients displayed an increase in striatal and cortical uptake of [^{18}F]FDG, consistent with an improvement of glucose metabolism in these two areas. After a further 24 months, both striatal and cortical metabolism was maintained at levels similar to the first follow-up. Motor and neuropsychological performances in recipients showed a less severe decline, compared with a group of non-transplanted manifest HD patients (Paganini et al., 2014).

A frequent problem experienced by transplantation recipients is the allo-immunization to donor's antigens. It has been hypothesized that the lack of clinical response observed in some cases may be partly due to this phenomenon. For example, a transplanted HD patient experienced a sudden clinical deterioration, that was reflected by a local, unilateral striatal reduction of glucose metabolism and the concomitant presence of anti-HLA-I antibodies in the blood (Krystkowiak et al., 2007). However, it has subsequently been demonstrated that both clinical response and imaging findings from transplantation were not correlated with the presence of class I and II HLA antibodies (Krebs et al., 2011).

3.3 Magnetic Resonance Spectroscopy

[^{1}H]MRS has been used in some studies as a biomarker in HD; patients with manifest and pre-manifest HD display lower levels of NAA and manifest HD show significantly increased levels of myo-inositol (Sturrock et al., 2010).

NAA is a molecule that is expressed only in mature neuronal cells and not in immature, fetal cells. Twenty-four HD patients who received graft transplantation were studied with [^{1}H]MRS imaging and showed 1-year after transplantation, a similar average amount of NAA compared to untreated adult patients, suggesting that this may reflect the maturation of grafted tissue inside the putamen (Ross et al., 1999). However, it was not reported whether this reflected in a favorable clinical outcome of the surgery.

3.4 Conclusions

Transplantation in HD has provided inconsistent and short-term results in the few patients who have received this experimental procedure. Neuroimaging techniques are able to track over time, the survival of grafting from a functional point of view, and the decline in regional function detected by PET imaging may reflect the degeneration to which grafted tissue may come across in the long term. This seems to be confirmed by the few neuropathological studies available, in which despite a good survival of the grafts after as long as 18 months (Freeman et al., 2000) there are conflicting reports about the capacity of grafted cells to survive in an already compromised local environment over the years (Cicchetti et al., 2009; Keene et al., 2007). This hinders the applicability of this surgical technique in HD patients in the design of randomized controlled trials.

4. TRANSPLANTATION IN OTHER MOVEMENT DISORDERS

Primary dystonia is an inherited condition characterized by the presence of generalized dystonia, alone or in combination with other neurological symptoms. A case study of an 18-year-old young man with DYT1 caused by mutations in the gene Tor1 was reported (Ren et al., 2016). He underwent bilateral neural fetal stem cell grafting in both the globi pallidi and was followed up clinically and by means of [^{18}F]FDG PET repeated serially up to 4 years after surgery. The patient sustained significant and progressive motor improvement and an increase in self-autonomy. [^{18}F]FDG PET imaging showed a progressive bilateral increase of glucose metabolism in the putamen and thalamus when compared to the levels of glucose metabolism before surgery (Ren et al., 2016). This case has not been followed by other reports and more studies are warranted in order to assess the feasibility and efficacy of this procedure in primary dystonia.

5. NEUROIMAGING OF ADULT-DERIVED MESENCHYMAL STEM CELLS IN MOVEMENT DISORDERS

Transplantation of fetal cells carries the major disadvantages of the difficulty in supplying the adequate amount of striatal tissue to achieve a pool of totipotent stem cells, and of ethical constraints about the collection of fetal cells from aborted embryos. Therefore, alternative ways to obtain stem cells have been developed. Stem cells can be also obtained from multiple adult sites, such as the placenta, the adipocytes, and bone marrow. Particularly, mesenchymal stem cells obtained from the bone marrow have been extensively investigated in animal studies and have yielded promising results in the animal model of PD (Cova et al., 2012; Fricker-Gates & Gates, 2010). Mesenchymal stem cells can be injected peripherally and subsequently they reach the target tissue, where they settle and grow. They have critical biological properties which could be beneficial for the treatment of movement disorders. First, by replacing the loss neurons in the target sites they can restore neuronal pools in areas of neurodegeneration; second, they can secrete growth factors in the extracellular milieu that may prevent further degeneration and promote the growth of native neurons (Yasuhara et al., 2005); and third, it has been noted that they display immunomodulatory and anti-inflammatory properties (Fibbe, Nauta, & Roelofs, 2007).

Few studies on human patients with degenerative movement disorders, such as multiple system atrophy, heredodegenerative ataxias, and progressive supranuclear palsy, have so far been performed using mesenchymal stem cells. PET imaging has been used, coupled with clinical assessments, as outcome measure for these trials.

5.1 Multiple System Atrophy and Degenerative Ataxias

Multiple system atrophy (MSA) is a progressive neurodegenerative disorder characterized clinically by autonomic dysfunction and by various degrees of severity of parkinsonism and cerebellar ataxia. The disease course is relentlessly progressive and aggressive, leading to death after a mean disease duration of about a decade. Studies conducted on cell replacement therapy in animal models of this disease have shown that grafted animals displayed an increased response to L-DOPA to tests of complex motor behaviors (Kollensperger et al., 2009).

A pilot trial compared 11 MSA patients who received intra-arterial bone marrow-derived mesenchymal stem cells injection to 18 MSA patients who received a placebo. Recipients of mesenchymal stem cells improved clinically on the Unified Multiple System Atrophy Rating Scale (UMSARS) score at each time point for up to 12 months (Lee et al., 2008). [^{18}F]FDG PET imaging also demonstrated an increase of glucose metabolism in the cerebellum and in the white matter, indicating improved levels of glucose metabolism following mesenchymal stem cells injection. This study was followed by a double-blind, placebo-controlled, randomized study that involved 34 MSA patients who received either mesenchymal stem cells injection or placebo. Transplantation recipients showed a slight but significant slowing of clinical progression, as assessed with the UMSARS, at a 12-month follow-up. In contrast to the previous study, the levels of cerebral glucose metabolism did not show any increase. However, the decline of [^{18}F]FDG PET uptake progressed slower in the transplanted group compared with the placebo group, particularly in the cerebellum and in frontal areas (Lee et al., 2012). Overall, the results are potentially interesting although safety issues were raised by the high incidence of post-operatory evidence, on diffusion-weighted MRI, of small ischemic lesions that, for given their appearance in both the treated and placebo arms, have been attributed to the arterial injection and not to the material injected. Although the results from this randomized controlled trial are promising, this safety concern is potentially critical and longer follow-up will need to establish long-term safety of this methodic. Moreover, the low number of patients does not allow conclusive data about the possible efficacy of this method in this devastating illness to be drawn.

Adipocyte-derived mesenchymal stem cells have been implanted in a Taiwan-based phase I/IIa open-label pilot study in six patients with spinocerebellar ataxia 3 (SCA3) and in one patient with multiple system atrophy cerebellar type (MSA-C). After a 12-month follow-up, no discernible modification could be detected using magnetic resonance spectroscopy or [^{18}F]FDG PET imaging, paralleled by a substantial stabilization in their Scale for Assessment and Score of Ataxia scores (Tsai et al., 2017).

5.2 Progressive Supranuclear Palsy

Progressive supranuclear palsy (PSP) is a heterogeneous neurodegenerative disorder characterized clinically by parkinsonism, supranuclear gaze palsy, and dementia which is unresponsive to L-DOPA therapy and is invariably progressive with unfortunate prognosis after diagnosis of only 3–4 years (dell'Aquila et al., 2013).

The history of cell transplantation in this disorder is still at its infancy. After a report of a single case from Korea of a 71-year-old patient with PSP who attained clinical benefit from a systemic and intrathecal infusion of autologous adipose tissue-derived mesenchymal stem cells (Choi, Park, Woo, Kang, & Ra, 2014), a phase I study for injection of mesenchymal stem cells in patients with PSP has recently been designed (Giordano et al., 2014). Here, five PSP patients received bone marrow-derived mesenchymal stem cells through arterial administration and were followed up for 1 year with $[^{123}I]$FP-CIT SPECT and $[^{18}F]$FDG PET imaging. Follow-up data were available only for three patients. After 12 months, motor performances, assessed with the UPDRS part III score, were stabilized, however imaging data showed further deterioration of striatal uptake (Canesi et al., 2016). Further studies on larger cohorts are warranted to assess the efficacy of this procedure in this neurodegenerative disorder.

6. CONCLUSIONS

Cell replacement therapy can potentially have an impact in the management of movement disorders and will likely continue be one of the major topics of research over the next decade. The past trials have taught us important lessons which will help better define selection criteria of candidate patients for this therapy and methodologies of administration and follow-up. However, many questions still need to be addressed, such as which patients are the best candidates to receive this therapy, what, between fetal cell and stem cell, is the best population, what is the optimal treatment paradigm, and what can be done to improve survival of transplanted cells. Second-generation trials are already underway in patients with degenerative movement disorders (Bachoud-Levi, 2017) and hopefully will provide considerable progress in this field.

Neuroimaging is an invaluable tool to the researcher studying the impact of this treatment approach in humans *in vivo*. So far, it has provided precious information on the viability and efficacy of graft transplantation, and the biological function of the grafted tissue in the host brain as well as on tracking outcomes over time. However, much is still unknown about the behavior of grafted tissue in terms of cell differentiation, maturation, and in the connections that it takes with neighboring and distant neuronal cells. Neuroimaging techniques will be of great help to further improve our knowledge about these topics. The use of novel radiotracers targeting the dopaminergic system may improve our knowledge in the study of this molecular pathway; in addition, radiotracers targeting unexplored neurotransmitter pathways and molecular functions may provide additional information about the molecular biology of grafts. New data may also come from innovative MRI techniques including diffusion tensor imaging, which permits to study the structural connectivity between distant brain regions and may be helpful to assess the reconstruction of connections with other structures by grafted tissues; arterial spin labeling techniques, which could help in the definition of the modifications in global and territorial cerebral perfusion induced by the grafts; and finally, susceptibility weighted imaging and neuromelanin-sensitive MR imaging are sensitive to iron deposition and may be of help in studying the conversion of grafted cells, into dopaminergic nigral cells abundant in iron content.

REFERENCES

Andrews, T. C., Weeks, R. A., Turjanski, N., Gunn, R. N., Watkins, L. H., Sahakian, B., et al. (1999). Huntington's disease progression. PET and clinical observations. *Brain*, *122*(Pt. 12), 2353–2363.

Antonini, A., Leenders, K. L., Spiegel, R., Meier, D., Vontobel, P., Weigell-Weber, M., et al. (1996). Striatal glucose metabolism and dopamine D2 receptor binding in asymptomatic gene carriers and patients with Huntington's disease. *Brain*, *119*(Pt. 6), 2085–2095.

Bachoud-Levi, A. C. (2017). From open to large-scale randomized cell transplantation trials in Huntington's disease: Lessons from the multicentric intracerebral grafting in Huntington's disease trial (MIG-HD) and previous pilot studies. *Progress in Brain Research*, *230*, 227–261. https://doi.org/10.1016/bs.pbr.2016.12.011.

Bachoud-Levi, A. C., Gaura, V., Brugieres, P., Lefaucheur, J. P., Boisse, M. F., Maison, P., et al. (2006). Effect of fetal neural transplants in patients with Huntington's disease 6 years after surgery: A long-term follow-up study. *Lancet Neurology*, *5*(4), 303–309. https://doi. org/10.1016/s1474-4422(06)70381-7.

Bachoud-Levi, A. C., Remy, P., Nguyen, J. P., Brugieres, P., Lefaucheur, J. P., Bourdet, C., et al. (2000). Motor and cognitive improvements in patients with Huntington's disease after neural transplantation. *Lancet*, *356*(9246), 1975–1979.

Backman, L., Robins-Wahlin, T. B., Lundin, A., Ginovart, N., & Farde, L. (1997). Cognitive deficits in Huntington's disease are predicted by dopaminergic PET markers and brain volumes. *Brain*, *120*(Pt. 12), 2207–2217.

Barker, R. A., Barrett, J., Mason, S. L., & Bjorklund, A. (2013). Fetal dopaminergic transplantation trials and the future of neural grafting in Parkinson's disease. *Lancet Neurology*, *12*(1), 84–91. https://doi.org/10.1016/s1474-4422(12)70295-8.

Barker, R. A., Mason, S. L., Harrower, T. P., Swain, R. A., Ho, A. K., Sahakian, B. J., et al. (2013). The long-term safety and efficacy of bilateral transplantation of human fetal striatal tissue in patients with mild to moderate Huntington's disease. *Journal of Neurology, Neurosurgery, and Psychiatry*, *84*(6), 657–665. https://doi.org/10.1136/jnnp-2012-302441.

Bjorklund, A., & Stenevi, U. (1979). Reconstruction of the nigrostriatal dopamine pathway by intracerebral nigral transplants. *Brain Research*, *177*(3), 555–560.

Bluml, S., Kopyov, O., Jacques, S., & Ross, B. D. (1999). Activation of neurotransplants in humans. *Experimental Neurology*, *158*(1), 121–125. https://doi.org/10.1006/exnr.1999.7073.

Breier, A., Su, T. P., Saunders, R., Carson, R. E., Kolachana, B. S., de Bartolomeis, A., et al. (1997). Schizophrenia is associated with elevated amphetamine-induced synaptic dopamine concentrations: Evidence from a novel positron emission tomography method. *Proceedings of the National Academy of Sciences of the United States of America*, *94*(6), 2569–2574.

Brooks, D. J., Ibanez, V., Sawle, G. V., Quinn, N., Lees, A. J., Mathias, C. J., et al. (1990). Differing patterns of striatal 18F-dopa uptake in Parkinson's disease, multiple system atrophy, and progressive supranuclear palsy. *Annals of Neurology*, *28*(4), 547–555. https://doi.org/10.1002/ana.410280412.

Broussolle, E., Dentresangle, C., Landais, P., Garcia-Larrea, L., Pollak, P., Croisile, B., et al. (1999). The relation of putamen and caudate nucleus 18F-Dopa uptake to motor and cognitive performances in Parkinson's disease. *Journal of the Neurological Sciences*, *166*(2), 141–151.

Brundin, P., Nilsson, O. G., Strecker, R. E., Lindvall, O., Astedt, B., & Bjorklund, A. (1986). Behavioural effects of human fetal dopamine neurons grafted in a rat model of Parkinson's disease. *Experimental Brain Research*, *65*(1), 235–240.

Brundin, P., Pogarell, O., Hagell, P., Piccini, P., Widner, H., Schrag, A., et al. (2000). Bilateral caudate and putamen grafts of embryonic mesencephalic tissue treated with lazaroids in Parkinson's disease. *Brain*, *123*(Pt. 7), 1380–1390.

Canesi, M., Giordano, R., Lazzari, L., Isalberti, M., Isaias, I. U., Benti, R., et al. (2016). Finding a new therapeutic approach for no-option Parkinsonisms: Mesenchymal stromal cells for progressive supranuclear palsy. *Journal of Translational Medicine*, *14*(1), 127. https://doi.org/10.1186/s12967-016-0880-2.

Carta, M., Carlsson, T., Kirik, D., & Bjorklund, A. (2007). Dopamine released from 5-HT terminals is the cause of L-DOPA-induced dyskinesia in parkinsonian rats. *Brain*, *130*(Pt. 7), 1819–1833. https://doi.org/10.1093/brain/awm082.

Ceballos-Baumann, A. O., Boecker, H., Bartenstein, P., von Falkenhayn, I., Riescher, H., Conrad, B., et al. (1999). A positron emission tomographic study of subthalamic nucleus stimulation in Parkinson disease: Enhanced movement-related activity of motor-association cortex and decreased motor cortex resting activity. *Archives of Neurology*, *56*(8), 997–1003.

Choi, S. W., Park, K. B., Woo, S. K., Kang, S. K., & Ra, J. C. (2014). Treatment of progressive supranuclear palsy with autologous adipose tissue-derived mesenchymal stem cells: A case report. *Journal of Medical Case Reports*, *8*, 87. https://doi.org/10.1186/1752-1947-8-87.

Chou, K. L., Hurtig, H. I., Stern, M. B., Colcher, A., Ravina, B., Newberg, A., et al. (2004). Diagnostic accuracy of [99mTc]TRODAT-1 SPECT imaging in early Parkinson's disease. *Parkinsonism & Related Disorders*, *10*(6), 375–379. https://doi.org/10.1016/j.parkreldis.2004.04.002.

Ciarmiello, A., Cannella, M., Lastoria, S., Simonelli, M., Frati, L., Rubinsztein, D. C., et al. (2006). Brain white-matter volume loss and glucose hypometabolism precede the clinical symptoms of Huntington's disease. *Journal of Nuclear Medicine*, *47*(2), 215–222.

Cicchetti, F., Saporta, S., Hauser, R. A., Parent, M., Saint-Pierre, M., Sanberg, P. R., et al. (2009). Neural transplants in patients with Huntington's disease undergo disease-like neuronal degeneration. *Proceedings of the National Academy of Sciences of the United States of America*, *106*(30), 12483–12488. https://doi.org/10.1073/pnas.0904239106.

Cochen, V., Ribeiro, M. J., Nguyen, J. P., Gurruchaga, J. M., Villafane, G., Loc'h, C., et al. (2003). Transplantation in Parkinson's disease: PET changes correlate with the amount of grafted tissue. *Movement Disorders*, *18*(8), 928–932. https://doi.org/10.1002/mds.10463.

Cooper, O., Astradsson, A., Hallett, P., Robertson, H., Mendez, I., & Isacson, O. (2009). Lack of functional relevance of isolated cell damage in transplants of Parkinson's disease patients. *Journal of Neurology*, *256*(Suppl. 3), 310–316. https://doi.org/10.1007/s00415-009-5242-z.

Cova, L., Bossolasco, P., Armentero, M. T., Diana, V., Zennaro, E., Mellone, M., et al. (2012). Neuroprotective effects of human mesenchymal stem cells on neural cultures exposed to 6-hydroxydopamine: Implications for reparative therapy in Parkinson's disease. *Apoptosis*, *17*(3), 289–304. https://doi.org/10.1007/s10495-011-0679-9.

Deacon, T. W., Pakzaban, P., & Isacson, O. (1994). The lateral ganglionic eminence is the origin of cells committed to striatal phenotypes: Neural transplantation and developmental evidence. *Brain Research*, *668*(1–2), 211–219.

Deckel, A. W., Robinson, R. G., Coyle, J. T., & Sanberg, P. R. (1983). Reversal of long-term locomotor abnormalities in the kainic acid model of Huntington's disease by day 18 fetal striatal implants. *European Journal of Pharmacology*, *93*(3–4), 287–288.

dell'Aquila, C., Zoccolella, S., Cardinali, V., de Mari, M., Iliceto, G., Tartaglione, B., et al. (2013). Predictors of survival in a series of clinically diagnosed progressive supranuclear palsy patients. *Parkinsonism & Related Disorders*, *19*(11), 980–985. https://doi.org/10.1016/j.parkreldis.2013.06.014.

Eshuis, S. A., Jager, P. L., Maguire, R. P., Jonkman, S., Dierckx, R. A., & Leenders, K. L. (2009). Direct comparison of FP-CIT SPECT and F-DOPA PET in patients with Parkinson's disease and healthy controls. *European Journal of Nuclear Medicine and Molecular Imaging*, *36*(3), 454–462. https://doi.org/10.1007/s00259-008-0989-5.

Fibbe, W. E., Nauta, A. J., & Roelofs, H. (2007). Modulation of immune responses by mesenchymal stem cells. *Annals of the New York Academy of Sciences*, *1106*, 272–278. https://doi.org/10.1196/annals.1392.025.

Filippi, L., Manni, C., Pierantozzi, M., Brusa, L., Danieli, R., Stanzione, P., et al. (2005). 123I-FP-CIT semi-quantitative SPECT detects preclinical bilateral dopaminergic deficit in early Parkinson's disease with unilateral symptoms. *Nuclear Medicine Communications*, *26*(5), 421–426.

Freed, C. R., Breeze, R. E., Rosenberg, N. L., Schneck, S. A., Kriek, E., Qi, J. X., et al. (1992). Survival of implanted fetal dopamine cells and neurologic improvement 12 to 46 months after transplantation for Parkinson's disease. *The New England Journal of Medicine*, *327*(22), 1549–1555. https://doi.org/10.1056/nejm199211263272202.

Freed, C. R., Greene, P. E., Breeze, R. E., Tsai, W. Y., DuMouchel, W., Kao, R., et al. (2001). Transplantation of embryonic dopamine neurons for severe Parkinson's disease. *The New England Journal of Medicine*, *344*(10), 710–719. https://doi.org/10.1056/nejm200103083441002.

Freeman, T. B., Cicchetti, F., Hauser, R. A., Deacon, T. W., Li, X. J., Hersch, S. M., et al. (2000). Transplanted fetal striatum in Huntington's disease: Phenotypic development and lack of pathology. *Proceedings of the National Academy of Sciences of the United States of America*, *97*(25), 13877–13882. https://doi.org/10.1073/pnas.97.25.13877.

Freeman, T. B., Olanow, C. W., Hauser, R. A., Nauert, G. M., Smith, D. A., Borlongan, C. V., et al. (1995). Bilateral fetal nigral transplantation into the postcommissural putamen in Parkinson's disease. *Annals of Neurology*, *38*(3), 379–388. https://doi.org/10.1002/ana.410380307.

Fricker-Gates, R. A., & Gates, M. A. (2010). Stem cell-derived dopamine neurons for brain repair in Parkinson's disease. *Regenerative Medicine*, *5*(2), 267–278. https://doi.org/10.2217/rme.10.3.

Fukuda, M., Mentis, M., Ghilardi, M. F., Dhawan, V., Antonini, A., Hammerstad, J., et al. (2001). Functional correlates of pallidal stimulation for Parkinson's disease. *Annals of Neurology*, *49*(2), 155–164.

Furtado, S., Sossi, V., Hauser, R. A., Samii, A., Schulzer, M., Murphy, C. B., et al. (2005). Positron emission tomography after fetal transplantation in Huntington's disease. *Annals of Neurology*, *58*(2), 331–337. https://doi.org/10.1002/ana.20564.

Gaura, V., Bachoud-Levi, A. C., Ribeiro, M. J., Nguyen, J. P., Frouin, V., Baudic, S., et al. (2004). Striatal neural grafting improves cortical metabolism in Huntington's disease patients. *Brain, 127*(Pt 1), 65–72. https://doi.org/10.1093/brain/awh003.

Gerhard, A., Pavese, N., Hotton, G., Turkheimer, F., Es, M., Hammers, A., et al. (2006). In vivo imaging of microglial activation with [11C](R)-PK11195 PET in idiopathic Parkinson's disease. *Neurobiology of Disease, 21*(2), 404–412. https://doi.org/10.1016/j.nbd.2005.08.002.

Giordano, R., Canesi, M., Isalberti, M., Isaias, I. U., Montemurro, T., Vigano, M., et al. (2014). Autologous mesenchymal stem cell therapy for progressive supranuclear palsy: Translation into a phase I controlled, randomized clinical study. *Journal of Translational Medicine, 12*, 14. https://doi.org/10.1186/1479-5876-12-14.

Grasbon-Frodl, E. M., Nakao, N., Lindvall, O., & Brundin, P. (1997). Developmental features of human striatal tissue transplanted in a rat model of Huntington's disease. *Neurobiology of Disease, 3*(4), 299–311. https://doi.org/10.1006/nbdi.1996.0124.

Guttman, M., Burkholder, J., Kish, S. J., Hussey, D., Wilson, A., DaSilva, J., et al. (1997). [11C]RTI-32 PET studies of the dopamine transporter in early dopa-naive Parkinson's disease: Implications for the symptomatic threshold. *Neurology, 48*(6), 1578–1583.

Hagell, P., & Brundin, P. (2001). Cell survival and clinical outcome following intrastriatal transplantation in Parkinson disease. *Journal of Neuropathology and Experimental Neurology, 60*(8), 741–752.

Hagell, P., & Cenci, M. A. (2005). Dyskinesias and dopamine cell replacement in Parkinson's disease: A clinical perspective. *Brain Research Bulletin, 68*(1–2), 4–15. https://doi.org/10.1016/j.brainresbull.2004.10.013.

Hagell, P., Piccini, P., Bjorklund, A., Brundin, P., Rehncrona, S., Widner, H., et al. (2002). Dyskinesias following neural transplantation in Parkinson's disease. *Nature Neuroscience, 5*(7), 627–628. https://doi.org/10.1038/nn863.

Hagell, P., Schrag, A., Piccini, P., Jahanshahi, M., Brown, R., Rehncrona, S., et al. (1999). Sequential bilateral transplantation in Parkinson's disease: Effects of the second graft. *Brain, 122*(Pt. 6), 1121–1132.

Haslinger, B., Erhard, P., Kampfe, N., Boecker, H., Rummeny, E., Schwaiger, M., et al. (2001). Event-related functional magnetic resonance imaging in Parkinson's disease before and after levodopa. *Brain, 124*(Pt. 3), 558–570.

Hauser, R. A., Freeman, T. B., Snow, B. J., Nauert, M., Gauger, L., Kordower, J. H., et al. (1999). Long-term evaluation of bilateral fetal nigral transplantation in Parkinson disease. *Archives of Neurology, 56*(2), 179–187.

Hauser, R. A., Furtado, S., Cimino, C. R., Delgado, H., Eichler, S., Schwartz, S., et al. (2002). Bilateral human fetal striatal transplantation in Huntington's disease. *Neurology, 58*(5), 687–695.

Hayden, M. R., Hewitt, J., Stoessl, A. J., Clark, C., Ammann, W., & Martin, W. R. (1987). The combined use of positron emission tomography and DNA polymorphisms for preclinical detection of Huntington's disease. *Neurology, 37*(9), 1441–1447.

Herben-Dekker, M., van Oostrom, J. C., Roos, R. A., Jurgens, C. K., Witjes-Ane, M. N., Kremer, H. P., et al. (2014). Striatal metabolism and psychomotor speed as predictors of motor onset in Huntington's disease. *Journal of Neurology, 261*(7), 1387–1397. https://doi.org/10.1007/s00415-014-7350-7.

Hoffer, B. J., Leenders, K. L., Young, D., Gerhardt, G., Zerbe, G. O., Bygdeman, M., et al. (1992). Eighteen-month course of two patients with grafts of fetal dopamine neurons for severe Parkinson's disease. *Experimental Neurology, 118*(3), 243–252.

Huang, Z., de la Fuente-Fernandez, R., Hauser, R. A., Freeman, T. B., Sossi, V., Olanow, C. W., et al. (2003). Dopaminergic alterationin Parkinson's patients with off-period dyskinesia following striatal embryonic mesencephalic transplant. *Neurology, 60*, A126.

Isacson, O., Bjorklund, L. M., & Schumacher, J. M. (2003). Toward full restoration of synaptic and terminal function of the dopaminergic system in Parkinson's disease by stem cells. *Annals of Neurology, 53*(Suppl. 3). https://doi.org/10.1002/ana.10482. S135-146; discussion S146-138.

Jacobson, K. B. (1959). Studies on the role of N-acetylaspartic acid in mammalian brain. *The Journal of General Physiology, 43*, 323–333.

Jahanshahi, M., Jenkins, I. H., Brown, R. G., Marsden, C. D., Passingham, R. E., & Brooks, D. J. (1995). Self-initiated versus externally triggered movements. I. An investigation using measurement of regional cerebral blood flow with PET and movement-related potentials in normal and Parkinson's disease subjects. *Brain, 118*(Pt. 4), 913–933.

Keene, C. D., Sonnen, J. A., Swanson, P. D., Kopyov, O., Leverenz, J. B., Bird, T. D., et al. (2007). Neural transplantation in Huntington disease: Long-term grafts in two patients. *Neurology, 68*(24), 2093–2098. https://doi.org/10.1212/01.wnl.0000264504.14301.f5.

Kefalopoulou, Z., Politis, M., Piccini, P., Mencacci, N., Bhatia, K., Jahanshahi, M., et al. (2014). Long-term clinical outcome of fetal cell transplantation for Parkinson disease: Two case reports. *JAMA Neurology, 71*(1), 83–87. https://doi.org/10.1001/jamaneurol.2013.4749.

Kollensperger, M., Stefanova, N., Pallua, A., Puschban, Z., Dechant, G., Hainzer, M., et al. (2009). Striatal transplantation in a rodent model of multiple system atrophy: Effects on L-Dopa response. *Journal of Neuroscience Research, 87*(7), 1679–1685. https://doi.org/10.1002/jnr.21972.

Kordower, J. H., Chu, Y., Hauser, R. A., Freeman, T. B., & Olanow, C. W. (2008). Lewy body-like pathology in long-term embryonic nigral transplants in Parkinson's disease. *Nature Medicine, 14*(5), 504–506. https://doi.org/10.1038/nm1747.

Kordower, J. H., Chu, Y., Hauser, R. A., Olanow, C. W., & Freeman, T. B. (2008). Transplanted dopaminergic neurons develop PD pathologic changes: A second case report. *Movement Disorders, 23*(16), 2303–2306. https://doi.org/10.1002/mds.22369.

Kordower, J. H., Freeman, T. B., Chen, E. Y., Mufson, E. J., Sanberg, P. R., Hauser, R. A., et al. (1998). Fetal nigral grafts survive and mediate clinical benefit in a patient with Parkinson's disease. *Movement Disorders, 13*(3), 383–393. https://doi.org/10.1002/mds.870130303.

Kordower, J. H., Freeman, T. B., Snow, B. J., Vingerhoets, F. J., Mufson, E. J., Sanberg, P. R., et al. (1995). Neuropathological evidence of graft survival and striatal reinnervation after the transplantation of fetal mesencephalic tissue in a patient with Parkinson's disease. *The New England Journal of Medicine, 332*(17), 1118–1124. https://doi.org/10.1056/nejm199504273321702.

Kordower, J. H., Rosenstein, J. M., Collier, T. J., Burke, M. A., Chen, E. Y., Li, J. M., et al. (1996). Functional fetal nigral grafts in a patient with Parkinson's disease: Chemoanatomic, ultrastructural, and metabolic studies. *The Journal of Comparative Neurology, 370*(2), 203–230. https://doi.org/10.1002/(SICI)1096-9861(19960624)370:2<203:AID-CNE6>3.0.CO;2-6.

Krebs, S. S., Trippel, M., Prokop, T., Omer, T. N., Landwehrmeyer, B., Weber, W. A., et al. (2011). Immune response after striatal engraftment of fetal neuronal cells in patients with Huntington's disease: Consequences for cerebral transplantation programs. *Clinical and Experimental Neuroimmunology, 2*(2), 25–32.

Kremer, B., Clark, C. M., Almqvist, E. W., Raymond, L. A., Graf, P., Jacova, C., et al. (1999). Influence of lamotrigine on progression of early Huntington disease: A randomized clinical trial. *Neurology, 53*(5), 1000–1011.

Krystkowiak, P., Gaura, V., Labalette, M., Rialland, A., Remy, P., Peschanski, M., et al. (2007). Alloimmunisation to donor antigens and immune rejection following foetal neural grafts to the brain in patients with Huntington's disease. *PLoS One, 2*(1), e166. https://doi.org/10.1371/journal.pone.0000166.

Kuwert, T., Lange, H. W., Langen, K. J., Herzog, H., Aulich, A., & Feinendegen, L. E. (1990). Cortical and subcortical glucose consumption measured by PET in patients with Huntington's disease. *Brain, 113*(Pt. 5), 1405–1423.

Langston, J. W., Widner, H., Goetz, C. G., Brooks, D., Fahn, S., Freeman, T., et al. (1992). Core assessment program for intracerebral transplantations (CAPIT). *Movement Disorders, 7*(1), 2–13. https://doi.org/10.1002/mds.870070103.

Lao-Kaim, N. P., Piccini, P., & Tai, Y. F. (2017). Restorative strategies in movement disorders: The contribution of imaging. *Current Neurology and Neuroscience Reports, 17*(12), 98. https://doi.org/10.1007/s11910-017-0807-1.

Lee, P. H., Kim, J. W., Bang, O. Y., Ahn, Y. H., Joo, I. S., & Huh, K. (2008). Autologous mesenchymal stem cell therapy delays the progression of neurological deficits in patients with multiple system atrophy. *Clinical Pharmacology and Therapeutics, 83*(5), 723–730. https://doi.org/10.1038/sj.clpt.6100386.

Lee, P. H., Lee, J. E., Kim, H. S., Song, S. K., Lee, H. S., Nam, H. S., et al. (2012). A randomized trial of mesenchymal stem cells in multiple system atrophy. *Annals of Neurology, 72*(1), 32–40. https://doi.org/10.1002/ana.23612.

Lee, C. S., Samii, A., Sossi, V., Ruth, T. J., Schulzer, M., Holden, J. E., et al. (2000). In vivo positron emission tomographic evidence for compensatory changes in presynaptic dopaminergic nerve terminals in Parkinson's disease. *Annals of Neurology, 47*(4), 493–503.

Leenders, K. L., Salmon, E. P., Tyrrell, P., Perani, D., Brooks, D. J., Sager, H., et al. (1990). The nigrostriatal dopaminergic system assessed in vivo by positron emission tomography in healthy volunteer subjects and patients with Parkinson's disease. *Archives of Neurology, 47*(12), 1290–1298.

Levivier, M., Dethy, S., Rodesch, F., Peschanski, M., Vandesteene, A., David, P., et al. (1997). Intracerebral transplantation of fetal ventral mesencephalon for patients with advanced Parkinson's disease. Methodology and 6-month to 1-year follow-up in 3 patients. *Stereotactic and Functional Neurosurgery, 69*(1–4 Pt. 2), 99–111. https://doi.org/10.1159/000099859.

Li, R. (2012). Stem cell transplantation for treating Parkinson's disease: Literature analysis based on the web of science. *Neural Regeneration Research, 7*(16), 1272–1279. https://doi.org/10.3969/j.issn.1673-5374.2012.16.010.

Li, J. Y., Englund, E., Holton, J. L., Soulet, D., Hagell, P., Lees, A. J., et al. (2008). Lewy bodies in grafted neurons in subjects with Parkinson's disease suggest host-to-graft disease propagation. *Nature Medicine, 14*(5), 501–503. https://doi.org/10.1038/nm1746.

Lindvall, O., Brundin, P., Widner, H., Rehncrona, S., Gustavii, B., Frackowiak, R., et al. (1990). Grafts of fetal dopamine neurons survive and improve motor function in Parkinson's disease. *Science, 247*(4942), 574–577.

Lindvall, O., & Hagell, P. (2000). Clinical observations after neural transplantation in Parkinson's disease. *Progress in Brain Research, 127*, 299–320.

Lindvall, O., & Kokaia, Z. (2006). Stem cells for the treatment of neurological disorders. *Nature, 441*(7097), 1094–1096. https://doi.org/10.1038/nature04960.

Lindvall, O., Rehncrona, S., Brundin, P., Gustavii, B., Astedt, B., Widner, H., et al. (1989). Human fetal dopamine neurons grafted into the striatum in two patients with severe Parkinson's disease. A detailed account of methodology and a 6-month follow-up. *Archives of Neurology, 46*(6), 615–631.

Lindvall, O., Rehncrona, S., Gustavii, B., Brundin, P., Astedt, B., Widner, H., et al. (1988). Fetal dopamine-rich mesencephalic grafts in Parkinson's disease. *Lancet, 2*(8626–8627), 1483–1484.

Ma, Y., Feigin, A., Dhawan, V., Fukuda, M., Shi, Q., Greene, P., et al. (2002). Dyskinesia after fetal cell transplantation for parkinsonism: A PET study. *Annals of Neurology, 52*(5), 628–634. https://doi.org/10.1002/ana.10359.

Ma, Y., Peng, S., Dhawan, V., & Eidelberg, D. (2011). Dopamine cell transplantation in Parkinson's disease: Challenge and perspective. *British Medical Bulletin, 100,* 173–189. https://doi.org/10.1093/bmb/ldr040.

Ma, Y., Tang, C., Chaly, T., Greene, P., Breeze, R., Fahn, S., et al. (2010). Dopamine cell implantation in Parkinson's disease: Long-term clinical and (18)F-FDOPA PET outcomes. *Journal of Nuclear Medicine, 51*(1), 7–15. https://doi.org/10.2967/jnumed.109.066811.

Madrazo, I., Franco-Bourland, R., Ostrosky-Solis, F., Aguilera, M., Cuevas, C., Zamorano, C., et al. (1990). Fetal homotransplants (ventral mesencephalon and adrenal tissue) to the striatum of parkinsonian subjects. *Archives of Neurology, 47*(12), 1281–1285.

McLeod, M. C., Kobayashi, N. R., Sen, A., Baghbaderani, B. A., Sadi, D., Ulalia, R., et al. (2013). Transplantation of GABAergic cells derived from bioreactor-expanded human neural precursor cells restores motor and cognitive behavioral deficits in a rodent model of Huntington's disease. *Cell Transplantation, 22*(12), 2237–2256. https://doi.org/10.3727/096368912x658809.

Mendez, I., Dagher, A., Hong, M., Gaudet, P., Weerasinghe, S., McAlister, V., et al. (2002). Simultaneous intrastriatal and intranigral fetal dopaminergic grafts in patients with Parkinson disease: A pilot study. Report of three cases. *Journal of Neurosurgery, 96*(3), 589–596. https://doi.org/10.3171/jns.2002.96.3.0589.

Mendez, I., Vinuela, A., Astradsson, A., Mukhida, K., Hallett, P., Robertson, H., et al. (2008). Dopamine neurons implanted into people with Parkinson's disease survive without pathology for 14 years. *Nature Medicine, 14*(5), 507–509. https://doi.org/10.1038/nm1752.

Morrish, P. K., Rakshi, J. S., Bailey, D. L., Sawle, G. V., & Brooks, D. J. (1998). Measuring the rate of progression and estimating the preclinical period of Parkinson's disease with [18F]dopa PET. *Journal of Neurology, Neurosurgery, and Psychiatry, 64*(3), 314–319.

Morrish, P. K., Sawle, G. V., & Brooks, D. J. (1996). Regional changes in [18F]dopa metabolism in the striatum in Parkinson's disease. *Brain, 119*(Pt. 6), 2097–2103.

Nakamura, T., Dhawan, V., Chaly, T., Fukuda, M., Ma, Y., Breeze, R., et al. (2001). Blinded positron emission tomography study of dopamine cell implantation for Parkinson's disease. *Annals of Neurology, 50*(2), 181–187.

Niccolini, F., & Politis, M. (2014). Neuroimaging in Huntington's disease. *World Journal of Radiology, 6*(6), 301–312. https://doi.org/10.4329/wjr.v6.i6.301.

Olanow, C. W., Goetz, C. G., Kordower, J. H., Stoessl, A. J., Sossi, V., Brin, M. F., et al. (2003). A double-blind controlled trial of bilateral fetal nigral transplantation in Parkinson's disease. *Annals of Neurology, 54*(3), 403–414. https://doi.org/10.1002/ana.10720.

Olsson, M., Campbell, K., Wictorin, K., & Bjorklund, A. (1995). Projection neurons in fetal striatal transplants are predominantly derived from the lateral ganglionic eminence. *Neuroscience, 69*(4), 1169–1182.

Ouchi, Y., Yoshikawa, E., Sekine, Y., Futatsubashi, M., Kanno, T., Ogusu, T., et al. (2005). Microglial activation and dopamine terminal loss in early Parkinson's disease. *Annals of Neurology, 57*(2), 168–175. https://doi.org/10.1002/ana.20338.

Paganini, M., Biggeri, A., Romoli, A. M., Mechi, C., Ghelli, E., Berti, V., et al. (2014). Fetal striatal grafting slows motor and cognitive decline of Huntington's disease. *Journal of Neurology, Neurosurgery, and Psychiatry, 85*(9), 974–981. https://doi.org/10.1136/jnnp-2013 306533.

Pagano, G., Niccolini, F., Fusar-Poli, P., & Politis, M. (2017). Serotonin transporter in Parkinson's disease: A meta-analysis of positron emission tomography studies. *Annals of Neurology, 81*(2), 171–180. https://doi.org/10.1002/ana.24859.

Pagano, G., Niccolini, F., & Politis, M. (2016). Imaging in Parkinson's disease. *Clinical Medicine (London, England), 16*(4), 371–375. https://doi.org/10.7861/clinmedicine.16-4-371.

Perlow, M. J., Freed, W. J., Hoffer, B. J., Seiger, A., Olson, L., & Wyatt, R. J. (1979). Brain grafts reduce motor abnormalities produced by destruction of nigrostriatal dopamine system. *Science, 204*(4393), 643–647.

Peschanski, M., Defer, G., N'Guyen, J. P., Ricolfi, F., Monfort, J. C., Remy, P., et al. (1994). Bilateral motor improvement and alteration of L-dopa effect in two patients with Parkinson's disease following intrastriatal transplantation of foetal ventral mesencephalon. *Brain, 117*(Pt. 3), 487–499.

Piccini, P., Brooks, D. J., Bjorklund, A., Gunn, R. N., Grasby, P. M., Rimoldi, O., et al. (1999). Dopamine release from nigral transplants visualized in vivo in a Parkinson's patient. *Nature Neuroscience, 2*(12), 1137–1140. https://doi.org/10.1038/16060.

Piccini, P., Lindvall, O., Bjorklund, A., Brundin, P., Hagell, P., Ceravolo, R., et al. (2000). Delayed recovery of movement-related cortical function in Parkinson's disease after striatal dopaminergic grafts. *Annals of Neurology, 48*(5), 689–695.

Piccini, P., Pavese, N., & Brooks, D. J. (2003). Endogenous dopamine release after pharmacological challenges in Parkinson's disease. *Annals of Neurology, 53*(5), 647–653. https://doi.org/10.1002/ana.10526.

Piccini, P., Pavese, N., Hagell, P., Reimer, J., Bjorklund, A., Oertel, W. H., et al. (2005). Factors affecting the clinical outcome after neural transplantation in Parkinson's disease. *Brain, 128*(Pt. 12), 2977–2986. https://doi.org/10.1093/brain/awh649.

Pogarell, O., Koch, W., Gildehaus, F. J., Kupsch, A., Lindvall, O., Oertel, W. H., et al. (2006). Long-term assessment of striatal dopamine transporters in parkinsonian patients with intrastriatal embryonic mesencephalic grafts. *European Journal of Nuclear Medicine and Molecular Imaging, 33*(4), 407–411. https://doi.org/10.1007/s00259-005-0032-z.

Politis, M. (2010). Dyskinesias after neural transplantation in Parkinson's disease: What do we know and what is next? *BMC Medicine, 8*, 80. https://doi.org/10.1186/1741-7015-8-80.

Politis, M. (2014). Neuroimaging in Parkinson disease: From research setting to clinical practice. *Nature Reviews Neurology, 10*, 708–722.

Politis, M., & Lindvall, O. (2012). Clinical application of stem cell therapy in Parkinson's disease. *BMC Medicine, 10*, 1. https://doi.org/10.1186/1741-7015-10-1.

Politis, M., & Loane, C. (2011). Serotonergic dysfunction in Parkinson's disease and its relevance to disability. *ScientificWorldJournal, 11*, 1726–1734. https://doi.org/10.1100/2011/172893.

Politis, M., & Niccolini, F. (2015). Serotonin in Parkinson's disease. *Behavioural Brain Research, 277*, 136–145. https://doi.org/10.1016/j.bbr.2014.07.037.

Politis, M., Oertel, W. H., Wu, K., Quinn, N. P., Pogarell, O., Brooks, D. J., et al. (2011). Graft-induced dyskinesias in Parkinson's disease: High striatal serotonin/dopamine transporter ratio. *Movement Disorders, 26*(11), 1997–2003. https://doi.org/10.1002/mds.23743.

Politis, M., & Piccini, P. (2010). Brain imaging after neural transplantation. *Progress in Brain Research, 184*, 193–203. https://doi.org/10.1016/s0079-6123(10)84010-5.

Politis, M., & Piccini, P. (2012). In vivo imaging of the integration and function of nigral grafts in clinical trials. *Progress in Brain Research, 200*, 199–220. https://doi.org/10.1016/b978-0-444-59575-1.00009-0.

Politis, M., Wu, K., Loane, C., Brooks, D. J., Kiferle, L., Turkheimer, F. E., et al. (2014). Serotonergic mechanisms responsible for levodopa-induced dyskinesias in Parkinson's disease patients. *The Journal of Clinical Investigation, 124*(3), 1340–1349. https://doi.org/10.1172/jci71640.

Politis, M., Wu, K., Loane, C., Quinn, N. P., Brooks, D. J., Oertel, W. H., et al. (2012). Serotonin neuron loss and nonmotor symptoms continue in Parkinson's patients treated with dopamine grafts. *Science Translational Medicine, 4*(128), 128ra141. https://doi.org/10.1126/scitranslmed.3003391.

Politis, M., Wu, K., Loane, C., Quinn, N. P., Brooks, D. J., Rehncrona, S., et al. (2010). Serotonergic neurons mediate dyskinesia side effects in Parkinson's patients with neural transplants. *Science Translational Medicine*, *2*(38), 38ra46. https://doi.org/10.1126/scitranslmed.3000976.

Precious, S. V., Zietlow, R., Dunnett, S. B., Kelly, C. M., & Rosser, A. E. (2017). Is there a place for human fetal-derived stem cells for cell replacement therapy in Huntington's disease? *Neurochemistry International*, *106*, 114–121. https://doi.org/10.1016/j.neuint.2017.01.016.

Pundt, L. L., Kondoh, T., & Low, W. C. (1994). NADPH-diaphorase histochemistry and functional analysis of human fetal striatal brain tissue transplanted into a rodent model of Huntington's disease. *Transplantation Proceedings*, *26*(6), 3295.

Quinn, N., Brown, R., Craufurd, D., Goldman, S., Hodges, J., Kieburtz, K., et al. (1996). Core assessment program for intracerebral transplantation in Huntington's disease (CAPIT-HD). *Movement Disorders*, *11*(2), 143–150. https://doi.org/10.1002/mds.870110205.

Rascol, O., Sabatini, U., Chollet, F., Celsis, P., Montastruc, J. L., Marc-Vergnes, J. P., et al. (1992). Supplementary and primary sensory motor area activity in Parkinson's disease. Regional cerebral blood flow changes during finger movements and effects of apomorphine. *Archives of Neurology*, *49*(2), 144–148.

Rascol, O., Sabatini, U., Chollet, F., Fabre, N., Senard, J. M., Montastruc, J. L., et al. (1994). Normal activation of the supplementary motor area in patients with Parkinson's disease undergoing long-term treatment with levodopa. *Journal of Neurology, Neurosurgery, and Psychiatry*, *57*(5), 567–571.

Remy, P., Samson, Y., Hantraye, P., Fontaine, A., Defer, G., Mangin, J. F., et al. (1995). Clinical correlates of [18F]fluorodopa uptake in five grafted parkinsonian patients. *Annals of Neurology*, *38*(4), 580–588. https://doi.org/10.1002/ana.410380406.

Ren, W. Q., Yin, F., Zhang, J. N., Lu, W. S., Liang, Y. K., Adlerberth, J., et al. (2016). Neural stem cell transplantation for the treatment of primary torsion dystonia: A case report. *Experimental and Therapeutic Medicine*, *12*(2), 661–666. https://doi.org/10.3892/etm.2016.3392.

Reuter, I., Tai, Y. F., Pavese, N., Chaudhuri, K. R., Mason, S., Polkey, C. E., et al. (2008). Long-term clinical and positron emission tomography outcome of fetal striatal transplantation in Huntington's disease. *Journal of Neurology, Neurosurgery, and Psychiatry*, *79*(8), 948–951. https://doi.org/10.1136/jnnp.2007.142380.

Ross, B. D., Hoang, T. Q., Bluml, S., Dubowitz, D., Kopyov, O. V., Jacques, D. B., et al. (1999). In vivo magnetic resonance spectroscopy of human fetal neural transplants. *NMR in Biomedicine*, *12*(4), 221–236.

Rosser, A. E., Barker, R. A., Harrower, T., Watts, C., Farrington, M., Ho, A. K., et al. (2002). Unilateral transplantation of human primary fetal tissue in four patients with Huntington's disease: NEST-UK safety report ISRCTN no 36485475. *Journal of Neurology, Neurosurgery, and Psychiatry*, *73*(6), 678–685.

Roussakis, A. A., Politis, M., Towey, D., & Piccini, P. (2016). Serotonin-to-dopamine transporter ratios in Parkinson disease: Relevance for dyskinesias. *Neurology*, *86*(12), 1152–1158. https://doi.org/10.1212/wnl.0000000000002494.

Rowe, J., Stephan, K. E., Friston, K., Frackowiak, R., Lees, A., & Passingham, R. (2002). Attention to action in Parkinson's disease: Impaired effective connectivity among frontal cortical regions. *Brain*, *125*(Pt. 2), 276–289.

Sawle, G. V., Bloomfield, P. M., Bjorklund, A., Brooks, D. J., Brundin, P., Leenders, K. L., et al. (1992). Transplantation of fetal dopamine neurons in Parkinson's disease: PET [18F]6-L-fluorodopa studies in two patients with putaminal implants. *Annals of Neurology*, *31*(2), 166–173. https://doi.org/10.1002/ana.410310207.

Schackel, S., Pauly, M. C., Piroth, T., Nikkhah, G., & Dobrossy, M. D. (2013). Donor age dependent graft development and recovery in a rat model of Huntington's disease: Histological and behavioral analysis. *Behavioural Brain Research, 256*, 56–63. https://doi.org/10.1016/j.bbr.2013.07.053.

Shin, E., Garcia, J., Winkler, C., Bjorklund, A., & Carta, M. (2012). Serotonergic and dopaminergic mechanisms in graft-induced dyskinesia in a rat model of Parkinson's disease. *Neurobiology of Disease, 47*(3), 393–406. https://doi.org/10.1016/j.nbd.2012.03.038.

Snow, B. J., Tooyama, I., McGeer, E. G., Yamada, T., Calne, D. B., Takahashi, H., et al. (1993). Human positron emission tomographic [18F]fluorodopa studies correlate with dopamine cell counts and levels. *Annals of Neurology, 34*(3), 324–330. https://doi.org/10.1002/ana.410340304.

Spencer, D. D., Robbins, R. J., Naftolin, F., Marek, K. L., Vollmer, T., Leranth, C., et al. (1992). Unilateral transplantation of human fetal mesencephalic tissue into the caudate nucleus of patients with Parkinson's disease. *The New England Journal of Medicine, 327*(22), 1541–1548. https://doi.org/10.1056/nejm199211263272201.

Squitieri, F., Orobello, S., Cannella, M., Martino, T., Romanelli, P., Giovacchini, G., et al. (2009). Riluzole protects Huntington disease patients from brain glucose hypometabolism and grey matter volume loss and increases production of neurotrophins. *European Journal of Nuclear Medicine and Molecular Imaging, 36*(7), 1113–1120. https://doi.org/10.1007/s00259-009-1103-3.

Stromberg, I., Bygdeman, M., Goldstein, M., Seiger, A., & Olson, L. (1986). Human fetal substantia nigra grafted to the dopamine-denervated striatum of immunosuppressed rats: Evidence for functional reinnervation. *Neuroscience Letters, 71*(3), 271–276.

Sturrock, A., Laule, C., Decolongon, J., Dar Santos, R., Coleman, A. J., Creighton, S., et al. (2010). Magnetic resonance spectroscopy biomarkers in premanifest and early Huntington disease. *Neurology, 75*(19), 1702–1710. https://doi.org/10.1212/WNL.0b013e3181fc27e4.

Sullivan, A. M., Pohl, J., & Blunt, S. B. (1998). Growth/differentiation factor 5 and glial cell line-derived neurotrophic factor enhance survival and function of dopaminergic grafts in a rat model of Parkinson's disease. *The European Journal of Neuroscience, 10*(12), 3681–3688.

Tessa, C., Diciotti, S., Lucetti, C., Baldacci, F., Cecchi, P., Giannelli, M., et al. (2013). fMRI changes in cortical activation during task performance with the unaffected hand partially reverse after ropinirole treatment in de novo Parkinson's disease. *Parkinsonism & Related Disorders, 19*(2), 265–268. https://doi.org/10.1016/j.parkreldis.2012.07.018.

Tsai, Y. A., Liu, R. S., Lirng, J. F., Yang, B. H., Chang, C. H., Wang, Y. C., et al. (2017). Treatment of spinocerebellar ataxia with mesenchymal stem cells: A phase I/IIa clinical study. *Cell Transplantation, 26*(3), 503–512. https://doi.org/10.3727/096368916x694373.

Tsukada, H., Nishiyama, S., Kakiuchi, T., Ohba, H., Sato, K., & Harada, N. (1999). Is synaptic dopamine concentration the exclusive factor which alters the in vivo binding of [11C]raclopride?: PET studies combined with microdialysis in conscious monkeys. *Brain Research, 841*(1–2), 160–169.

Vingerhoets, F. J., Schulzer, M., Calne, D. B., & Snow, B. J. (1997). Which clinical sign of Parkinson's disease best reflects the nigrostriatal lesion? *Annals of Neurology, 41*(1), 58–64. https://doi.org/10.1002/ana.410410111.

Watts, C., Brasted, P. J., & Dunnett, S. B. (2000). Embryonic donor age and dissection influences striatal graft development and functional integration in a rodent model of Huntington's disease. *Experimental Neurology, 163*(1), 85–97. https://doi.org/10.1006/exnr.1999.7341.

Weeks, R. A., Piccini, P., Harding, A. E., & Brooks, D. J. (1996). Striatal D1 and D2 dopamine receptor loss in asymptomatic mutation carriers of Huntington's disease. *Annals of Neurology, 40*(1), 49–54. https://doi.org/10.1002/ana.410400110.

Wenning, G. K., Odin, P., Morrish, P., Rehncrona, S., Widner, H., Brundin, P., et al. (1997). Short- and long-term survival and function of unilateral intrastriatal dopaminergic grafts in Parkinson's disease. *Annals of Neurology, 42*(1), 95–107. https://doi.org/10.1002/ana.410420115.

Widner, H., Tetrud, J., Rehncrona, S., Snow, B., Brundin, P., Gustavii, B., et al. (1992). Bilateral fetal mesencephalic grafting in two patients with parkinsonism induced by 1-methyl-4-phenyl-1,2,3,6-tetrahydropyridine (MPTP). *The New England Journal of Medicine, 327*(22), 1556–1563. https://doi.org/10.1056/nejm199211263272203.

Wilson, H., De Micco, R., Niccolini, F., & Politis, M. (2017). Molecular imaging markers to track Huntington's disease pathology. *Frontiers in Neurology, 8*, 11. https://doi.org/10.3389/fneur.2017.00011.

Yasuhara, T., Shingo, T., Muraoka, K., Kobayashi, K., Takeuchi, A., Yano, A., et al. (2005). Early transplantation of an encapsulated glial cell line-derived neurotrophic factor-producing cell demonstrating strong neuroprotective effects in a rat model of Parkinson disease. *Journal of Neurosurgery, 102*(1), 80–89. https://doi.org/10.3171/jns.2005.102.1.0080.

Yhnell, E., Dunnett, S. B., & Brooks, S. P. (2016). A longitudinal motor characterisation of the HdhQ111 mouse model of Huntington's disease. *The Journal of Huntington's Disease, 5*(2), 149–161. https://doi.org/10.3233/jhd-160191.

Young, A. B., Penney, J. B., Starosta-Rubinstein, S., Markel, D. S., Berent, S., Giordani, B., et al. (1986). PET scan investigations of Huntington's disease: Cerebral metabolic correlates of neurological features and functional decline. *Annals of Neurology, 20*(3), 296–303. https://doi.org/10.1002/ana.410200305.

FURTHER READING

Politis, M., Pagano, G., & Niccolini, F. (2017). Imaging in Parkinson's Disease. *International Review of Neurobiology, 132*, 233–274. https://doi.org/10.1016/bs.irn.2017.02.015.

Roussakis, A. A., & Piccini, P. (2015). PET imaging in Huntington's disease. *The Journal of Huntington's Disease, 4*(4), 287–296. https://doi.org/10.3233/jhd-150171.

CPI Antony Rowe

Chippenham, UK

2019-01-17 23:08